MicroComputed Tomography
Methodology and Applications

MicroComputed Tomography

Methodology and Applications

Stuart R. Stock

CRC Press
Taylor & Francis Group
Boca Raton London New York

CRC Press is an imprint of the
Taylor & Francis Group, an **informa** business

CRC Press
Taylor & Francis Group
6000 Broken Sound Parkway NW, Suite 300
Boca Raton, FL 33487-2742

© 2009 by Taylor & Francis Group, LLC
CRC Press is an imprint of Taylor & Francis Group, an Informa business

No claim to original U.S. Government works
Printed in the United States of America on acid-free paper
10 9 8 7 6 5 4 3 2 1

International Standard Book Number-13: 978-1-4200-5876-5 (Hardcover)

This book contains information obtained from authentic and highly regarded sources. Reasonable efforts have been made to publish reliable data and information, but the author and publisher cannot assume responsibility for the validity of all materials or the consequences of their use. The authors and publishers have attempted to trace the copyright holders of all material reproduced in this publication and apologize to copyright holders if permission to publish in this form has not been obtained. If any copyright material has not been acknowledged please write and let us know so we may rectify in any future reprint.

Except as permitted under U.S. Copyright Law, no part of this book may be reprinted, reproduced, transmitted, or utilized in any form by any electronic, mechanical, or other means, now known or hereafter invented, including photocopying, microfilming, and recording, or in any information storage or retrieval system, without written permission from the publishers.

For permission to photocopy or use material electronically from this work, please access www.copyright.com (http://www.copyright.com/) or contact the Copyright Clearance Center, Inc. (CCC), 222 Rosewood Drive, Danvers, MA 01923, 978-750-8400. CCC is a not-for-profit organization that provides licenses and registration for a variety of users. For organizations that have been granted a photocopy license by the CCC, a separate system of payment has been arranged.

Trademark Notice: Product or corporate names may be trademarks or registered trademarks, and are used only for identification and explanation without intent to infringe.

Library of Congress Cataloging-in-Publication Data

Stock, Stuart R.
  MicroComputed tomography : methodology and applications / Stuart R. Stock.
      p. cm.
  Includes bibliographical references and index.
  ISBN 978-1-4200-5876-5 (hardcover : alk. paper)
  1. Tomography. I. Title.

RC78.7.T6S73 2009
616.07'57--dc22   2008042180

Visit the Taylor & Francis Web site at
http://www.taylorandfrancis.com

and the CRC Press Web site at
http://www.crcpress.com

Dedication

In memory of

Kathryn L. S. Banner and Merlyn L. Stock

Deyr fé, deyja frændr,
deyr sjalfr it sama;
ek veit ei∧n at aldri deyr:
dómr of dâuðarhvern.

Hávamál, Poetic Edda
Stuff dies, kindred die,
No one lives forever.
The one thing that does live on
is one's good name and actions.

Contents

Preface xi
Acknowledgments xiii
Biography xv

1 Introduction 1
   References 6

2 Fundamentals 9
   2.1 X-Radiation 9
      2.1.1 Generation 9
      2.1.2 Interaction with Matter 13
   2.2 Imaging 15
   2.3 X-Ray Contrast and Imaging 17
   References 20

3 Reconstruction from Projections 21
   3.1 Basic Concepts 21
   3.2 Algebraic Reconstruction 23
   3.3 Back-Projection 24
   3.4 Fourier-Based Reconstruction 29
   3.5 Performance 31
   3.6 Sinograms 33
      3.6.1 Related Methods 33
   References 36

4 MicroCT Systems and Their Components 39
   4.1 Absorption MicroCT Methods 39
   4.2 X-Ray Sources 44
   4.3 Detectors 46
   4.4 Positioning Components 51
   4.5 Tube-Based Systems 52
   4.6 Synchrotron Radiation Systems 57
   4.7 NanoCT (Full-Field, Microscopy-Based) 61
   4.8 MicroCT with Phase, Fluorescence, or Scattering Contrast 62
      4.8.1 Phase Contrast MicroCT 62
      4.8.2 Fluorescence MicroCT 67
      4.8.3 Scatter MicroCT 69
   4.9 System Specification 69
   References 71

5 MicroCT in Practice 85
5.1 Reconstruction Artifacts 85
5.1.1 Motion Artifacts 85
5.1.2 Ring Artifacts 86
5.1.3 Reconstruction Center Errors 87
5.1.4 Mechanical Imperfections Including Rotation Stage Wobble 87
5.1.5 Undersampling 89
5.1.6 Beam Hardening 89
5.1.7 Streak Artifacts 91
5.1.8 Phase Contrast Artifacts 91
5.2 Performance: Precision and Accuracy 92
5.2.1 Correction for Nonidealities 93
5.2.2 Partial Volume Effects 93
5.2.3 Detection Limits for High-Contrast Features 94
5.2.4 Geometry 96
5.2.5 Linear Attenuation Coefficients 99
5.3 Contrast Enhancement 103
5.4 Data Acquisition Challenges 104
5.5 Speculations 106
References 108

6 Experimental Design, Data Analysis, Visualization 115
6.1 Experiment Design 115
6.2 Data Analysis 117
6.2.1 Segmentation 119
6.2.2 Distance Transform Method 122
6.2.3 Watershed Segmentation 122
6.2.4 Other Methods 122
6.2.5 Image Texture 127
6.2.6 Interpretation of Voxel Values 128
6.2.7 Tracking Evolving Structures 128
6.3 Data Representation 129
References 138

7 Simple Metrology and Microstructure Quantification 145
7.1 Distribution of Phases 146
7.1.1 Pharmaceuticals 146
7.1.2 Geological Materials 146
7.1.3 Two or More Phase Metals, Ceramics, and Polymers 148
7.1.4 Manufactured Composites 148
7.1.5 Biological Tissues as Phases 151
7.2 Metrology and Phylogeny 152
7.2.1 Industrial Metrology 153
7.2.2 Paleontology and Archeology 153

	7.2.3 Invertebrates and Micro-Organisms 154
	7.2.4 Vertebrates 159
	References 162

8 Cellular or Trabecular Solids 171
8.1 Cellular Solids 171
8.2 Static Cellular Structures 173
8.3 Temporally Evolving, Nonbiological Cellular Structures 176
8.4 Mineralized Tissue 181
    8.4.1 Echinoderm Stereom 182
    8.4.2 Cancellous Bone: Motivations for Study and the
          Older Literature 182
    8.4.3 Cancellous Bone: Growth and Aging 184
    8.4.4 Cancellous Bone: Deformation, Damage, and Modeling 191
8.5 Implants and Tissue Scaffolds 194
    8.5.1 Implants 194
    8.5.2 Scaffold Structures and Processing 195
    8.5.3 Bone Growth into Scaffolds 196
References 198

9 Networks 215
9.1 Engineered Network Solids 215
9.2 Networks of Pores 217
9.3 Circulatory System 223
9.4 Respiratory System 227
References 229

10 Evolution of Structures 237
10.1 Materials Processing 237
    10.1.1 Solidification 238
    10.1.2 Vapor Phase Processing 240
    10.1.3 Plastic Forming 243
    10.1.4 Particle Packing and Sintering 247
10.2 Environmental Interactions 248
    10.2.1 Geological Applications 249
    10.2.2 Construction Materials 251
    10.2.3 Degradation of Biological Structures 253
    10.2.4 Corrosion of Metals 257
10.3 Bone and Soft Tissue Adaptation 260
    10.3.1 Mineralized Tissue: Implants, Healing, Mineral
           Levels, and Remodeling 260
    10.3.2 Soft Tissue and Soft Tissue Interfaces 264
References 266

11 Mechanically Induced Damage, Deformation, and Cracking 281
11.1 Deformation Studies 281
11.2 Monolithic Materials: Crack Face Interactions and Crack Closure 283
11.3 Composite Systems 289
    11.3.1 Particle-Reinforced Composites 289
    11.3.2 Fiber-Reinforced Composites 294
References 300

12 Multimode Studies 307
12.1 Sea Urchin Teeth 307
12.2 Sulfate Ion Attack of Portland Cement 308
12.3 Fatigue Crack Path and Mesotexture 309
12.4 Creep Damage 309
12.5 Load Redistribution in Damaged Monofilament Composites 310
12.6 Bone 312
12.7 Networks 315
References 316

Index 319

Preface

MicroComputed Tomography (microCT) systems are high-resolution siblings of the medical CT scanners and are a bit more than one decade younger than the clinical CT scanners of the mid-1970s. MicroCT has developed at a slower rate than clinical CT for the obvious economic reason: much more expensive systems were and are more viable in hospitals than in the research realm where microCT finds its principal home. The number of microCT systems began to climb about the time that biomedical researchers began to emphasize use of small animal knockout models for the study of human diseases and that commercial microCT systems could be obtained at costs lower than many electron microscopes. In broad brush strokes, this corresponded to the mid- to late 1990s.

The increase in the number of x-ray microCT papers since the turn of the century amounts to an explosion (discussed in Chapter 1), and summarizing the literature to date was one motivation for the author to write this book. There have been any number of reviews covering microCT or, more recently, nanoCT, including two by the author (Stock, 1999, 2008), but, to the best of the author's knowledge, only one book has had x-ray microCT as its central focus (Baruchel et al., 2000), and this book was a collection of chapters by different authors on different materials science topics. There has been no detailed synthesis of biological and physical sciences and engineering approaches to microCT and analysis of its data, a lack this book was designed to address.

It is difficult for readers new to microCT to learn enough about the experimental side of microCT without a text starting at the beginning. Therefore, a second motivation for the author was filling this gap.

The author continues to be amazed by the many microCT papers he has reviewed that inadequately cite prior work. Whether this is at all unique to microCT (the author suspects not), it is certainly worth assembling a comprehensive report of the field to help improve the citation situation. The 10^3 references in this book certainly do not cover all of the literature, but these are a significant fraction of the papers, or at least a significant sampling of those employing microCT for diverse purposes. The citations in this book are biased toward those the author was able to obtain electronically through Northwestern University's libraries.

The nature of microCT as a subspecialty is responsible, in part, for some individuals' poor appreciation for prior work. Some authors of microCT studies use synonyms (including micro-CT, x-ray tomographic microscopy, computerized microtomography, the recent nanoCT, and even just tomography) in describing their studies, and this complicates the search for relevant papers. Furthermore, the same class of structure requiring similar analysis tools can occur in disciplines spanning the life sciences to art conservation to the physical sciences and engineering, and reports appear in a wide dispersion of journals and conference proceedings. One example is cellular solids with trabecular or spongy bone and bone growth scaffolds found in the biomedical literature and with metal foams in engineering publications. These two factors combine to hinder newcomers finding prior paradigms on which to base their analyses and to produce examples of unneeded (except perhaps in an existential sense) sweat expenditure via wheel reinvention.

Consider the following experiment (done at the end of 2006) in locating microCT papers relating to foams, a class of cellular solids described in Chapter 8. Over the preceding several years, the author desultorily collected nine papers on microCT of cellular solids (excluding those on trabecular bone) without any particular purpose beyond perhaps preparing a review. A literature search in Compendex, a database for engineering papers, on "microCT and foam" revealed one paper (one of the nine), a search on "microtomography and foam" produced 30 hits (three more of the nine), and a search on "tomography and foam" resulted in 139 hits (six of the nine). Separate searches on "cellular solid and tomography" or "wood and tomography" or "scaffolds and tomography" would be required to reveal the other three papers of the nine; note that the middle search yields 204 hits, most of which are irrelevant to microCT.

In summary, this book, covering the literature through the end of 2007, with only a very few exceptions, follows two principles. First, it gathers together fundamentals and applications into as integrated a treatment as possible without descending too deeply into details. The author hopes that presenting a few fundamentals (x-ray generation, instrumentation, etc.) will allow those with no background in x-ray imaging to achieve, relatively quickly, an appreciation for the literature and for how to design their own microCT studies. Second, given that different subdisciplines require similar analyses, the book gathers like structures together instead of grouping applications by subdiscipline. Learning from other fields seems only sensible. Although the mental exercise of reinventing the wheel has its own existential benefits, the author thinks it would be better to spend the effort on science rather than algorithm.

References

Baruchel, J., J.Y. Buffière, et al. (Eds.) (2000). *X-Ray Tomography in Materials Science*. Paris, Hermes Science.

Stock, S.R. (1999). Microtomography of materials. *Int Mater Rev* **44**: 141–164.

Stock, S.R. (2008). Recent advances in x-ray microtomography applied to materials. *Int Mater Rev* **58**: 129–181.

Acknowledgments

Over the years, many people helped the author reach the point where he could put this book in your hands. Here mention is restricted to those directly involved with the subject and topics of this book, lest the author go on for too long. The author feels extraordinarily lucky to have benefited from these individuals' expertise, advice, and guidance.

John Hilliard introduced the author to numerical analysis of microstructure, Mike Meshii to incorporation of microstructural data into numerical models, and Morris Fine to the notion of microstructure mapping of numerical quantities derived from different modalities. Howard Birnbaum and Hadyn Chen guided the author in his first foray into x-ray imaging of materials, and Keith Bowen introduced him to synchrotron radiation imaging. Jerry Cohen taught the author much about representing spatial structure by its periodicities. Ray Young helped the author refine his thinking about this subject.

During his Georgia Tech years, the author's graduate and undergraduate students taught him much about x-ray imaging and microstructure characterization, more than anyone else will appreciate. These students' work, among the very earliest in microCT, dictated, to a large extent, the future direction of this field. They should have gotten more credit as true innovators.

For many years, Zofia Rek collaborated with the author on synchrotron x-ray imaging studies. John Kinney collaborated with the author for some years, producing some wonderful microCT papers. Jim Elliott and his coworkers, Paul Anderson, Graham Davis, and Stephanie Dowker, have collaborated with the author for more than twenty years in a wide variety of microCT studies and have helped him enter the area of mineralized tissue research. Wah-Keat Lee and Kamel Fezzaa introduced the author to x-ray phase imaging. Since moving to the Chicago area, the author has collaborated with Francesco De Carlo in a large number of very productive synchrotron microCT studies of all manner of subjects biological and material.

The editors of this book, Marsha Pronin and Nora Konopka, were very patient, and the author thanks them for their help.

Finally, the author's wife, Chris, and children, Michala, Sebastian, and Meredith, need to be thanked for their support and patience during many periods of synchrotron beam time and during inconveniences during the several periods in which this material was compiled.

Biography

Stuart R. Stock

Dr. Stock completed his undergraduate and master's degrees in materials science and engineering at Northwestern University, where he later was a post-doc in the same fields. His PhD was in metallurgical engineering at the University of Illinois Urbana-Champaign. He was on the materials science faculty at Georgia Tech for more than sixteen years, rising to the rank of professor. In 2001, he returned to Northwestern University, this time to the medical school.

Dr. Stock has used x-ray diffraction for materials characterization for more than thirty years and revised Cullity's classic text *Elements of X-ray Diffraction*. He has employed x-ray imaging for the same length of time. His first synchrotron radiation experiments were twenty-five years ago, and he currently travels to the Advanced Photon Source six or more times a year to collect data. He has published results of microCT studies of inorganic materials and composites and of mineralized tissue throughout the last twenty years.

1

Introduction

X-ray computed tomography (CT) is an imaging method where individual projections (radiographs) recorded from different viewing directions are used to reconstruct the internal structure of the object of interest. It offers the additional advantage of being noninvasive and nondestructive; that is, the same component can be reinstalled after inspection or the same sample can be interrogated multiple times during the course of mechanical or other testing.

X-ray CT is quite familiar in its medical manifestations (CT- or CAT-scans), but it is less known as an imaging modality for components or materials. Computed tomography provides an accurate map of the variation of x-ray absorption within an object, regardless of whether there is a well-defined substructure of different phases or there are slowly varying density gradients. High-resolution x-ray CT is also termed microComputed Tomography (microCT) or microtomography and reconstructs samples' interiors with the spatial and contrast resolution required for many problems of interest. The application of microCT to biological, physical science, or engineering problems is the subject at hand. The division between conventional CT and microCT is, of course, an artificial distinction, but here microCT is taken to include results obtained with at least 50–100 µm spatial resolution. The actual resolution needed for a particular application depends on the microstructural features of interest and their shapes.

No sooner had Röntgen discovered x-radiation than he applied the new-found penetrating radiation to radiological imaging: the first widely disseminated publication of the discovery of x-rays featured a radiological image (Röntgen, 1898). Within 20 years this new medical tool was in use across the battlefields of World War I (Hildebrandt, 1992). Locating projectiles and shrapnel and checking the reduction of fractures noninvasively was a true breakthrough. One of the advantages of radiological images or radiographs, simplicity, can also be a severe limitation: these images are nothing more than two-dimensional projections of the variation of x-ray absorptivity within the object under study. Although recording stereo pairs allows precise three-dimensional location of high-contrast objects, this approach is impractical when a large number of similar objects produce a confusing array of overlapping images or when there are no sharp charges of contrast on which one can orient.

A strategy for recovering three-dimensional internal structure evolved prior to digital computers (Webb, 1990). It involves translating the patient (or object to be imaged) together with the detection medium (film or other two-dimensional detector) in such a way that only one narrow slice parallel to the translation plane remains in focus. This approach is termed *laminography* or *focal plane tomography*.* The contrast from features outside this slice are out of focus and are blurred to the point where they disappear from the image. Sharp images are difficult to obtain, in part because of the thickness of each slice, and the smearing of images outside the imaging "plane" across the image of the plane of interest seriously degrades contrast within the slice. Laminography continues to be used as an inspection tool for objects whose geometry is impractical for CT, for example, relatively planar objects such as printed wiring boards.

Computed tomography, an approach superior to laminography for most applications, became possible with the development of digital computers. Radon established the mathematics underlying computed tomography in 1917 (Radon, 1917; 1986), and in 1963 Cormack demonstrated the feasibility of using x-rays and a finite number of radiographic viewing directions to reconstruct the distribution of x-ray absorptivity within a cross-section of an object (Cormack, 1963). Early in the 1970s, Hounsfield developed a commercial CT system for medical imaging (Hounsfield, 1968–1972, 1971–1973, 1973), and the number of medical systems is now virtually uncountable.

In CT of patients there were several constraints that affected the way in which apparatus were developed. First, the dose of x-rays received by the patient must be kept to a minimum. Second, the duration of data collection must be limited to several seconds to prevent involuntary patient movement from blurring the image. These considerations do not apply in general to imaging of inanimate objects, and longer data collection times can be used to improve the signal-to-noise ratio in the data. Early on, engineering components and assemblies were characterized by medical CT scanners, but, because the medical systems were optimized for the range of contrast encountered in the body and not for objects of technological interest, systems for nondestructive evaluation and materials characterization were soon marketed.

Industrial CT has gained a measure of acceptance (Bossi and Knutson, 1994; Copely et al., 1994), but the high cost of the instrumentation means that it will not replace x-radiography in many nondestructive evaluation applications. Applications where x-ray CT offers significant economic advantages include five areas (Bossi and Knutson, 1994): new product development, process control, noninvasive metrology, materials performance prediction, and failure analysis.

---

* The word *tomography* arises from the Greek *tomos* for slice, section, or cut, as in common medical usage such as appendectomy, plus -*graphy* (Compact Edition of the Oxford English Dictionary, 1987), and appears in print as early as 1935 (Grossman 1935a,b).

Introduction

The information CT provides can drastically shorten the iterative cycles of prototype manufacture and testing required to bring new manufacturing processes under control. Evaluation of castings by radiography is very time consuming because of widely varying thicknesses of these components, whereas CT allows relatively inexpensive inspection; with accurate three-dimensional tomographic measurements, castings with critical flaws can be eliminated before subsequent costly manufacturing steps and those with anomalies such as voids that can be demonstrated to be noncritical can be retained and not scrapped. Integrating CT data of as-manufactured components into structural analysis programs seems very promising, particularly for anisotropic materials such as metal matrix composites (Bonse and Busch, 1996). Final assembly verification, for example, in small jet engines, is a third area where CT appears to be cost effective.

The extreme sensitivity of CT to density charges can be exploited to follow damage propagation in polymeric matrix composites (Bathias and Cagnasso, 1992), even when the microcracks produced cannot be resolved by the most sensitive x-ray imaging techniques. X-ray CT can be performed with portable units and offers considerable promise for studying the processes active in growing trees and for milling lumber: the environmental effects on a forest of a nearby chemical or power plant can be assessed over a number years on the same set of trees, daily and seasonal changes in the cross-sectional distribution of water can be obtained, and luck can be removed from the process of obtaining large wooden panels with beautiful ring patterns and without knots or decay (Onoe et al., 1984; Habermehl and Ridder, 1997). Pyrometric cones used for furnace temperature calibration are produced in the millions annually and in at least 100 compositions, and CT has been applied to understand why certain powder compositions for these dry-pressed, self-supporting cones produce large density gradients in dies and rejection rates (due to fracture) several times higher than most other compositions (Phillips and Lannuti, 1993). Of interest also is comparative work using magnetic resonance imaging and x-ray CT to study ceramics (Ellingson et al., 1989).

As with any other imaging modality, new applications required resolution of even smaller features, and this became the goal of one branch of workers in CT. If instead of resolving features with dimensions barely smaller than millimeters, as is typical of industrial or medical computed tomography equipment, one were able to image features on the scale of one to ten micrometers, then many microstructural features in engineering materials could be studied nondestructively. This size scale is also important in biological structural materials such as calcified tissue. Areas in which microCT has been employed profitably include damage accumulation in composites, fatigue crack closure in metals, and densification of ceramics.

One can view the advances in microCT imaging since the 1980s as a by-product of the demand for improved area detectors for consumer electronics. Before considering the mathematics and physics of computed tomography and the hardware requirements for microCT, it is constructive to review the early chronology of microCT (or at least the author's perspective of how this developed). The first realization of microCT seems to have been in 1982 (Elliott and Dover, 1982); until recently, this group has used a microfocus x-ray source and a pinhole collimator to collect high resolution data.* In work published in 1983, Grodzins (1983a,b) suggested how using the tunability of synchrotron x-radiation would allow one to obtain enhanced contrast from a particular element within a sample imaged with CT, viz. by comparing a reconstruction from data collected at a wavelength below that of the absorption edge of the element in question with a reconstruction from a wavelength above the edge.

In 1984, Thompson et al. (1984) published low-resolution CT results using synchrotron radiation and the approach advocated by Grodzins. Within a few years, multiple groups had demonstrated microCT using synchrotron radiation (Bonse et al., 1986; Flannery et al., 1987; Flannery and Roberge, 1987; Hirano et al., 1987; Spanne and Rivers, 1987; Ueda et al., 1987; Kinney et al., 1988; Sakamoto et al., 1988; Suzuki et al., 1988; Engelke et al., 1989a,b), and others applied x-ray tube-based microCT (Burstein et al., 1984; Feldkamp et al., 1984; Seguin et al., 1985; Feldkamp and Jesion, 1986; Feldkamp et al., 1988). It is important to emphasize the shift from collecting single slices to collecting volumetric (i.e., simultaneously collected multiple adjoining slices) data.

Dedicated microCT instruments at third-generation synchrotron x-radiation sources (e.g., APS, ESRF, SPring-8) and at other storage rings have multiplied opportunities for 3D imaging at highest spatial resolution and contrast sensitivity, but daily access is not an option. Multiple manufacturers now offer affordable, turnkey microCT systems for routine, day-to-day laboratory characterization à la SEM (scanning electron microscopy). Recently, commercial nanoCT systems (spatial resolutions substantially below one micrometer) and in vivo microCT systems (for small animal models of human diseases) began to appear in research laboratories.

Many microCT papers are being added to the literature each year. Quantifying the rate of increase of publication of microCT papers is problematic because of artificial issues such as the division between microCT and conventional tomography (here the author arbitrarily takes the definition that microCT describes tomographic imaging with ~50-μm voxels, i.e., volume elements). Nonetheless, a feel for the increase can be gained by considering the results of use of two different search engines for the scientific literature: Compendex, which covers engineering subjects, and

---

* Sato et al. (1981) presented reconstructions of an optical fiber, claiming 20-μm spatial resolution, but, due to the noise in the image and the unfortunate sample geometry, it appears to the author that the slices are dominated by reconstruction artifacts.

Introduction

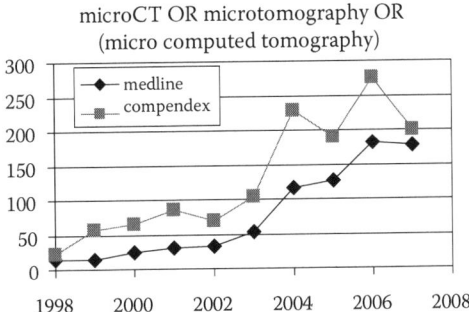

FIGURE 1.1
Annual number of citations for "microCT" OR "microtomography" for Compendex and for Medline for the years 1998–2007.

Medline, which covers biological and medical research areas. Figure 1.1 shows the annual number of microCT papers found by each index over the years 1998 through 2007. The number of papers has increased eightfold or more over the decade. Most surprising to the author was the very small fraction of papers appearing in both indices, less than 15 percent. The author expects the total for 2008 to surpass 400 papers, a daunting number of abstracts to scan, let alone papers to read.

Figure 1.2 shows the distribution of subjects of microCT papers found in Compendex and Medline for 2007. Setting aside the 111 papers on nonbiological subjects, the other 239 papers fall into a number of topical areas. There are, for example, almost as many papers on bone (99) as there are in the nonbiological literature. There are nearly half as many microCT papers on tissue engineering scaffolds as there are in the nonbiology subject areas.

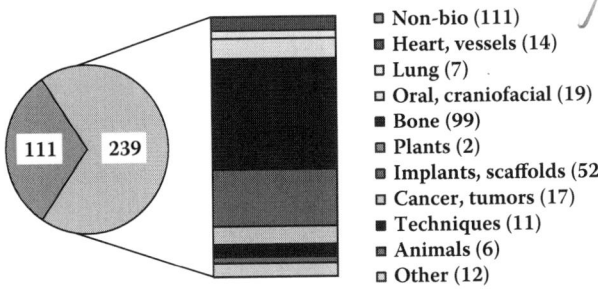

FIGURE 1.2
Distribution of citations for 2007 by subject areas identified by the author. The citations are from Compendex and Medline.

At the time of writing, there were on the order of 500 microCT systems operating worldwide, the majority of these commercial tube-based systems. An organized description of the fundamentals and applications of x-ray microCT should be helpful, therefore, to a number of individuals. Many aspects are common to both commercial laboratory and synchrotron research systems, and comparing and contrasting the two realms is rarely assayed, perhaps because only a few individuals work to a significant degree in both realms. Such a synthesis is particularly valuable (and efficient) in light of the recent explosion of publications based on microCT, literature that is not covered by any single electronic database. The common thread through the book is the presentation of microCT as a single modality, each study discussed employing one set of capabilities or experimental designs from a continuum of possibilities.

The chapters that follow fall into two categories: methodology, which is covered in Chapters 2–6, and applications, which constitute the bulk of the book (Chapters 7–12). Chapter 2 briefly reviews fundamentals of x-radiation and imaging, and Chapter 3 discusses reconstruction from projections. Experimental methods used to perform microCT are the subject of Chapter 4, and Chapter 5 covers microCT in practice. Data analysis and visualization are the subjects of Chapter 6. Chapter 7 introduces simple microstructure quantification and metrology. More complex analyses required for quantifying microstructure in cellular solids and in network specimens are discussed in Chapters 8 and 9, respectively. Microstructural evolution is an important area of microCT research, described in Chapter 10. The subject of Chapter 11 is mechanically induced damage and deformation and of Chapter 12 is multimode studies, where microCT is integrated with another characterization technique such as x-ray diffraction.

References

Bathias, C. and A. Cagnasso (1992). Application of x-ray tomography to the nondestructive testing of high-performance polymer composites. *Damage Detection in Composite Materials*. J. E. Master. West Conshocken, PA, ASTM. ASTM STP **1128**: 35–54.

Bonse, U. and F. Busch (1996). X-ray computed microtomography (µCT) using synchrotron radiation. *Prog Biophys Molec Biol* **65**: 133–169.

Bonse, U., Q. Johnson, M. Nichols, R. Nusshardt, S. Krasnicki, and J. Kinney (1986). High resolution tomography with chemical specificity. *Nucl Instrum Meth* **A246**: 644–648.

Bossi, R.H. and B.W. Knutson (1994). The advanced development of X-ray computed tomography applications. *United States Air Force Wright Laboratory Publication* WL-TR-93-4016.

Burstein, P., P.J. Bjorkholm, R.C. Chase, and F.H. Seguin (1984). The largest and smallest x-ray computed tomography systems. *Nucl Instrum Meth* **221**: 207–212.

Compact Edition of the Oxford English Dictionary (1987). Oxford: Clarendon Press.

Copely, D.C., J.W. Eberhard, and G.A. Mohr (1994). Computed tomography Part I: Introduction and industrial applications. *J Metals* 14–26.

Cormack, A.M. (1963). Representation of a function by its line integrals, with some radiological applications. *J Appl Phys* **34**: 2722–2727.

Ellingson, W.A., P.E. Engel, T.I. Hertea, K. Goplan, P.S. Wang, S.L. Dieckman, and N. Gopalsani (1989). Characterization of ceramics by NMR and x-ray CT. *Industrial Computed Tomography*. Columbus (OH), ASNT: 10–14.

Elliott, J.C. and S.D. Dover (1982). X-ray microtomography. *J Microsc* **126**: 211–213.

Engelke, K., M. Lohmann, W.R. Dix, and W. Graeff (1989a). Quantitative microtomography. *Rev Sci Instrum* **60**: 2486–2489.

Engelke, K., M. Lohmann, W.R. Dix, and W. Graeff (1989b). A system for dual energy microtomography of bones. *Nucl Instrum Meth* **A274**: 380–389.

Feldkamp, L.A. and G. Jesion (1986). 3-D x-ray computed tomography. *Rev Prog Quant NDE* **5A**: 555–566.

Feldkamp, L.A., L.C. Davis, and J.W. Kress (1984). Practical cone-beam algorithm. *J Opt Soc Am* **A1**: 612–619.

Feldkamp, L.A., D.J. Kubinski, and G. Jesion (1988). Application of high magnification to 3D x-ray computed tomography. *Rev Prog Quant NDE* **7A**: 381–388.

Flannery, B.P. and W.G. Roberge (1987). Observational strategies for three-dimensional synchrotron microtomography. *J Appl Phys* **62**: 4668–4674.

Flannery, B.P., H.W. Deckman, W.G. Roberge, and K.L. D'Amico (1987). Three-dimensional x-ray microtomography. *Science* **237**: 1439–1444.

Grodzins, L. (1983a). Critical absorption tomography of small samples: Proposed applications of synchrotron radiation to computerized tomography II. *Nucl Instrum Meth* **206**: 547–552.

Grodzins, L. (1983b). Optimum energies for x-ray transmission tomography of small samples: Applications of synchrotron radiation to computerized tomography I. *Nucl Instrum Meth* **206**: 541–545.

Grossman, G., (1935a). Lung tomography. *Brit J Radiol* **8**: 733-751.

Grossman, G., (1935b). Tomographie 1; röntgenographische darstellung von körperschnitten (x-ray imaging of body sections). *Fortschr Geb Röntgenstr* **51**: 61-80.

Habermehl, A. and H. W. Ridder (1997). γ-Ray tomography in forest and tree sciences. *Developments in X-ray Tomography*. U. Bonse (Ed.). Bellingham, WA, SPIE. *SPIE Vol.* **3149**: 234–244.

Hildebrandt, G. (1992). Paul P. Ewald, the German period. *P.P. Ewald and His Dynamical Theory of X-ray Diffraction*. D.W.T. Cruickshank, H.J. Juretschke, and N. Kato (Eds.). Oxford, Int. Union Cryst.: 27–34.

Hirano, T., K. Usami, K. Sakamoto, and Y. Suziki (1987). High resolution tomography employing an x-ray sensing pickup tube, photon factory. Japanese National Laboratory for High Energy Physics, KEK 187.

Hounsfield, G.N. (1968–1972). A method of and apparatus for examination of a body by radiation such as X or gamma radiation. UK Patent 1, 283,915.

Hounsfield, G.N. (1971–1973). Method and apparatus for measuring x- and γ-radiation absorption or transmission at plural angles and analyzing the data. US Patent 3,778,614.

Hounsfield, G.N. (1973). Computerized transverse axial scanning (tomography): Part I description of system. *Brit J Radiol* **46**: 1016–1022.

Kinney, J.H., Q.C. Johnson, U. Bonse, M.C. Nichols, R.A. Saroyan, R. Nusshardt, R. Pahl, and J.M. Brase (1988). Three-dimensional x-ray computed tomography in materials science. *MRS Bull* (January): 13–17.

Onoe, M., J.W. Tsao, H. Yamada, H. Nakamura, J. Kogure, H. Kawamura, and M. Yoshimatsu (1984). Computed tomography for measuring the annual rings of a live tree. *Nucl Instrum Meth* **221**: 213–230.

Phillips, D.H. and J.J. Lannuti (1993). X-ray computed tomography for testing and evaluation of ceramic processes. *Am Ceram Soc Bull* **72**: 69–75.

Radon, J. (1917). Über die Bestimmung von Functionen durch ihre Integralwerte längs gewisser Mannigfaltigkeiten. *Berichte der Sächsischen Akademie der Wissenschaft* **69**: 262–277.

Radon, J. (1986). On the determination of functions from their integral values along certain manifolds (translated by P.C. Parks). *IEEE Trans Med Imaging* **MI-5 (4)**: 170–175.

Röntgen, W. (1898). Über eine neue Art von Strahlen (Concerning a new type of radiation). *Ann Phys Chem New Series* **64**: 1–37.

Sakamoto, K., Y. Suzuki, T. Hirano, and K. Usami (1988). Improvement of spatial resolution of monochromatic x-ray CT using synchrotron radiation. *J Appl Phys* **27**: 127–132.

Sato, T. et al. (1981). X-ray tomography for microstructural objects. *Appl Optics* **20**: 3880-3883.

Seguin, F.H., P. Burstein, P.J. Bjorkholm, F. Homburger, and R.A. Adams (1985). X-ray computed tomography with 50-μm resolution. *Appl Optics* **24**: 4117–4123.

Spanne, P. and M.L. Rivers (1987). Computerized microtomography using synchrotron radiation from the NSLS. *Nucl Instrum Meth* **B24/25**: 1063–1067.

Suzuki, Y., K. Usami, K. Sakamoto, H. Kozaka, T. Hirano, H. Shiono, and H. Kohno (1988). X-ray computerized tomography using monochromated synchrotron radiation. *Japan J Appl Phys* **27**: L461–L464.

Thompson, A.C., J. Llacer, L.C. Finman, E.B. Hughes, J.N. Otis, S. Wilson, and H.D. Zeman (1984). Computed tomography using synchrotron radiation. *Nucl Instrum Meth* **222**: 319–323.

Ueda, K., K. Umetani, R. Suzuki, and H. Yokouchi (1987). A high-speed subtraction angiography system for phantom and small animal studies. *Photon Factory*, Japanese National Laboratory for High Energy Physics, KEK: 186.

Webb, S. (1990). *From the Watching of Shadows: The Origins of Radiological Tomography*. Bristol, UK: Adam Hilger.

2

Fundamentals

A certain amount of background is required before one can appreciate the constraints on microCT performance. Limits exist to the rate at which data can be collected and the spatial resolution and contrast sensitivity that can be obtained for a given specimen. Understanding the fundamentals underlying experimental trade-offs is essential to using microCT effectively and efficiently.

2.1 X-Radiation

The details of x-ray generation and interaction with matter are covered very briefly in this section under the assumption that most readers will have encountered this subject before. Considerably more detail can be found in texts on x-ray diffraction analysis of materials, for example, Cullity and Stock (2001), or on nondestructive evaluation, for example, Halmshaw (1991).

2.1.1 Generation

X-rays are generated when charged particles are accelerated or when electrons change shells within an atom. Figure 2.1 shows a schematic of an x-ray tube, the source used in lab microCT systems. Electrons flow through a filament (generally W) at a potential kVp relative to the target (generally a metal such as Cu, Mo, Ag, or W). Electrons are emitted from the filament and accelerate toward the target under the effect of the potential. Upon striking the target, the electrons decelerate, producing (a) the Bremsstrahlung or continuous spectrum or (b) characteristic radiation (Figure 2.2). Characteristic radiation arises from electronic transitions generated by the incident electrons, and the high-intensity peaks have a very narrow energy range. As the potential across the tube increases, the intensities of both types of radiation increase and the peak of the continuous spectrum shifts to higher energies. Only a very small fraction of the energy of the electron beam is converted to x-radiation; most of the energy is released as heat. Further limiting the x-ray intensity that is available for

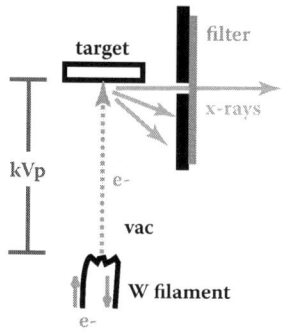

FIGURE 2.1
Schematic of an x-ray tube with electrons e- flowing through a W filament, thermionically emitted from the filament and accelerated within a vacumn (vac) by potential kVp. As drawn, x-rays pass through an aperture and a filter.

imaging is the fact that the x-rays are emitted in all directions (Figure 2.1) and most never encounter the specimen.

Electrons (or positrons) accelerating in storage rings such as synchrotrons are another source of x-radiation (Figure 2.3). If the electrons are moving at relativistic velocities and are deflected by a magnetic field, a continuous spectrum of electromagnetic radiation results, spanning from microwaves to very hard x-rays (Figure 2.4). Several factors give synchrotron radiation an advantage over tube sources for x-ray imaging. First, the intensity of x-rays delivered to a specimen is much greater, and synchrotron radiation can be tuned to a very narrow energy range of wavelength most advantageous for examining a given sample. The relativistic character of the synchrotron radiation emission process confines the resulting radiation to directions very close to the plane of the electron orbit, and the divergence of the beam is very small. Thus, synchrotron radiation not only possesses very high flux, but it also has much higher brightness and spectral brightness (intensity

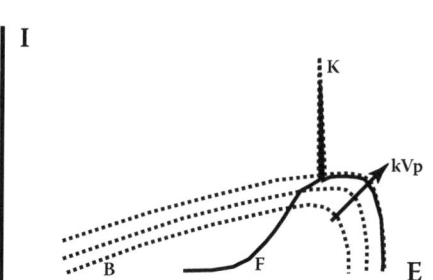

FIGURE 2.2
Schematic of x-ray tube spectra (intensity I as a function of photon energy E) consisting of continuous spectrum B and characteristic lines (K, only one shown). Each curve (dashed line) is the spectrum at a particular potential kVp, and the arrow labeled with kVp shows the shift of the continuous spectrum with increasing kVp. The solid gray curve labeled F shows the effect of the filter (eliminating the softer, lower-energy radiation) decreasing the energy range of the x-rays incident on the specimen. The presence of the filter is particularly important in x-ray imaging: the softer radiation would not pass through a specimen but would saturate the detector where the beam passes around the specimen and would increase scatter, degrading contrast. Note that the x-ray (photon) energy E is in keV, whereas the x-ray tube voltage kVp is in kV.

Fundamentals

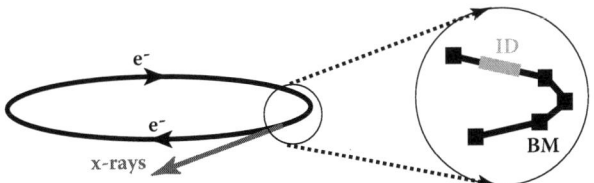

FIGURE 2.3
Schematic of a synchrotron storage ring showing electron bunches e⁻ circulating around the ring. The bunches travel at relativistic velocities (the electron energy is 7 GeV at the APS), and, where the bunches are deflected by bending magnets BM or insertion devices ID, x-rays are emitted. In a sense, the x-ray beam is a highly collimated searchlight blinking on and off.

per unit area of source and intensity per unit area per unit solid angle per unit energy bandwidth, respectively).

Several types of devices produce the intense magnetic fields required to produce synchrotron radiation (Figure 2.3). Bending magnets situated periodically around the storage ring deflect the electrons and force them to circulate within the ring (which is a polygon and not a circle). Insertion devices placed between the bending magnets and consisting of a number of closely spaced magnets are another way synchrotron radiation is delivered to experiments. Each bending magnet or insertion device line at a given synchrotron is optimized for certain operating characteristics, and it is beyond the scope of this section to discuss specifics of x-ray imaging stations at a particular ring. Storage rings for synchrotron radiation are typically very large facilities and are found around the world. The characteristics of each differ markedly and change from time to time, the interested reader is advised to do an Internet search for further details. Recent development of tabletop synchrotron radiation sources offers another option for x-ray brightness imaging (Hirai et al., 2006).

Figure 2.4 compares synchrotron radiation brilliance versus photon energy for different storage rings (bending magnet, insertion devices) and for x-ray tube sources. The APS (Advanced Photon Source) bending magnet (center right side of Figure 2.4), for example, produces at least three and one-half orders of magnitude higher brilliance than is obtained from an x-ray tube with a Mo target (lower right side of Figure 2.4), that is, at the energy of the Mo Kα characteristic line, 17.44 keV. At higher energies, the difference between x-ray tubes and the APS bending magnets is much greater. The highest brilliance at a storage ring, however, is not produced by bending magnets but, rather, by insertion devices. In general, synchrotron radiation with energies up to 25–30 keV can be obtained at many stations and many storage rings. Sources for energies above 30 keV are rarer, and those above 60 keV are rarer still.

Discussion of how this radiation is conditioned for use in imaging is postponed until later in this chapter. Approaches differ for x-ray tubes

12 *MicroComputed Tomography: Methodology and Applications*

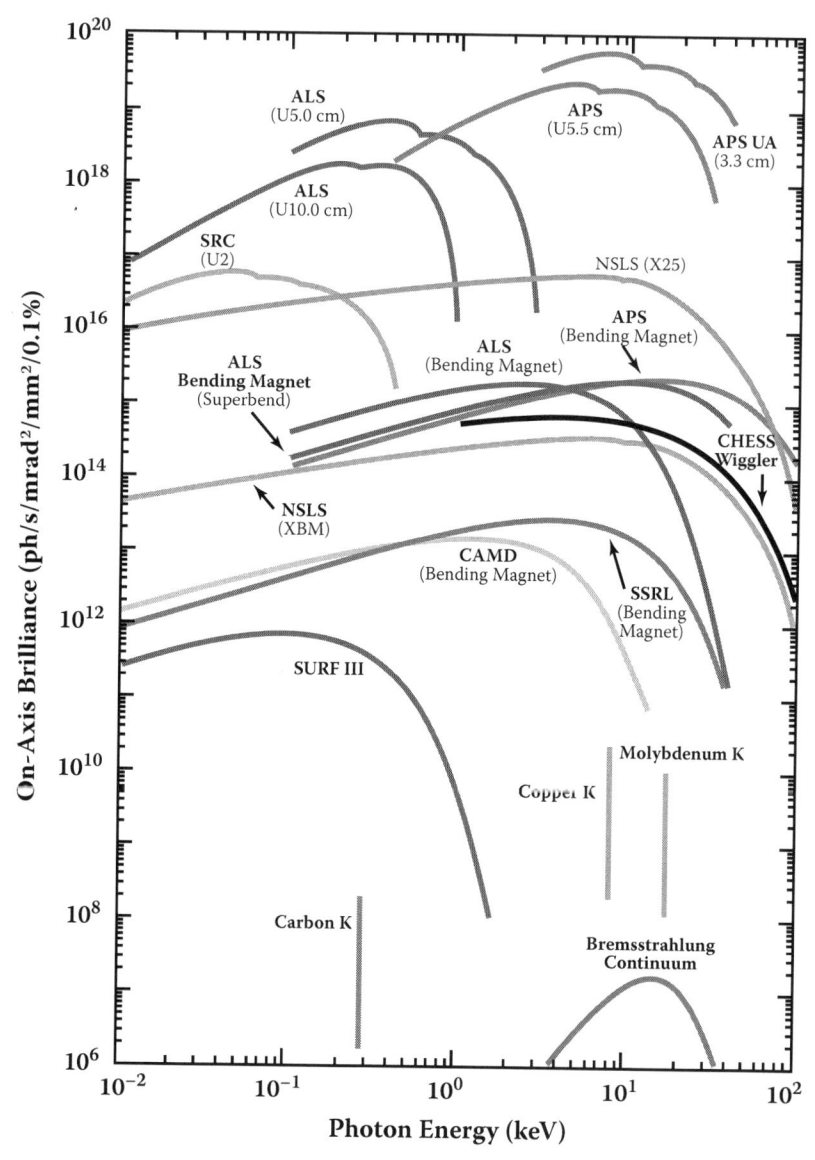

FIGURE 2.4 (SEE COLOR INSERT FOLLOWING PAGE 144.)
Relative intensities of x-ray tubes and of synchrotron radiation sources in the United States as a function of photon energy. The vertical axis plots brilliance, that is, intensity per unit time per unit area per unit solid angle per unit bandpass. (The figure was produced by Argonne National Laboratory, managed and operated by UChicago Argonne LLC for the U.S. Department of Energy under Contract No. DE-AC02-06CH11357, and is used with permission.)

Fundamentals

2.1.2 Interaction with Matter

As discovered by Röntgen (Röntgen, 1898), the attenuation of x-rays of wavelength λ is given for a homogeneous object by the familiar equation:

$$I = I_o \exp(-\mu x), \tag{2.1}$$

where I_o is the intensity of the unattenuated x-ray beam, and I is the beam's intensity after it traverses a thickness of material x characterized by a linear attenuation coefficient μ (Cullity and Stock, 2001). Typically, one finds μ given in cm^{-1}. Rewriting Equation (2.1) in terms of the mass attenuation coefficient μ/ρ (units cm^2/g) and the density ρ (units g/cm^3) explicitly recognizes that the fundamental basis of the amount of attenuation is the number of atoms encountered by the x-ray beam:

$$I = I_o \exp[(-\mu/\rho) \rho x]. \tag{2.2}$$

Mass attenuation coefficients are a materials property and are a strong function of the atomic number of the absorber Z as well as the x-ray wavelength λ (the inverse of energy). Over much of the energy range used most frequently for microCT and except at absorption edges, mass attenuation coefficients can be described by the relationship $\mu/\rho \sim Z^m \lambda^n$, where m equals three or four and n equals (approximately) three. Figure 2.5 is a log–log plot of μ/ρ as a function of x-ray photon energy for several materials of interest (two elemental solids and two multielement solids). In Figure 2.5, two processes produce the curves shown. The curve's linear sections are from photoelectric absorption, and the flattening of the plots at higher energies results from a change in the dominant absorption mechanism to Compton scattering. More details on these absorption mechanisms appear in texts on nondestructive evaluation, for example, Halmshaw (1991). In Figure 2.5, the absorption edges for bone (calcium) and titanium are indicated by arrows, and the dashed rectangle roughly indicates the energies most frequently used in microCT and, on the other hand, the range of mass attenuation coefficients allowing practical imaging.

Two of the substances plotted in Figure 2.5 (cortical bone and polyethylene) consist of more than one element. One is often not fortunate enough to find values for a specific mixture already tabulated. Values of the linear attenuation coefficient of any mixture or compound can be calculated for a particular energy from first principles using mass attenuation coefficients for the elements and densities for the phases present or the density of the compound in question. Specifically, the linear attenuation coefficient of the mixture or compound equals

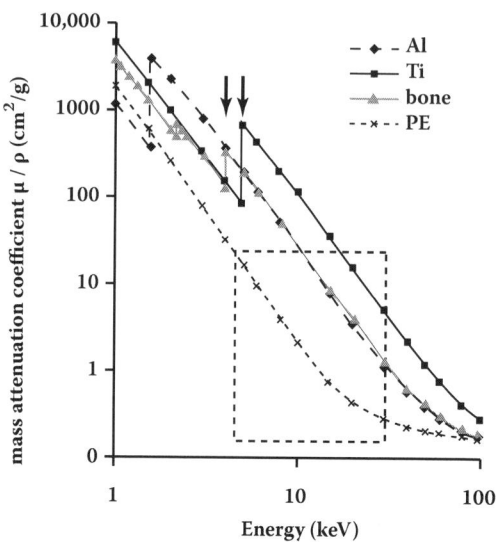

FIGURE 2.5
Mass attenuation coefficient μ/ρ as a function of photon energy for four materials: Al, Ti, cortical bone, and polyethylene (PE). The arrows identify absorption edges for Ti and Ca, the principle absorber in bone. The rectangle bounded by the dashed line very roughly indicates the ranges of the two variables most often encountered in microCT. (Plotted from NIST tabulations; Hubbel and Seltzer (2006).)

$$<\mu> = \Sigma \, (\mu/\rho)_i \, \rho_i, \qquad (2.3)$$

that is, the weight-fraction-weighted average.

X-rays scatter from the atoms in objects and from very small objects (submicrometer dimensions). Incoherent scattering is generally folded into the linear attenuation coefficients described above. Reinforced scattering can lead to peaks of intensity in certain directions: small angle scattering from various sources including identically sized, but nonperiodically dispersed, submicrometer objects and diffraction from periodic arrays of atoms in crystalline solids. This latter phenomenon is utilized to produce monochromatic radiation for x-ray imaging and improving contrast sensitivity, and is taken up in Section 2.3.

In strict terms, Equations (2.1) and (2.2) are incomplete descriptions of x-ray attenuation, but they suffice for most applications. These equations do not consider that x-rays are ever so slightly refracted when passing through solids (indices of refraction differ from one by a few parts per million), enough so that the x-ray wavefronts distort when passing through regions of different electron density (see Fitzgerald (2000) for an introduction). One situation where a more complete description is needed is imaging with a beam possessing significant spatial coherence. Such

Fundamentals

coherence can be achieved using (a) an x-ray tube with a very small focal spot or with slits providing a small virtual source size or (b) synchrotron radiation from some (but not all) storage rings. Aside from a brief mention of focusing optics in Section 2.3, further discussion is postponed until Chapter 4.

2.2 Imaging

Various features' visibilities within an object depend on the spatial resolution with which they can be imaged and on the contrast the features have relative to their surroundings. The interplay of contrast sensitivity and spatial resolution defines what can be achieved with CT.

Contrast is a measure of how well a feature can be distinguished from the neighboring background. Frosty the Snowman's eyes of coal show high contrast, whereas writing with a yellow highlighter marker on a white sheet of paper provides little contrast. Figure 2.6 uses a variable background to illustrate this effect on feature visibility. The more closely spaced pairs of black disks gradually disappear as the background becomes darker.

It is important to be able to quantify the amount of contrast present in an object's image because the smallest change in contrast that can

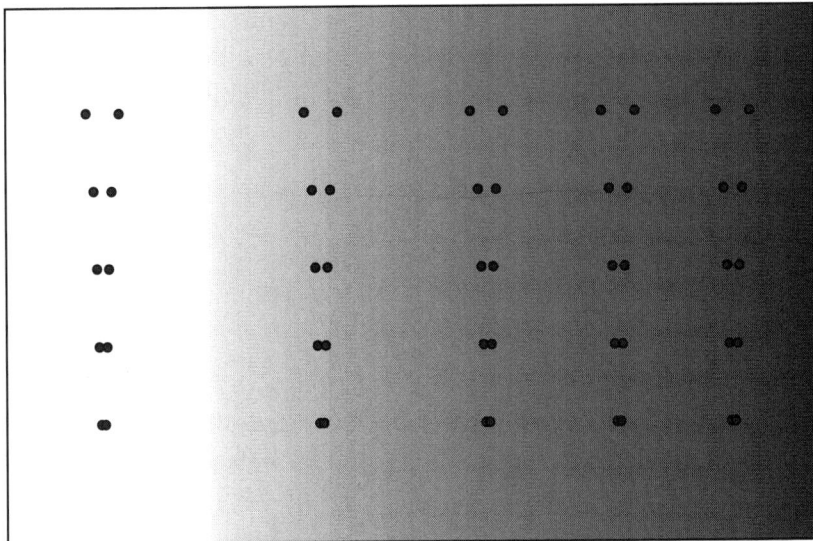

FIGURE 2.6
Influence of contrast on resolution of closely spaced features. The 1D gradient of background gray levels makes it more difficult to resolve the pairs of disks the farther to the right one goes.

be reliably discerned by the imaging system dictates quantities such as detection limits. Contrast is often defined in terms of the ratio of the difference in signal between feature and background to the signal from the background. Thus, the fractional contrast is given by

$$\text{contrast} = (|sig_f - sig_b|) / sig_b, \qquad (2.4)$$

where *sig* is the signal observed from the object and whose point-by-point variation makes up the image and the subscripts *f* and *b* denote feature and background, respectively.

Spatial resolution describes how well small details can be imaged or small features can be located with respect to some reference point. Figure 2.7 shows two pairs of black disks with different spacing. When imaged with a pixel (picture element) size of 8 units, there is one open pixel between the pixels containing the disks of the more widely spaced features (Figure 2.7a), and it is clear that this pair of features can be resolved. The more closely spaced pair of disks, vertically oriented in Figure 2.7a, occupies adjacent pixels and cannot be resolved. The situation changes, however, when the specimen is imaged with a pixel size of four units (Figure 2.7b): both disk pairs have at least one open pixel of separation, and the individual disks are resolved based on this simplistic criterion. As implicitly noted above in Figure 2.6, a pair of high-contrast features can be differentiated at a smaller separation than the same-sized low-contrast features. One generally quantifies spatial resolution in terms of the smallest separation at which two points can be perceived as discrete entities.

The presence of noise and inherent imaging imperfection means that quantities such as apparatus performance and image fidelity must be measured in probabilistic terms for a given set of imaging conditions. The point spread function (PSF), for example, describes how the system responds to a point input (i.e., how it images a point), and the modulation transfer function (MTF) represents the interaction of the system (PSF) with multiple features of the object being imaged (i.e., the convolution of all these factors; see the following paragraph). The number of line pairs per millimeter that can be resolved is an often-used simplification. A more accurate approach, reflecting the fact that features must be both detected and resolved, is plotting the contrast required for 50 percent discrimination of pairs of features as a function of their diameters in pixels; this is

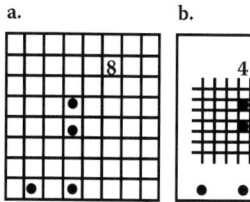

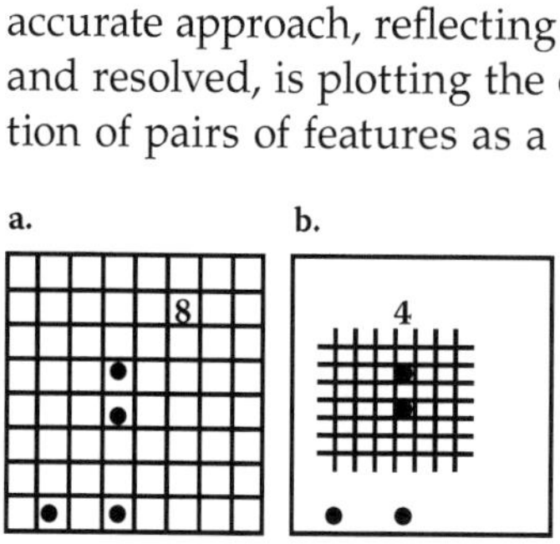

FIGURE 2.7
Two pairs of disks with different spacings. (a) One pair of disks is resolved with pixels of size 8 units. (b) Both pairs of disks resolved when the pixel size is 4 units.

Fundamentals

termed the *contrast–detail–dose curve* (1996). The reader is directed to texts on microscopy for further details.

Convolution is the mathematical operation of smearing one feature over another feature and is written in one dimension as

$$g(x) = f(x) * h(x) \tag{2.5}$$

$$= \int f(x) \, h(x-u) \, du \tag{2.6}$$

and in two dimensions as

$$g(x,y) = f(x,y) * h(x,y) \tag{2.7}$$

$$= \iint f(x,y) \, h(x-u, y-v) \, du \, dv, \tag{2.8}$$

where the limits of integration are ±∞. As mentioned above, the convolution operation is widely used when considering the PSF and MTF of imaging systems including microCT; it is also the basis of the most widely used reconstruction method, the filtered back-projection algorithm, described in Section 3.3.

In closing this section, two geometric effects should be mentioned that pertain to tube-based x-ray imaging. As mentioned in Section 2.1, x-radiation from a tube diverges from the area on the target where the electron beam is incident. If this, the x-ray source size, is small enough, then geometric magnification can be used to see smaller features than would otherwise be the case (Figure 2.8); the spreading beam can match the feature size to the detector resolution. Source sizes in x-ray tubes used for high-resolution imaging are generally quite small, a few micrometers or larger in diameter, and this, through the effect of penumbral blurring (Figure 2.8), affects the resolving power of a tube-based microCT system.

2.3 X-Ray Contrast and Imaging

Two factors dictate the optimum sample thickness for x-ray imaging (i.e., for greatest contrast). If the specimen is too thick, no x-rays pass through it and no contrast can be seen. If the specimen is very, very thin, no measurable contrast is produced (more precisely, the intensity transmitted through the specimen cannot be distinguished from that passing to either side of the object). The quantity defining x-ray attenuation is the product μx (see Equation (2.1)), and optimum imaging in microCT is found when

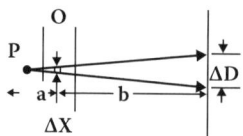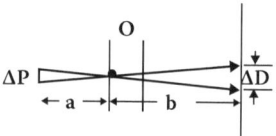

FIGURE 2.8
Illustration of geometrical magnification (left) and penumbral blurring (right). In geometrical magnification, a divergent beam from point source P spreads sample feature at O with width Δx to width ΔD on the detector or film plane. The amount of magnification depends on the ratio of the object to detector separation b to the source to object separation a. Penumbral blurring occurs when there is a finite source size ΔP at a distance a from a pointlike feature O in the sample. The crossfire from the source spreads the contrast from O over ΔD on the detector or film b from the object. In both cases trigonometry is all that is required to calculate the amount of magnification or the level of blurring. (Reproduced from Stock (1999).)

$\mu x < 2$ for the longest path length through the sample (Grodzins, 1983), that is, greater than 13–14 percent transmission through the specimen.

Implicit in the previous paragraph and in the application of Equations (2.1) and (2.2) is that the radiation is monochromatic; that is, the x-ray photons have a single energy. X-rays both from tubes and from synchrotrons are polychromatic before any treatment of the beam is performed. As indicated above in Section 2.1.2, attenuation coefficients are a strong function of x-ray wavelength (energy), and the presence of more than one wavelength complicates analysis of x-ray images. Use of a thin filter between the target and specimen in tube-based imaging (Figure 2.1) hardens the beam considerably (Figure 2.2), that is, it preferentially removes the softer radiation and substantially decreases the range of x-ray energies present. Effects such as beam hardening (Section 5.1.6) persist in CT and microCT and can lead the unwary astray. As discussed below, monochromators (crystal or multilayer) can also be used, but this cuts beam intensities too much for practical microCT with x-ray tubes.

Scattering from periodic arrays of atoms or molecules reinforces for certain combinations of x-ray wavelength and angle of incidence; this relationship is described by Bragg's law:

$$\lambda = 2d \sin \theta, \tag{2.9}$$

where d is the period of scatterers and θ is the angle of incidence (Figure 2.9a). The reinforced scattering is termed diffraction and is a basic tool of materials characterization covered in undergraduate texts, for example, Cullity and Stock (2001). X-rays with wavelength λ but incident at an angle θ' not satisfying Bragg's law will not reinforce; x-rays with differing wavelength λ' but incident at the same angle θ will also not reinforce. Atoms within a large single crystal are the basis of a crystal monochromator, and d is the

Fundamentals

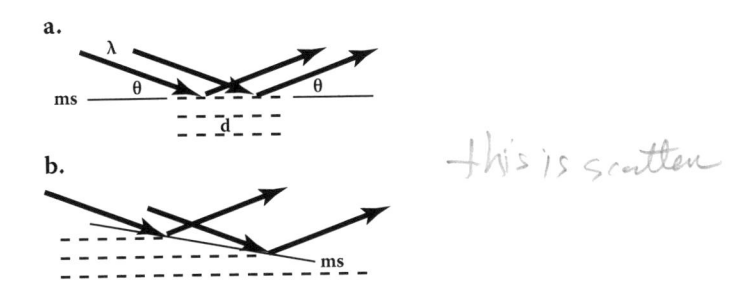

$\lambda = 2 \cdot d \cdot \sin\theta$

FIGURE 2.9
Monochromators. (a) X-rays with wavelength λ incident at angle θ to the direction of periodicity d produce reinforced scattering (diffraction) along the direction shown. Here the periodicity is normal to the monochromator surface ms. (b) When the monochromator surface is inclined relative to the period of the scatterers, the beam is spread spatially. The orientation and spacing of the scatterers and the width and orientation of the incident beam are the same as in (a), but the orientation of ms has changed. The brightness (intensity per unit area) of the spread beam is decreased as a result.

this is scatter

Bragg plane spacing. Multilayer monochromators are another option and consist of alternating low Z and high Z layers whose period d is produced by a technique such as vapor deposition. Because of the high brilliance of synchrotron x-radiation, both types of monochromators are very effective for conditioning beams used in synchrotron microCT. When extremely rapid data acquisition is needed, synchrotron microCT is performed with polychromatic radiation, that is, with a "pink" beam.

Until the last couple of decades, it was a truism that x-ray focusing was largely ineffective; that is, x-ray optics except of the crudest sort were impractical. One exception was the use of asymmetrically cut crystal monochromators as beam spreaders (Figure 2.9b). If such a beam spreader were placed in a synchrotron (parallel) x-ray beam that had passed through a specimen, then the image could be enlarged before the detector was reached. Development of Fresnel optics (zone plates) and of refractive optics has proceeded to the point where they can be used for x-rays hard enough for small specimens of some of the materials discussed in Chapters 7 through 12. X-ray optics, although of great interest, are only a small aspect of micro- and nanoCT and are not examined in depth.

The previous paragraph considered ways of extending spatial resolution in x-ray imaging, and some options exist for extending the range of contrast that can be discriminated. As mentioned above, imaging with monochromatic radiation is preferable to imaging with polychromatic radiation if high-contrast sensitivity is required: values of the linear attenuation coefficient returned by the reconstruction algorithm are not convoluted with the spread of wavelengths in the monochromatic case. Assuming reasonable mechanical and source stability, increased counting can improve the signal-to-noise ratio in an image, and this is discussed in Chapter 4. Imaging the same specimen at energies above and below the absorption edge of an element within the specimen (and numerically comparing the images) is particularly useful for enhancing detection limits for that particular element when it is in low concentration or when other phases are present that have similar linear attenuation coefficients and lack the element in question. As seen in Figure 2.5, the difference in μ/ρ can be greater than five across the absorption edge.

References

(1996). E 1441 — 95 Standard guide for computed tomography (CT) imaging. *1996 Annual Book of ASTM Standards*. Philadelphia: ASTM. 03.03: 704–733.

Cullity, B.D. and S.R. Stock (2001). *Elements of X-ray Diffraction*. Upper Saddle River, NJ: Prentice-Hall.

Fitzgerald, R. (2000). Phase sensitive x-ray imaging. *Phys Today* **53**: 23–26.

Grodzins, L. (1983). Optimum energies for x-ray transmission tomography of small samples: Applications of synchrotron radiation to computerized tomography. I. *Nucl Instrum Meth* **206**: 541–545.

Halmshaw, R. (1991). *Non-Destructive Testing*. London: Edward Arnold.

Hirai, T., H. Yamada, M. Sasaki, D. Hasegawa, M. Morita, Y. Oda, J. Takaku, T. Hanashima, N. Nitta, M. Takahashi, and K. Murata (2006). Refraction contrast 11x-magnified x-ray imaging of large objects by MIRRORCLE-type table-top synchrotron. *J Synchrotron Rad* **13**: 397–402.

Hubbel, J.H. and S.M. Seltzer (2006). Tables of x ray mass attenuation coefficients and mass energy-absorption coefficients. *NISTIR* 5632. Retrieved Jan 22, 2007, from http://physics.nist.gov/PhysRefData?XrayMassCoef/cover.html.

Röntgen, W. (1898). Über eine neue Art von Strahlen (Concerning a new type of radiation). *Ann Phys Chem* New Series **64**: 1–37.

Stock, S.R. (1999). Microtomography of materials. *Int Mater Rev* **44**: 141–164.

3

Reconstruction from Projections

Understanding the principles of tomographic reconstruction is essential to understanding what CT and microCT can and cannot do and what causes certain artifacts. The treatment here is limited to absorption tomography; reconstruction with x-ray phase contrast is covered separately in Section 4.8. The basic concepts of reconstruction from x-ray projections are the subject of the first section of this chapter. The second section covers the algebraic reconstruction method that uses an iterative approach to reconstruction, which can be understood without resorting to mathematics but which is used infrequently. The third section examines the convolution back-projection method; this discussion focuses on physical explanations with only a small amount of mathematics. Section 3.4 introduces Fourier-based reconstruction and requires some application of mathematics; mathematical depth beyond that presented can be found elsewhere (Newton and Potts, 1981; Kak and Slaney, 2001; Natterer, 2001). Section 3.5 discusses performance limits for tomographic reconstruction, and the last section introduces alternative methods for 3D mapping of structure.

$$d \ln I = \frac{1}{I} \cdot \partial I = -(\mu/\rho)\rho\, dx$$

3.1 Basic Concepts

Equations (2.1) and (2.2) reveal what is observed after attenuation is complete, and writing the differential form of these equations focuses attention on what occurs within each small thickness element dx:

$$dI / I = -(\mu/\rho)\,\rho\, dx. \tag{3.1}$$

The size of dx into which the path can be divided varies from instrument to instrument and sample to sample, but, on the scale of the minimum physically realistic thickness element dx, $(\mu/\rho)\rho$ is regarded as a constant and is written simply as μ. Figure 3.1 illustrates how each voxel (volume element) with attenuation coefficient μ_i along path s contributes to the total absorption. Adding the increments of the attenuation along the direction of x-ray propagation yields the more general form

$$I = I_o \exp[-\int \mu(s)\, ds\,], \tag{3.2}$$

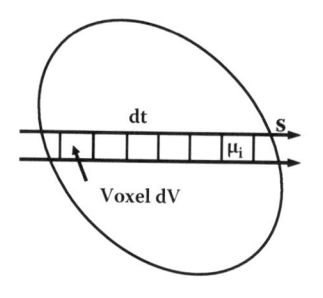

FIGURE 3.1
Contribution of each voxel dV (dimension $dt \times dt \times dt$) with linear attenuation coefficient μ_i to the total x-ray absorption along rays.

where $\mu(s)$ is the linear absorption coefficient at position s along ray **s**. Assigning the correct value of μ to each position along this ray (and along all the other rays traversing the sample), knowing only the values of the line integral for the various orientations of **s**, that is,

$$\int \mu(s)\, d\mathbf{s} = \ln(I_o/I) = p_s(s), \quad (3.3)$$

is the central problem of computed tomography.

Locating and defining the different contributions to attenuation requires measuring I/I_o for many different ray directions **s**. Measuring I/I_o for many different positions for a given **s** is also required: a radiograph measures exactly this quantity, the variation of I/I_o as a function of position for a given projection or ray direction. Thus, a set of high-resolution radiographs collected at enough well-chosen directions **s** can be used to reconstruct the volume through which the x-rays traverse.

The reality of being able to reconstruct volumes can be illustrated simply by considering how the profile or projection of x-ray attenuation $P(\mathbf{s})$ from a simple object changes with viewing direction.* The low-absorption rectangle within the slice pictured in Figure 3.2 casts a spatially narrow but deep "shadow" in the attenuation profile seen along one viewing direction and a spatially wide and shallow "shadow" along the second viewing direction. For the views parallel to the sides of the rectangle, the corresponding changes in the profile are quite sharp, but for views oriented between the two pictured in Figure 3.2, the changes in the profile would be more complex. Note that paths through the cylinder are relatively short near its edges and are much larger at or near its center; the result is the curved profile outside the "shadows" which, for simplicity, is not reproduced within the shadowed area. For this simple case, the two correctly chosen views suffice to define the location of the rectangle and the change of μ between the cylinder and rectangle.

In general, the internal structure of the specimen is much more complex than that shown in Figure 3.2, and a more general reconstruction approach will be required. At one extreme, the images of an array of many small objects embedded in a matrix overlap to such an extent that they cannot be distinguished from view to view. At the other extreme, the linear attenuation coefficient can vary continuously across the specimen, with no sharp internal interfaces present. Sections 3.2–3.4 describe

---

* Note that $P(\mathbf{s})$ consists of the set of individual projection rays $p_s(s)$, that is, the values of attenuation at each position along the profile.

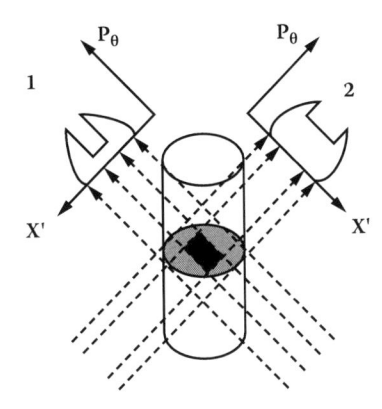

Reconstruction from Projections

FIGURE 3.2
Illustration of how internal structure of objects can be determined from projections. For simplicity, only a single plane and parallel x-radiation are pictured, and the rotation axis for collecting views (absorption profiles P_θ) along different directions θ is vertical and in the center of the cylindrical sample. (Stock, 1990; © ASME.)

approaches for reconstructions with parallel beam. As is described in Chapter 4, most tube-based microCT systems collect data in a fan-beam or cone beam geometry; reconstruction with these geometries is briefly covered in Chapter 4.

3.2 Algebraic Reconstruction

The algebraic reconstruction technique (sometimes abbreviated ART) is an iterative approach to reconstruction, that is, a mathematical trial-and-error approach. The algebraic method is best illustrated with an example of a 2 × 2 voxel object and projections along six rays and four directions (Figure 3.3). The numbers within each voxel in Figure 3.3a are summed along the four directions; these sums are the projections. For example, in Figure 3.3a, the top row sums to 14, and the upper left to lower right diagonal sums to 9. For the first guess (not required to be particularly good), choose constant values in each column such that each column yields the correct sum (Figure 3.3b). The arrows to the right of the object show the resulting (incorrect) sum for each row. The top row of Figure 3.3b is too low by two units, and the bottom row is too high by two units (left side of Figure 3.3c). Figure 3.3c shows the correction of the rows: in the top row 1 is added to the value of each voxel (i.e., one-half of the total difference between measured value and the value calculated for the previous iteration), and so on. At this point, the sums of the columns and of the rows are correct, but the values of the diagonals are incorrect (12 instead of 15 from upper right to lower left, 12 instead of 9 from upper left to lower right). Figure 3.3d shows correction along one diagonal: the value of each voxel is increased by one-half of the difference between the measured value and the value in the previous iteration. Figure 3.3e shows the correction of the second diagonal, and the four voxels' values exactly match those in the object.

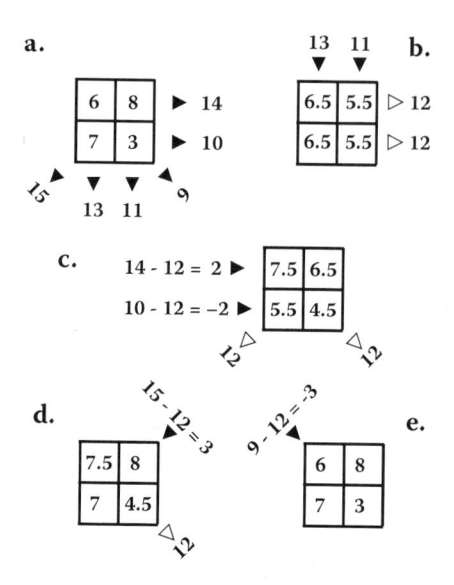

FIGURE 3.3
Algebraic reconstruction: (a) 2 × 2 voxel object. The sum of the voxel values is indicated along the directions shown. These values are the projections from which the unknown voxel values are determined in (b)–(e). See the text for the details.

The example of Figure 3.3 is quite simplified but should give an idea of the approach. One area where an iterative reconstruction technique can be quite useful is when projections are missing (e.g., a certain range of angles is missing, perhaps because they are being blocked by an opaque object outside the field of view of interest or the aspect ratio of a platelike object prevents some radiographs from being collected). Because the algebraic method is rarely used at present, no further details are given, and the discussion proceeds to the back-projection method.

3.3 Back-Projection

The principles behind reconstruction via the back-projection method are illustrated in Figures 3.4 and 3.5. Figure 3.4 shows an array of circular features, one of which is enlarged at the lower left. These features have a lower density at their centers than at their outer borders; the monotonic density variation was chosen for the convenience of producing a uniform, constant absorption projection along any viewing direction. Views along directions *i*, *ii*, and *iii* produce the profiles shown. In view *i*, for example, the projection consists of peaks, from left to right, of 1, 3, 3, 4, and 2 units

Reconstruction from Projections 25

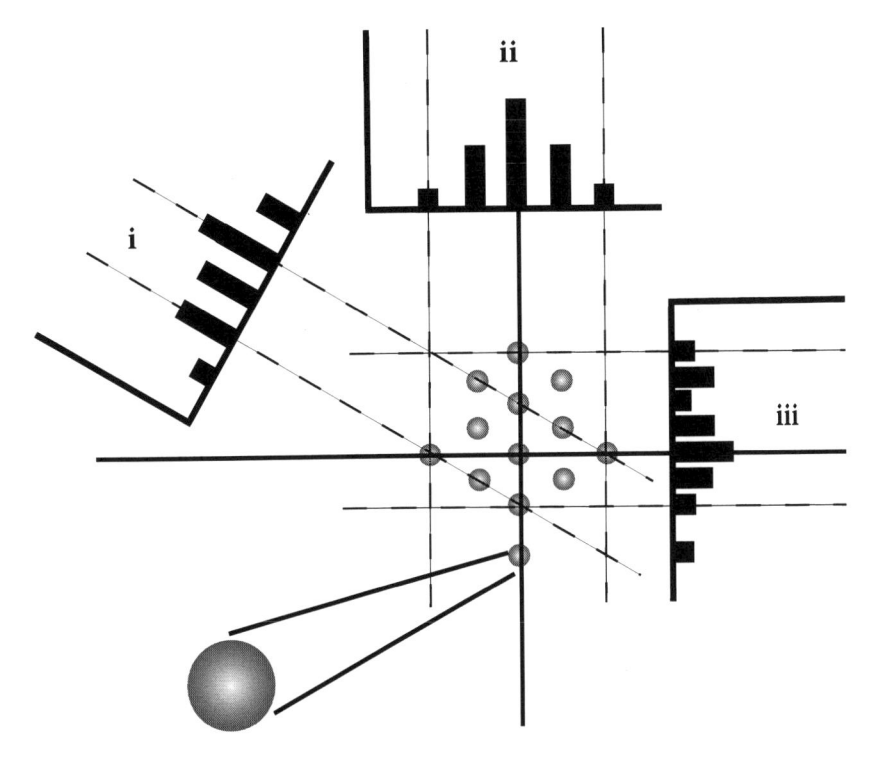

FIGURE 3.4
Absorption profiles of an array of circular objects along three projection directions *i–iii*. These projections are used in Figure 3.5 to illustrate back-projection.

of absorption matching the number of circular features aligned along the viewing direction.

The three profiles in Figure 3.4 can be used to demonstrate how reconstruction via back-projection is performed. In Figure 3.5a, profile *i* is back-projected along the viewing direction along which it was obtained. The number of thin lines is used to indicate the amount of absorption along each ray. Figure 3.5b adds projection *ii* to projection *i*, and Figure 3.5c adds projection *iii* to *ii* and *i*. The correct orientations are maintained in Figure 3.5b and c. Positions where three rays intersect are possible positions for the circular features, and these positions are labeled in Figure 3.5d with two kinds of disks: solid disks, where the features actually are, and open disks, where the presence of features cannot be excluded based on the only three projections available. Incorporation of additional data (projections recorded at different angles) would eliminate the open disks.

In the cartoon representation of Figures 3.4 and 3.5, the specimen and viewing directions were selected for a simple illustration of back-projection and reconstruction. The circular features within the specimen were positioned in well-defined rows and columns, and the views in Figure 3.4 were

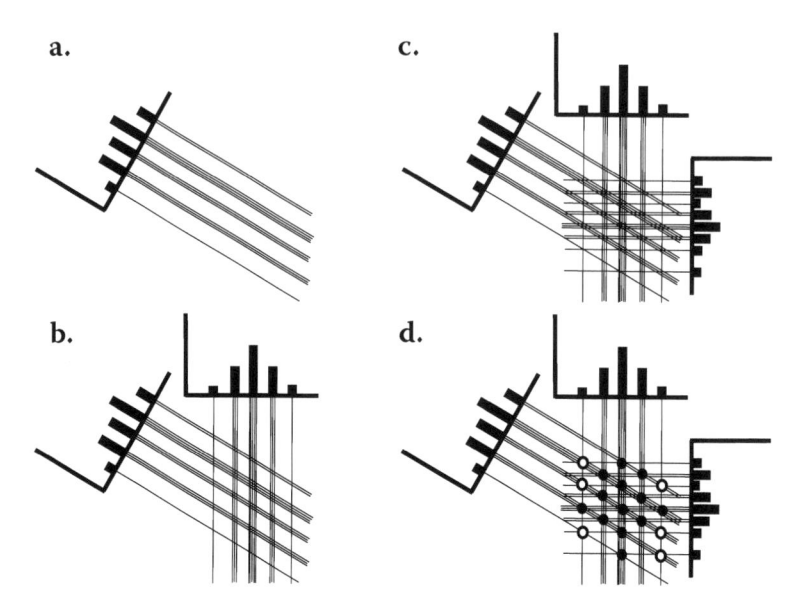

FIGURE 3.5
Back-projection illustrated using the "data" of Figure 3.4. (a) Back-projection of radiograph *i*. The number of lines in each ray represents the amount of absorption. (b) Back-projection of *ii* added to *i*. (c) Back projection of *i–iii*. (d) Identification of possible object positions (circles) based on the three projections. The open circles would be eliminated based on additional views.

chosen along directions where all of the circular features aligned into discrete peaks in the projection profiles. Views at an angle between *i* and *ii*, for example, will not show well-separated peaks. Displacement of a few of the circular features will similarly blend the peaks in the absorption profiles for projections *i*, *ii*, and *iii*. Real specimens will rarely be this convenient.

In the general case, the attenuating mass in each projection is back-projected onto a grid and the contributions added within the space covered by the projections. Different positions along the profile P have different levels of absorption; that amount of attenuation is added to each voxel within the object space that lies along the ray direction in question (represented by the number of thin closely spaced lines in Figure 3.5). Figure 3.6 shows a reconstruction grid, the cells of which have been filled with the numerical values for two projections from Figure 3.4. Each value is the sum of the value along the column and the value along the row. Larger values result where both projections have nonzero values.

The careful observer will have noted in both Figures 3.5 and 3.6 that mass has been distributed across regions where, in fact, no mass is present. Use of filtered back-projection alleviates this shortcoming and has become a standard algorithm. Figure 3.7 illustrates the filtering process graphically. In Figure 3.7a, the blurred reconstruction of a point is shown,

Reconstruction from Projections 27

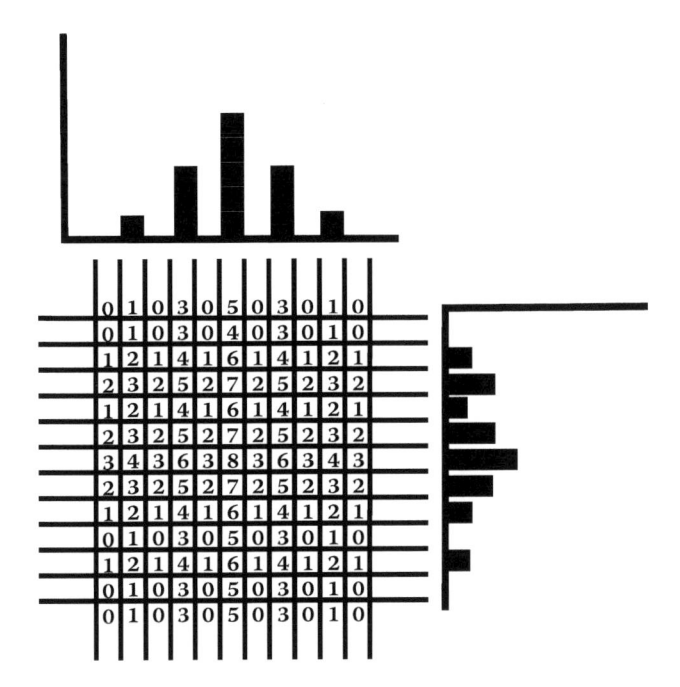

FIGURE 3.6
Back-projection onto a grid (see text).

that is, amplitude *A* as a function of distance *d* from the center of the point mass. Correction is accomplished using a sharpening filter, and this is done mathematically by convoluting the object's projection with the filtering function. Setting aside the mathematics for a moment, the application of a filter to the blurred profile produces the distance versus amplitude plot shown in Figure 3.7b; when back-projected, the negative tails cancel the blur shown in Figure 3.7a. Consider how this result extends to a profile of several peaks (Figure 3.7c); in some positions the negative tails overlap (Figure 3.7d). The enlargement of two closely spaced peaks (Figure 3.7e) shows how overlapping negative tails add to make the total profile. Note that filtering is applied to each absorption profile.

Mathematically, filtered projections are produced by convoluting the filter function with the projection in question. Although the choice of the filter function is extremely important in microCT, it is ignored here in favor of a more general (brief) description of filtered back-projection. Generally, reconstruction is done in polar coordinates, with *x* replaced by *t* in a coordinate system **t** rotated by angle *q* from **x** (left-hand side of Figure 3.8), using Equations (2.5) and (2.6) instead of (2.7) and (2.8).

Now consider how the convolution operation applies to the reconstruction problem. The object is given by $\mu(x,y)$ (see Equation (3.3) for the relationship between μ and I/I_0 in polar coordinates); the function *f(t)* in

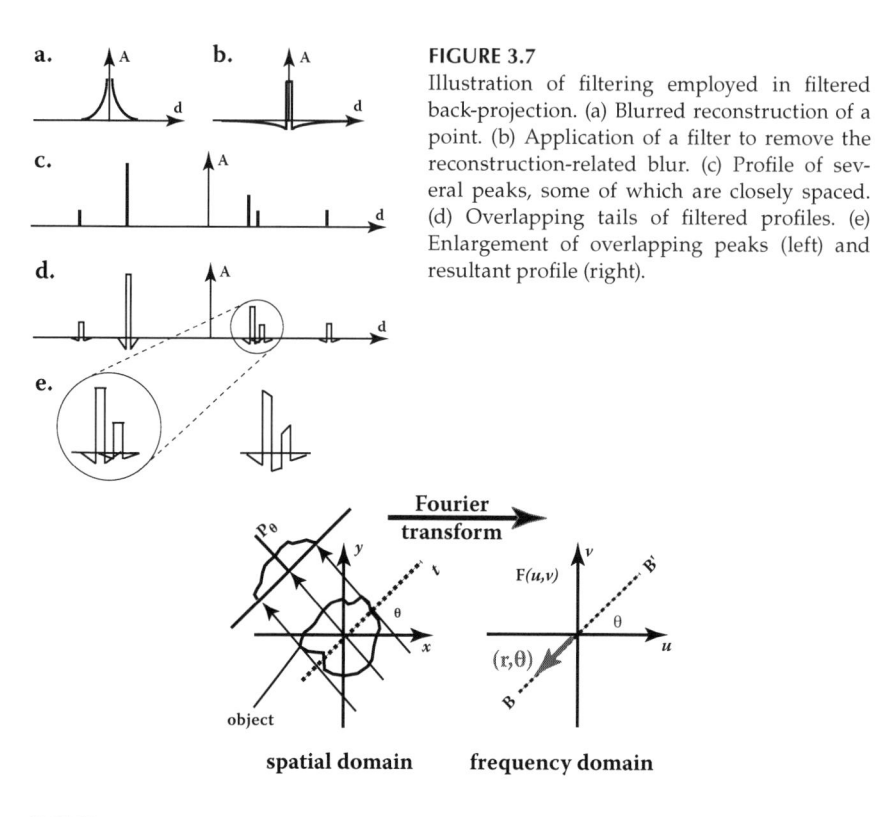

FIGURE 3.7
Illustration of filtering employed in filtered back-projection. (a) Blurred reconstruction of a point. (b) Application of a filter to remove the reconstruction-related blur. (c) Profile of several peaks, some of which are closely spaced. (d) Overlapping tails of filtered profiles. (e) Enlargement of overlapping peaks (left) and resultant profile (right).

FIGURE 3.8
Relationship between the profile measured in the spatial domain and the corresponding representation in the frequency domain (see text).

Equation (2.6) is then the projected ray $p_q(t)$ of Equation (3.3) and the projection P_q is made up of the individual $p_q(t)$. One obtains the map of µ by performing several steps:

1. Calculating the Fourier transform* S_q of measured projection P_q for each angle q;

2. Multiplying S_q by the value of the weighting function (the convolution of two functions is equivalent to simple multiplication of their Fourier transforms, so the weighting function is the transform of the filter) to obtain S_q'; and

3. Calculating the inverse Fourier transform of S_q' and summing over the image plane (direct space or spatial domain) which is the back-projection process.

---

* The Fourier transform of a function and the inverse transform are defined mathematically in the following section.

Reconstruction from Projections

Note that the summation of the smeared projections is conducted in direct space, as are any interpolations; this differs from the algorithm presented in the following section and relies on the Fourier slice theorem described elsewhere (Kak and Slaney, 2001).

3.4 Fourier-Based Reconstruction

Reconstruction can be performed in direct space, that is, in the 3D space in which we move and live; this approach was just described in Section 3.3. Other spaces can be more convenient (more efficient, more robust) for certain operations including reconstruction. In x-ray or electron diffraction, for example, reciprocal space representations are often more instructive than the corresponding direct space data of the single crystal or polycrystalline specimens (Cullity and Stock, 2001).

Figure 3.8 indicates schematically how Fourier transforms of absorption profiles P_q (in the spatial domain) can be used to populate the frequency domain. The Fourier components (frequencies and amplitudes along line t, projected along s in the spatial domain) of the absorption profile provide points along line B–B' in the frequency domain. Each frequency is plotted at radius r along the line shown, and the amplitude of that frequency component provides the numerical value for that point. Data from projections at different angles q from 0° to 180° are used to populate frequency space. Because the frequency space representation is as valid as the direct space version of the object (provided, of course, that it is adequately populated with observations), Fourier transformation of the frequency data will produce a valid reconstruction of the object in the spatial domain. As the amplitude–frequency representation of an arbitrary profile (or, in fact, any curve) may not be familiar to all readers, a brief digression is appropriate at this point.

Consider for a moment a square wave (Figure 3.9); this might represent the absorption profile of the rectangle in Figure 3.2. In terms of Fourier components (i.e., of sine waves of different frequencies and amplitudes), the expression for a square wave is

$$f(x) = 4/\pi \{\sin(\pi[x/L]) + (1/3)\sin(3\pi[x/L]) + (1/5)\sin(5\pi[x/L]) + \ldots\}, \quad (3.4)$$

where L is the period and the amplitude is one. Figure 3.9a plots the amplitudes of the different components (areas of the disks) as a function of frequency. Figure 3.9b pictures the idealized square wave and the first term of this series, which, one can see, is hardly an accurate representation. Figure 3.9c, however, shows the sum of the first seven terms of the series;

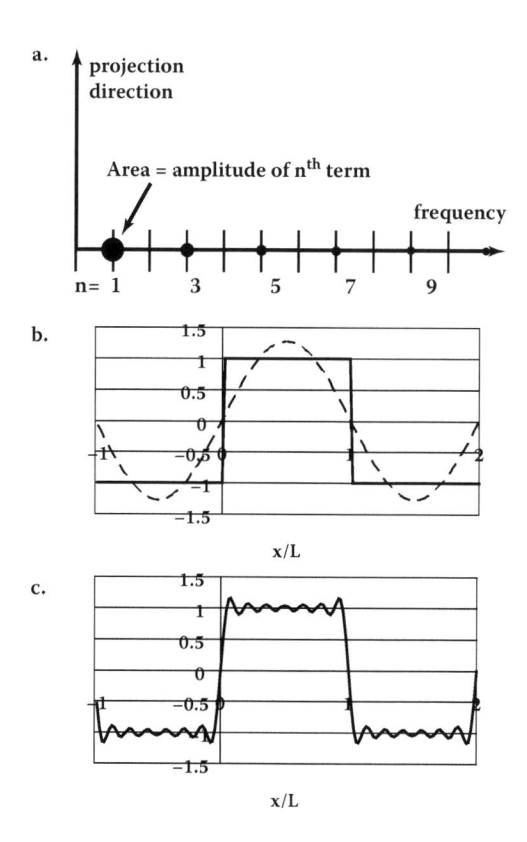

FIGURE 3.9
Illustration of how frequency/amplitude can represent a specific absorption profile, here a square wave. (a) Amplitudes (represented by areas of the disks) as a function of frequency of the terms of the square wave. (b) Idealized square wave, first term of this series. (c) Sum of the first seven terms of the series.

the square wave is evident, although more terms would be required to damp out the small oscillations (Weisstein, 2007).

The two-dimensional Fourier transform $F(u,v)$ of an object function $f(x,y)$ is

$$F(u,v) = \int\int f(x,y)\exp(-2\pi i [ux + vy])\, dx\, dy, \qquad (3.5)$$

where the limits of integration are $\pm \infty$. A projection $P_q(t)$ and its transform $S_q(w)$ are related by a similar equation, and, for parallel projections, some mathematical manipulation yields the relationship

$$F(u,0) = S_{q=0}(u), \qquad (3.6)$$

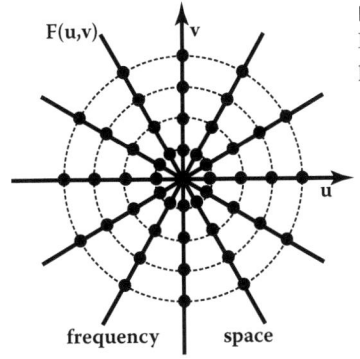

Reconstruction from Projections

FIGURE 3.10
Frequency space filled with data from many different projections.

a result that is independent of orientation between the object and coordinate system. This is a form of the Fourier slice theorem which can be stated as:

> The Fourier transform of a parallel projection of an image $f(x,y)$ taken at an angle q gives a slice of a two-dimensional transform $F(u,v)$ subtending an angle q with the u-axis. In other words, the Fourier transform of $P_q(t)$ gives values of $F(u,v)$ along line B–B' in Figure 3.8. (Kak and Slaney, 2001)

Collecting projections at many angles fills the frequency domain as shown in Figure 3.10, and the inverse Fourier transform

$$f(x,y) = \int \int F(u,v)\exp(2\pi i\,[ux + vy])\,du\,dv, \tag{3.7}$$

with integration limits again at $\pm\infty$, can be used to recover the object function $f(x,y)$. Typically, the fast Fourier transform algorithm is used for these operations, and interpolation between the spokes of data in the frequency domain is required. Inspection of Figure 3.10 shows sparser coverage farther from the origin of the frequency space; therefore, the high-frequency components are more subject to error than the lower spatial frequencies.

3.5 Performance

In understanding the various experimental approaches to microCT (described in the following chapter), it helps first to consider the requirements for reconstructing an $M \times M$ object (i.e., a planar slice through an object consisting of M voxels in one direction and M voxels along a second direction perpendicular to the first). A set of systematically sampled line integrals $\ln(I_o/I)$ must be measured over the entire cross-section of interest such that the geometrical relationship between these measurements is precisely defined. The quality of reconstruction depends on how finely the object is sampled (i.e., the spatial frequencies resolved in the profiles $P(s)$ and the number of viewing directions), on how accurately individual measurements of $\ln(I_o/I)$ are made (i.e., the levels of random and systematic errors) and on how precisely each measurement can be related to a common frame of reference.

The number of samples per projection and the number of views needed depend on the reconstruction method and on the size of features one wishes to resolve in the reconstruction. For an $M \times M$ slice, a minimum of $(\pi/4) M^2$ independent measurements are required if the data is noise-free, but faithful reconstruction can still be obtained with sampling approaching this minimum, even in the presence of noise (1996). Features down to one-tenth of the reconstructed voxels can be seen if contrast is high enough (Breunig et al., 1992, 1993), and metrology algorithms can measure dimensions to about one-tenth of a pixel with a three-sigma confidence level (1996). The number of samples per view is generally more important than the number of views, errors in I/I_o of 10^{-3} are significant, and both place important constraints on detectors for CT and microCT. The details of the various reconstruction algorithms lie outside the scope of this review (see, e.g., Newton and Potts (1981) and Kak and Slaney (2001)).

The precision with which the linear attenuation coefficients can be determined can be expressed in terms of its variance

$$\sigma^2_0 = \text{const } v \, (M_{proj} \langle N_0 \rangle)^{-1}, \qquad (3.8)$$

where v is the spatial sampling frequency, M_{proj} is the number of views, and $\langle N_0 \rangle$ is the mean number of photons transmitted through the center of the specimen (Kak and Slaney, 2001). To be strict, Equation (3.8) applies only to the center voxel of the specimen, but this equation provides important guidance in terms of how changes in several parameters affect reconstructed data.

For example, consider the mean value of the linear attenuation coefficient $\langle \mu \rangle$ for a region encompassing a significant number of voxels. Reconstructions produced from projections recorded for time t_0 (with N_0 counts per voxel) would have a standard deviation of the linear attenuation coefficient σ', whereas those recorded for time $4t_0$ ($4N_0$ counts) would be expected to have a standard deviation equal to $0.5\sigma'$, if counting statistics were the sole contribution to the variance. Similarly, for the same counting time and x-ray source, if one were to collect data with two sampling dimensions (voxel sizes) v' and $v'/2$, one would expect the standard deviation of the latter measurement to be substantially larger. Other contributions to broadened distributions of linear attenuation coefficients can be substantial and should not be ignored; these include partial volume effects (voxels partly occupied by two very different phases giving rise to an intermediate value of μ), which are discussed in Chapter 5.

All of the "exact" reconstruction algorithms require a full 180° set of views, although approximate reconstructions can be obtained where views are missing, for example, where opacity and sample size limit the directions along which useful views may be obtained (Tonner et al., 1989; Haddad and Trebes, 1997); the cost is a degraded quality reconstruction. Another approximate data collection approach is spiral tomography (Kalender et al., 1997; Wang et al., 1997), and it has received considerable attention because it affords increased speed and lower patient x-ray dosage. Only those details important in a particular data collection strategy and those reconstruction artifacts important in the examples are discussed.

3.6 Sinograms

One of the methods of representing projections for reconstructing a slice is the sinogram. Figure 3.11 schematically illustrates the information contained in a sinogram, a plot of intensity within the projection of a slice as a function of rotation angle. Essentially, the sinogram is the plot of the absorption data for the reconstruction of a single slice, and it gets its name from the fact that projections of objects follow sinusoidal paths as they rotate around the rotation axis. Note that sinograms appear fairly infrequently in the literature but are quite useful in illustrating certain aspects of the projection data. For example, a method of correcting for mechanical imperfections in a microCT rotation stage relies on properties of the sinogram to refine reconstructions (Section 5.2.1).

3.6.1 Related Methods

Recording radiographic stereo pairs is often used to precisely triangulate sharply defined features: that is, views of the same sample are recorded along two view directions separated by a precisely known angle, typically between 5 and 10°. This very rapid approach to three-dimensional inspection is of little use and gives way to CT when there are so many similar overlapping objects that individuals cannot be distinguished, when contrast does not vary sharply within the sample, or when the features to be imaged are so anisotropic that they produce significant contrast only along certain viewing directions that cannot be determined a priori (e.g., a crack).

Region-of-interest or local tomography is an approach where portions of the specimen pass out of the field of view (FOV) during rotation (Figure 3.12). The effect of the missing mass can be corrected by stitching together lower-resolution data for the missing areas of the project or by using known sample composition and geometry and calculating

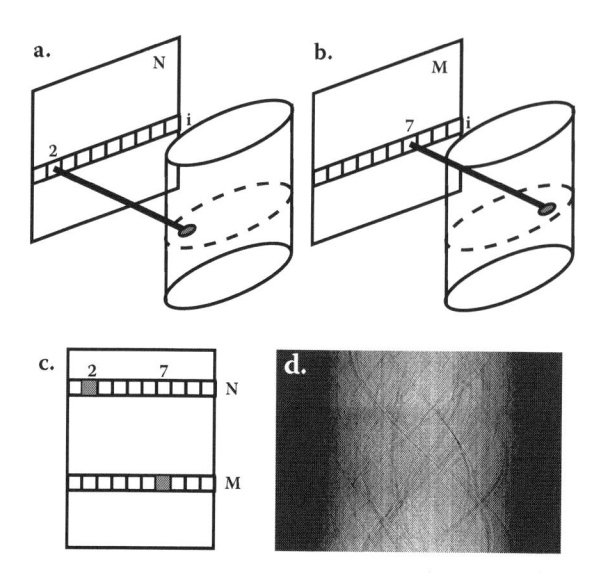

FIGURE 3.11
Illustration of the information contained in a sinogram. The gray object in the cylindrical sample projects onto the second column of the *i*th row of the digital radiograph N in (a). (b) After rotation about the vertical axis (the cylinder's axis, not shown), the gray object now projects onto the seventh column of the *i*th row of the digital radiograph M. (c) The rows of the sinogram for slice *i* (i.e., the slice to be reconstructed from the *i*th row of the radiograph) consist of the successive absorption profiles for the *i*th row derived from radiographs ..., N, ..., M, ... The plot is termed a sinogram because the gray object (and all others) trace a sinusoidal path in this representation. (d) Experimental sinogram from an Al corrosion specimen. (From Rivers and Wang (2006) but with the original linear grayscale altered for visibility.)

corrected views (see Lewitt and Bates (1978) and Nalcioglu et al. (1979)). Uncorrected local tomography reconstructions are necessarily approximate, but the extent to which their fidelity is degraded (geometry, linear attenuation coefficient values) depends on many factors. Errors will become more important as more mass remains longer out of the FOV, and one expects a priori that specimens with anisotropic cross-sections will provide the greatest problems. In general terms, the internal geometries in local reconstructions will be reproduced with good fidelity, but, if there is significant mass outside the FOV, dynamic range may be suppressed or linear attenuation coefficients affected (Xiao et al., 2007). For specimens with complex, highly anisotropic cross-sections or with high-frequency, anisotropic internal structure, it is essential to check for the presence of artifacts (Kalukin et al., 1999).

A number of groups and facilities routinely use local tomography. In a custom-built lab microCT system, local tomographic reconstruction compared well with reconstruction with the complete FOV (Jorgensen et al., 1998). Local tomography is routinely used at ESRF, so much so that it is

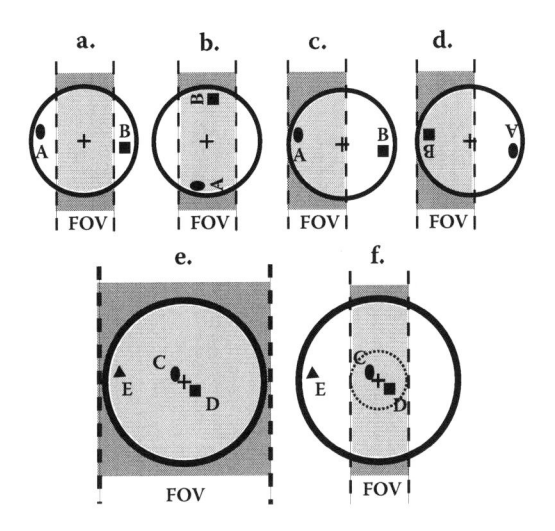

Reconstruction from Projections

FIGURE 3.12
Field of view (FOV) and specimen diameter. The x-ray beam illuminates the area shaded gray. (a), (b) Points A and B rotate into and out of the FOV. (c), (d) Illustration of how placing the center of rotation to one side of the FOV and rotating through 360° provides data missing in (a) and (b). (e) The entire specimen diameter is within the FOV, but the smallest voxel size is limited by the number of detector elements. Points C, D, and E remain in the FOV. (f) Local or region-of-interest tomography where the FOV is much smaller; points C and D remain in the FOV, whereas E moves in and out and only the region within the dotted line is reconstructed. Here the voxel size (region diameter divided by the number of detector elements) can be much smaller than in (e) (Stock, 2008).

sometimes only mentioned in passing, for example, Peyrin et al. (1999). Local tomography is particularly effective in specimens with relatively low absorption, such as foams; it has been applied to good effect to study deformation of an Al foam (Ohgaki et al., 2006).

Partial view reconstruction, where an angular range of projections is unavailable, is related to local tomography in the sense that information is missing. Interpolation of the missing views from the existing data seems to produce tolerable reconstructions (Brunetti et al., 2001), but this sort of approximation should be avoided if at all possible. If only one or two adjacent projections are interpolated within an otherwise complete set of views, one is hard-pressed to see the effect of the missing data.

Laminography, also termed tomosynthesis, is an alternative approach and is particularly valuable for specimens whose aspect ratios are impractical for conventional microCT (e.g., platelike specimens). Recent digital methods have been reviewed (Dobbins and Godfrey, 2003), albeit from a clinical and not a microimaging perspective, and Figure 3.13 illustrates one method of determining 3D positions from a series of views limited to one side of the specimen. There is a cost in terms of degraded contrast by methods such as the shift and add algorithm, illustrated in Figure 3.13.

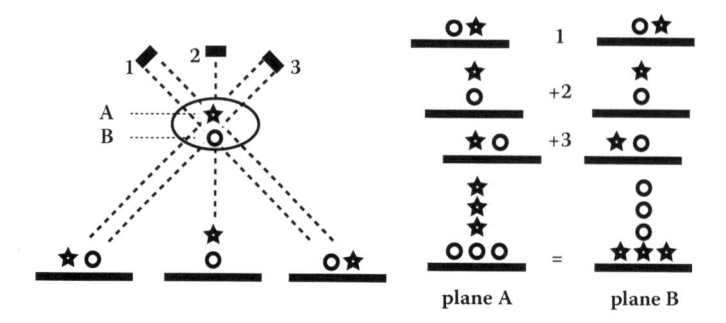

FIGURE 3.13
Illustration of tomosynthesis via the add and shift method for parallel rays. (Left) Image positions of features (on planes A and B) on the detector plane are shown for source positions 1, 2, and 3 relative to the object. (Right) Images shifted to reinforce objects in plane A (star) or to reinforce those in plane B (circle). The features out of the plane of interest are smeared out across the detector, and sharp images occur only for the focal plane. In this illustration, the amount of shift depends on experimental quantities such as the separation between specimen plane and detector and the angle of incidence of the x-ray beam (Stock, 2008).

Tomosynthesis has been applied in the microscopic imaging regime in recent studies of perfusion (Nett et al., 2004), of integrated circuits (Helfen et al., 2006), and of NDE (nondestructive evaluation) of long objects (Huang et al., 2004). In situations where displacement of well-defined features can be followed versus rotation, the relative translations of each resolvable point can be converted in depth from one of the specimen surfaces. In this approach, termed *stereometry*, use of eight to ten views allows a feature's depth to be determined to higher precision than in simple two-view triangulation; the 3D fatigue crack surface positions determined with stereometry were in excellent agreement with conventional microCT (Ignatiev, 2004; Ignatiev et al., 2005).

References

(1996). E 1441-95 Standard guide for computed tomography (CT) imaging. *1996 Annual Book of ASTM Standards*. Philadelphia, ASTM. **03.03**: 704–733, and E 1570-95a Standard practice for computed tomographic (CT) examination. *1996 Annual Book of ASTM Standards*. Philadelphia, ASTM. **03.03**: 784–795.

Breunig, T.M., J.C. Elliott, S.R. Stock, P. Anderson, G.R. Davis, and A. Guvenilir (1992). Quantitative characterization of damage in a composite material using x-ray tomographic microscopy. In *X-ray Microscopy III*. A.G. Michette, G. R. Morrison, and C. J. Buckley (Eds.). New York, Springer: 465–468.

Breunig, T.M., S.R. Stock, A. Guvenilir, J.C. Elliott, P. Anderson, and G.R. Davis (1993). Damage in aligned fibre SiC/Al quantified using a laboratory x-ray tomographic microscope. *Composites* **24**: 209–213.

Brunetti, A., B. Golosio, R. Cesareo, and C.C. Borlino (2001). Computer tomographic reconstruction from partial-view projections. In *Developments in X-ray Tomography III*. U. Bonse (Ed.). Bellingham, WA, SPIE. *SPIE Proc Vol* **4503**: 330–337.

Cullity, B.D. and S.R. Stock (2001). *Elements of X-ray Diffraction*. Upper Saddle River, NJ: Prentice-Hall.

Dobbins III, J.T. and D.J. Godfrey (2003). Digital x-ray tomosynthesis: Current state of the art and clinical potential. *Phys Med Biol* **48**: R65-R106.

Haddad, W.S. and J.E. Trebes (1997). Developments in limited data image reconstruction techniques for ultrahigh-resolution x-ray tomographic imaging of microchips. In *Developments in X-ray Tomography*. U. Bonse (Ed.). Bellingham, WA, SPIE. *SPIE Proc Vol*. **3149**: 222–231.

Helfen, L., T. Baumbach, P. Pernot, P. Mikulik, M. Di Michiel, and J. Baruchel (2006). High resolution three-dimensional imaging by synchrotron radiation computed laminography. In *Developments in X-Ray Tomography V*. U. Bonse (Ed.). Bellingham,WA, SPIE. *SPIE Proc Vol* **6318**: 63180N–1–9.

Huang, A., Z. Li, and K. Kang (2004). The application of digital tomosynthesis to the CT nondestructive testing of long large objects. In *Developments in X-Ray Tomography IV*. U. Bonse (Ed.). Bellingham, WA, SPIE. *SPIE Proc Vol* **5535**: 514–521.

Ignatiev, K.I. (2004). *Development of X-Ray Phase Contrast and Microtomography Methods for the 3D Study of Fatigue Cracks*. Ph.D. Thesis. Atlanta, Georgia Institute of Technology.

Ignatiev, K.I., W.K. Lee, K. Fezzaa, and S.R. Stock (2005). Phase contrast stereometry: Fatigue crack mapping in 3D. *Phil Mag* **83**: 3273–3300.

Jorgensen, S.M., O. Demirkaya, and E.L. Ritman (1998). Three-dimensional imaging of vasculature and parenchyma in intact rodent organs with x-ray microCT. *Am J Physiol* 275 (*Heart Circ Physiol* 44) **275**: H1103–H1114.

Kak, A.C. and M. Slaney (2001). *Principles of Computerized Tomographic Imaging*. Philadelphia: SIAM (Soc. Industrial Appl. Math.).

Kalender, W.A., K. Engelke, and S. Schaller (1997). Spiral CT: Medical use and potential industrial applications. In *Developments in X-ray Tomography*. U. Bonse (Ed.). Bellingham, WA, SPIE. *SPIE Proc Vol* **3149**: 188–202.

Kalukin, A.R., D.T. Keane, and W.G. Roberge (1999). Region-of-interest microtomography for component inspection. *IEEE Trans Nucl Sci* **46**: 36–41.

Lewitt, R.M. and R.H.T. Bates (1978). Image reconstruction from projections: III: Projection completion methods (theory) and IV: Projection completion methods (computational examples). *Optik* **59**: 189–204 and 269–278.

Nalcioglu, O., Z.H. Cho, and R.Y. Lou (1979). Limited field of view reconstruction in computerized tomography. *IEEE Trans Nucl Sci* **NS-26**: 546–551.

Natterer, F. (2001). *The Mathematics of Computerized Tomography*. Philadelphia: SIAM (Soc. Industrial Appl. Math.).

Nett, B.E., G.H. Chen, M.S. Van Lysel, T. Betts, M. Speidel, H.A. Rowley, B.A. Kienitz, and C.A. Mistretta (2004). Investigation of tomosynthetic perfusion measurements using the scanning-beam digital x-ray (SBDX). In *Developments in X-Ray Tomography IV*. U. Bonse (Ed.). Bellingham, WA, SPIE. *SPIE Proc Vol* **5535**: 89–100.

Newton, T.H. and D.G. Potts (1981). *Radiology of the Skull and Brain: Technical Aspects of Computed Tomography*. St. Louis: Mosby.

Ohgaki, T., H. Toda, M. Kobayashi, K. Uesugi, M. Niinom, T. Akahori, T. Kobayashi, K. Makii, and Y. Aruga (2006). In situ observations of compressive behaviour of aluminium foams by local tomography using high-resolution tomography. *Phil Mag* **86**: 4417–4438.

Peyrin, F., S. Bonnet, W. Ludwig, and J. Baruchel (1999). Local reconstruction in 3D synchrotron radiation microtomography. In *Developments in X-ray Tomography II*. U. Bonse (Ed.). Bellingham,WA, SPIE. *SPIE Proc Vol* **3772**: 128–137.

Rivers, M.L. and Y. Wang (2006). Recent developments in microtomography at GeoSoilEnviroCARS. In *Developments in X-Ray Tomography V*. U. Bonse (Ed.). Bellingham, WA, SPIE. *SPIE Proc Vol* **6318**: 63180J–1–15.

Stock, S.R. (1990). X-ray methods for mapping deformation and damage. In *Micromechanics — Experimental Techniques*. J.W.N. Sharpe (Ed.). ASME. *AMD* **102**: 147–162.

Stock, S.R. (2008). Recent advances in x-ray microtomography applied to materials. *Int Mater Rev* **58**: 129–181.

Tonner, P.D., B.D. Sawicka, G. Tosello, and T. Romaniszyn (1989). Region-of-interest tomography imaging for product and material characterization. In *Industrial Computerized Tomography*. Columbus, OH, ASNT: 160–165.

Wang, G., P. Cheng, and M. W. Vannier (1997). Spiral CT: Current status and future directions. In *Developments in X-ray Tomography*. U. Bonse (Ed.). Bellingham, WA, SPIE. *SPIE Proc Vol* **3149**: 203–212.

Weisstein, E.W. (2007). Fourier Series-Square Wave, *MathWorld* — A Wolfram Web Resource. Retrieved August 10, 2007. http://mathworld.wolfram.com/fourierseriessquarewave.html.

Xiao, X., F. De Carlo, and S. Stock (2007). Practical error estimation in zoom-in and truncated tomography reconstructions. *Rev Sci Instrum* **78**: 063705–1–7.

4
MicroCT Systems and Their Components

This chapter begins with a brief description of different absorption microCT methods. Section 4.2 describes x-ray source characteristics that affect the performance of microCT systems, and Section 4.3 covers detectors. Discussion of the third important component of systems, sample positioning and rotation subsystems, appears in Section 4.4. Tube-based microCT systems are covered in Section 4.5 and synchrotron microCT systems in Section 4.6. Full-field nanoCT and lens-based nanoCT are discussed in Section 4.7. MicroCT with contrast mechanisms other than absorption (mainly phase contrast but also fluorescence and scatter microCT) is the subject of Section 4.8. The final section of this chapter is intended mainly for those new to microCT and discusses how to determine which commercial system or synchrotron microCT facility is best suited for the intended applications. Topics including artifacts found in actual systems, precision and accuracy of reconstructions, and challenges and speculations for the future are covered in Chapter 5.

4.1 Absorption MicroCT Methods

Most microCT systems employ one of four geometries shown in Figure 4.1. Although two arrangements are the same as two of the four generations of scanners into which the CT literature classifies apparatus, the other two are different.

In first-generation or pencil beam systems (Figure 4.1a), a pinhole collimator C and a pointlike source P produce a narrow, pencil-like beam that is scanned across the object O along x_1 to produce each view; successive views are obtained by rotation about x_2. Only a simple zero-dimensional x-ray detector D is required, perhaps with some scatter shielding S. Energy-sensitive detectors are readily available and, if used instead of gas proportional or scintillation detectors, allow reconstruction with very accurate values of linear attenuation coefficients. Successive views are obtained by rotating the sample and repeating the translation. Obtaining volumetric data (i.e., a set of adjacent slices) borders on infeasible because of the long scan times required, but this is balanced by the inherent simplicity and flexibility of such apparatus and by a relatively

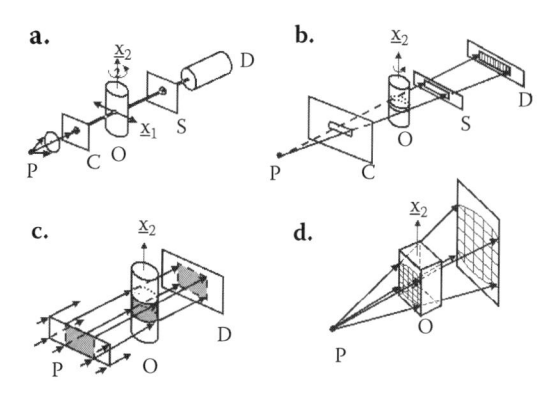

FIGURE 4.1
Four experimental methods for microCT: (a) pencil, (b) fan, (c) parallel, and (d) cone beam geometries. P is the x-ray source, C the collimator, O the object being imaged, x_2 the specimen rotation axis, x_1 a translation axis perpendicular to the x-ray beam and the rotation axis, S the scatter slit, and D the x-ray detector. (Reproduced from Stock (1999).)

greater immunity to degradation of contrast due to scatter. Pencil beam microCT continues to be used with laboratory x-ray sources (Elliott and Dover, 1982, 1984, 1985; Borodin et al., 1986; Bowen et al., 1986; Breunig et al., 1990, 1992, 1993; Elliott et al., 1994a,b; Stock et al., 1989, 1994; Mummery et al., 1995; Davis and Wong, 1996), and very high spatial resolution has been achieved in small samples using synchrotron radiation (Spanne and Rivers, 1987; Connor et al., 1990; Ferrero et al., 1993).

Fan beam systems (Figure 4.1b, i.e., third-generation apparatus) use a rotate-only geometry: a flat fan of x-rays defined by collimator C and spanning the sample originates at the pointlike source P, passes through the sample and scatter shield S, and is collected by the one-dimensional x-ray detector. These systems are often used with laboratory microfocus-generated x-radiation. This detector consists of an array of discrete elements that allows the entire view to be collected simultaneously. One to two thousand detectors are typically in the array, making fan beam systems much more rapid than pencil beam systems, but data for only one slice is recorded at a time. Incorporating a linear or area detector makes the system much more susceptible to scatter (than pencil beam systems), that is, the redirection of photons from the detector element on a line-of-sight from the x-ray source into another detector element. In severe cases, this greatly affects the fidelity of a reconstruction. Furthermore, it is necessary to normalize the response of the different detector elements; even with careful correction, ring artifacts from various nonuniformities can still appear in reconstructions.

When examining slices from fan beam systems, it is not only important to note the dimensions of the voxels in the plane of reconstruction, but it is also important to ascertain the thickness of the slice: systems collecting data for one slice at a time often are used with a detector width (perpendicular to the reconstruction plane) and slice thickness substantially larger than the voxels' dimensions in the reconstruction plane. This certainly improves the signal-to-noise ratio in the reconstruction and is very effective when imaging samples with slowly varying structure along the axis perpendicular to the reconstruction plane (London et al., 1990). This approach sacrifices sensitivity to defects much smaller than the slice thickness.

In situations where spatially wide, parallel beams of x-rays are available, the parallel-beam geometry (Figure 4.1c) allows straightforward and very rapid data collection for multiple slices (i.e., a volume) simultaneously. A parallel beam from a source P (with a certain cross-sectional area) shines through the sample and is collected by a two-dimensional detector array. Because the x-ray beam is parallel, the projection of each slice of the object O on the detector D (i.e., each row of the array) is independent of all other slices. In practice, this must be done at storage rings optimized for the production of the hard synchrotron x-radiation (Flannery et al., 1987; Kinney et al., 1988). High-performance area detectors are required, but there is an enormous increase in data collection rates over the geometries described above (see Section 5.4). Because most area detectors consist of square detector elements, slices are generally, but not always, reconstructed with isotropic voxels (i.e., the voxel dimensions within the reconstruction plane equal the slice thickness).

The cone beam geometry (Figure 4.1d), the three-dimensional analogue of the two-dimensional fan beam arrangement, is a fourth option; it is especially well suited for volumetric CT employing microfocus tube sources (Feldkamp et al., 1984, 1988; Feldkamp and Jesion, 1986). The x-rays diverge from the source, pass through the sample, and are recorded on the area detector. In this geometry each detector row, except the central row, receives contributions from more than one slice, and the effect becomes greater the farther one goes from the plane perpendicular to the rotation axis.

The cone beam reconstruction algorithm is an approximation, however, and some blurring is to be expected in the axial direction for features that do not have significant extent along this direction. Nonetheless, reconstructions of the same 8-mm cube of trabecular bone from data collected with orthogonal rotation axes show only minor differences when the same numerical sections are compared (Feldkamp et al., 1988). With an x-ray source size of 5 µm or smaller, system resolution is limited by that of the x-ray detector array and by penumbral blurring (Figure 2.8) and can be considerably better than 20 µm. Only the portions of the sample that remain in the beam throughout the entire rotation can be reconstructed exactly. As noted at the end of this section, the greater the cone angle is, the larger the reconstruction errors, particularly along the direction parallel to the specimen rotation axis.

Fan beam and cone beam apparatus generally employ point x-ray sources, and this allows geometrical magnification to match the desired sample voxel size to detector pixel size (Figure 2.8). The incorporation of time delay integration (mechanically coupled scanning of sample and detector) into a microCT system (Davis and Elliott, 1997) has allowed reconstruction of specimens larger than the detector imaging area or the x-ray beam; directional correlation of noise in large-aspect ratio samples remains a problem because this introduces streak artifacts (Davis, 1997). Time delay integration has been quite successful in eliminating ring artifacts caused by nonuniform response of the individual detector elements (Davis and Elliott, 1997). Multiple frame acquisition with detector translations is a hardware-based approach used to reduce ring artifacts, for example, in tube-based systems manufactured by Skyscan.

Use of an asymmetrically cut crystal (see Figure 2.9), positioned between the sample and x-ray area detector and set to diffract the monochromatic synchrotron radiation incident on the sample, has been demonstrated to reject scatter and improve sensitivity as well as to provide, through beam spreading, magnification of the x-ray beam prior to its sampling by the x-ray detector (Sakamoto et al., 1988; Suzuki et al., 1988; Kinney et al., 1993). This is an adaptation of a commonly used method in x-ray diffraction topography that allows one to overcome limitations of the detector, that is, to approach resolutions inherent to the x-ray source.

NanoCT, that is, tomography with systems designed to produce voxel sizes substantially below one micrometer, requires much higher precision and accuracy of the various components as well as much longer counting times or much brighter x-ray sources than in microCT (operating in the one micrometer or greater voxel-size regime). Parallel beam and cone beam geometries are used, the latter employing zone plate or other x-ray optics. To a first approximation, the level of mechanical performance required of a nanoCT apparatus scale with the voxel size is used, and the details appear in Section 4.7.

Before leaving the discussion of the generalities of microCT systems and considering the characteristics of systems' individual components, a brief digression on reconstructions with fan and cone beam is useful. Explaining this earlier in the text, say, in the chapter on reconstructions, would have been confusing prior to describing the experimental geometries.

Consider first the fan beam geometry and the rays projected from the point source through the specimen and onto the detector (Figure 4.2a). Because each ray follows a different, albeit predictable, angle relative to the reference direction, the spatial frequencies within each projection cannot be placed along a single spoke in frequency space as they would in a parallel beam projection. Each voxel within the physical slice irradiated by the x-ray beam does, however, contribute to each projection (unlike the case of the cone beam; see below), but, with each ray describing a

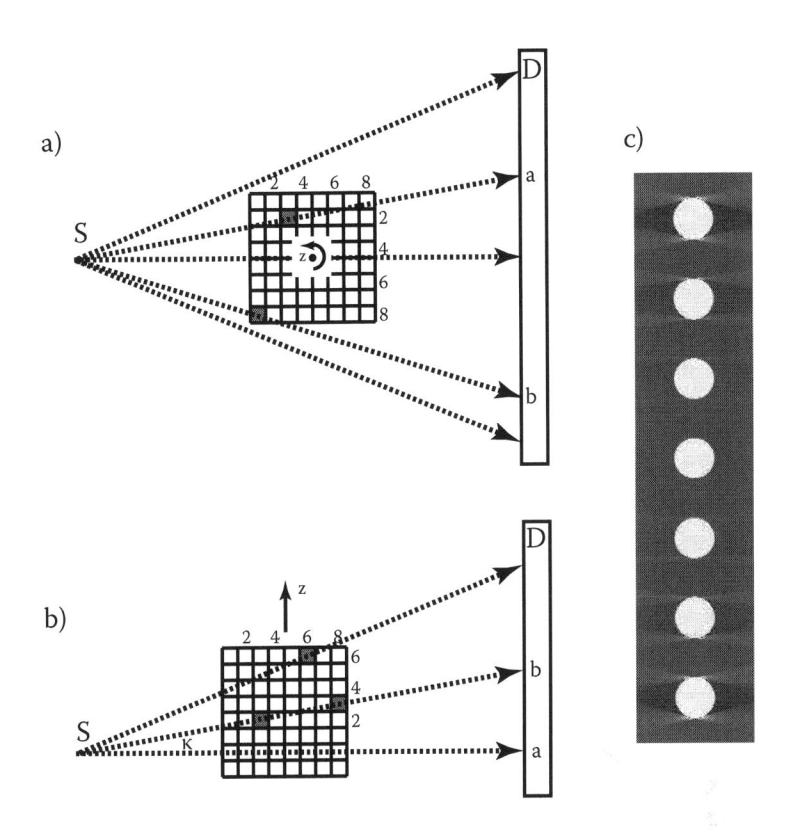

MicroCT Systems and Their Components 43

FIGURE 4.2
Illustration of fan beam and cone beam reconstruction: (a) fan beam geometry. The source S, detector D, and rotation axis z are shown, and all of the small squares within the large square object represent voxels within one physical slice of the object. The voxel position is denoted by the numbers outside the specimen (rows, columns). The voxels shown in gray (third column, second row; first column, eighth row) project onto the detector at positions a and b, and all of the voxels within the physical slice contribute to the projected profile of the slice recorded on the detector; that is, all voxels within the physical slices are sampled in each projection. (b) Cone beam geometry showing a plane perpendicular to the slice plane, that is, parallel to the rotation axis z. One-half of the specimen and rays are shown; the other half (below the central ray a) are not because they add no additional information to the figure. Each row of squares in the object represents the voxels within a physical section of the specimen, a section that will be reconstructed into a slice. As pictured, the ray striking the detector at point b contains contributions from voxels in rows 2 and 3 (e.g., the gray voxels in column 3, row 2, and column 8, row 3). Note that the ray reaching point a is within the central plane, a special plane in cone beam reconstructions: this slice can be reconstructed exactly because it is identical to the fan beam situation pictured in (a). The central task of cone beam reconstruction algorithms is the correct reapportionment of absorption within rays like b to the correct slices; the greater the cone angle κ is, the greater the potential error. (c) Errors (arrows) at larger cone angles illustrated using a simulated reconstruction of stack of seven solid balls. (Courtesy of Tom Case, Xradia Inc.)

different projection direction, a coordinate transformation or other procedure is required to transform the projections into a form convenient for reconstruction. A simple trigonometric relationship between the two coordinate systems can be used, although the mathematics required to propagate this transformation through the basic reconstruction integrals is somewhat involved. The interested reader is directed elsewhere for these details (Kak and Slaney, 2001).

The situation in the cone beam geometry is more complicated than that with fan beam data. Here, one must differentiate the situation on the central plane, on the one hand, and those of the planes above and below, on the other. In the central plane, details of the reconstruction follow those of the fan beam. Off this plane, rays pass from one slice plane to another, and voxels from more than one "slice" contribute to the projection at points such as "b" in Figure 4.2b. The farther a given plane is from the central plane (i.e., the larger the cone angle κ), the greater the potential errors in the reconstruction. Figure 4.2c shows a simulation of a cone beam reconstruction of several spheres stacked along the rotation axis; here a numerical section perpendicular to the slice planes shows the increasing blurring with distance away from the central plane. Structures with periods along the specimen rotation axis are particularly prone to errors, but other structures reconstruct accurately.

4.2 X-Ray Sources

Most microCT with tube sources of x-radiation has been performed using the entire spectrum, Bremsstrahlung, and characteristic radiation, because the cost in data collection times is prohibitive in this photon-starved environment. Exceptions include studies done with pencil beam systems. One group used an energy-sensitive detector to correct for polychromaticity (Elliott et al., 1994a) and another used a channel cut monochromator (Kirby et al., 1997). In a tube-based microCT system, the investigator can vary the tube potential and the tube current to affect imaging conditions. Increasing the tube current produces a linear increase in x-ray intensity but no change in the distribution of x-ray energies. With increased electron flux incident on the target, an unintended consequence may be spreading of the x-ray focal spot, with adverse effect on resolution (i.e., increased penumbral blurring). Altering the tube voltage (kVp) changes the spectrum of x-ray energies emitted, and this can be used to optimize contrast in reconstructions. Increased voltage allows more absorbing specimens to be studied. Lower voltages are used to enhance contrast between low absorption phases such as different soft tissue types. Changes in tube voltage can also alter the focal spot on the target.

In synchrotron-based microCT, the cost of discarding most of the x-ray spectrum during monochromatization is insignificant compared to the times required for sample movement, detector readout, and so on: in most cases the resulting monochromatic beam is intense enough for a view to be collected in a fraction of a second. Most synchrotron microCT facilities allow the user to select the x-ray energy used to image the specimen (of course, there are limits based on the optics available); a little thought and a bit of trial and error will allow the user to obtain the optimum available contrast for a given type of specimen. In some applications (e.g., transient phases), a pink (i.e., polychromatic) beam can be used for very rapid data acquisition. As mentioned earlier (Sections 2.1.2 and 2.2.3), collecting microCT data above and below the absorption edge of an element of interest and comparing the two reconstructions provide considerably improved sensitivity (Nusshardt et al., 1991; Dilmanian et al., 1997).

Recent development of tabletop synchrotron radiation sources offers another option for a source for x-ray microCT (Hirai et al., 2006). Essentially, these are super x-ray tubes with the accelerated electron beam striking a target, and performance can be quite good. Other types of electron accelerators can also be used with targets to generate x-radiation for imaging, but use of these types of sources, because of their complexity and cost, remains a rarity.

The size of the x-ray source affects the spatial resolution that can be obtained. This is normally not a consideration with synchrotron radiation, given the minuscule intrinsic divergence and the typical source-to-sample distances employed (>10 m). With x-ray tubes, using a small (5–10 μm) diameter x-ray source limits the loading of the target (i.e., the amount of energy that may be deposited) and the resulting x-ray intensity.* This must be balanced against use of a larger spot size where penumbral blurring would prevent small features from being seen. As discussed in the context of quantification of crack openings (Chapter 11), there are situations where much can be done, even in the presence of significant penumbral blurring.

In closing this section, the reader should note that electron optics and modern filament materials (e.g., LaB_6) such as are used in SEMs can produce beam diameters and x-ray sources sizes substantially smaller than 5 μm. It is not surprising, therefore, that SEM-based nanoCT systems have been developed and commercially marketed.

---

* At the time of writing, still smaller sources were becoming available. The reader should check with system manufacturers for what they reckon their spot size is on the target, whether it is measured and reported for each tube, and how much it changes with time (due to pitting at the target surface, etc.). Values quoted should be taken with a grain of salt, not only because we all want to put our best foot forward, but also because good measurements of x-ray-sourced sizes are nontrivial and many factors can influence actual values.

4.3 Detectors

The characteristics of the x-ray detector array used have important consequences for the performance of a given microCT apparatus. Most groups engaged in microCT based their apparatus on linear or area detectors because volumetric work is largely impractical with pencil beam systems. Details typical of pencil beam systems appear elsewhere (Elliott et al., 1994a), and the discussion here, therefore, focuses on one- and two-dimensional detectors. It is worth mentioning that mechanical or electronic shutters are needed to separate different frames recorded, but further development of this subject exceeds the intended level of detail.

Most one- or two-dimensional detector arrays are based on semiconductor devices (e.g., photodiode arrays, charge injection devices, and charge coupled devices (CCDs)), which work efficiently with optical photons and which are not suitable as direct x-ray detectors.* These detectors are quite transparent at photon energies above 10 keV and quickly suffer radiation damage. Instead of detecting the x-ray photons directly, the array images light given off by an x-ray scintillator chosen for the x-ray energies of interest. These x-ray camera systems couple the scintillator to the visible light detector typically through a lens system or fiber-optic channel plate (Figure 4.3). The former is generally used with synchrotron radiation, and the latter is often employed with tube-based systems. Note that electronics are needed to read the linear or area detectors, and these are normally integral to the camera system delivered by the vendor.

One-dimensional photodiode arrays (Reticon devices, typically 1,024 elements) have been successfully used in several single-slice microCT apparatus (Burstein et al., 1984; Seguin et al., 1985; Armistead, 1988; Suzuki et al., 1988; Engelke et al., 1989a,b; London et al., 1990). In these systems, optical fibers are typically used to couple the detector with the phosphor. Some systems have $Gd_2O_2S{:}Tb$-based phosphors coated directly on the end of the fibers, whereas another system couples to a 650-µm-thick transparent layer in which 4-µm diameter particles of $Gd_2O_2S{:}Tb$ are embedded (Engelke et al., 1989a,b). As is noted elsewhere (Kinney and Nichols, 1992), photodiode arrays are quite noisy and suffer from significant nonlinearities that can lead to very serious ring artifacts in reconstruction.

Some two-dimensional detector systems have been based on vidicons (Feldkamp et al., 1984, 1988; Feldkamp and Jesion, 1986; Sasov, 1987a,b, 1989; Goulet et al., 1994), but most are based on two-dimensional CCDs. Within each device (metal oxide silicon or MOS) of the CCD array, an absorbed photon creates a charge pair. The electrons are accumulated in

---

* A new generation of highly absorbing, and in some cases energy-sensitive, area detectors are under development but are not widely available. It will be interesting to track the future use of these detectors and the development of still larger area arrays.

MicroCT Systems and Their Components

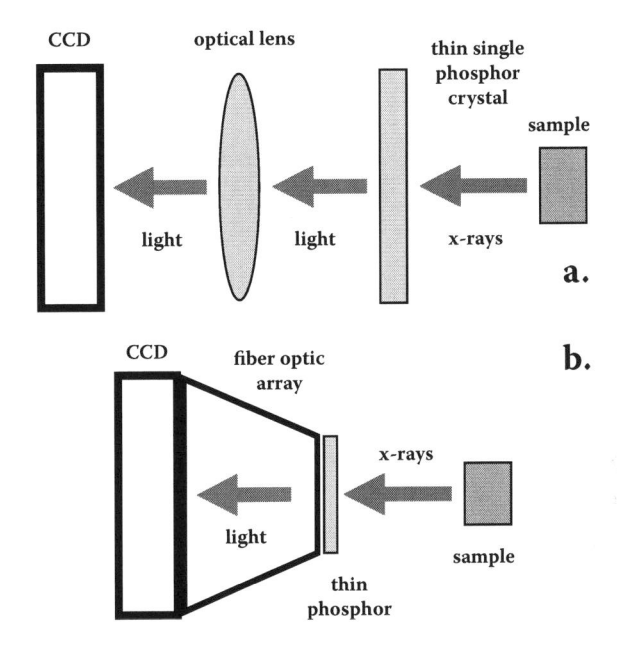

FIGURE 4.3
Scintillator–CCD coupling: (a) thin crystal scintillator and optical lens; (b) scintillator screen and fiber optic taper, where the individual light guides direct the light to the CCD detector elements.

each detector well during exposure, and the individual detector elements are read digitally by transferring charge from pixel to pixel until all columns of pixels have reached the readout register and been stored in the computer system. The larger the detector element area is, the greater the number of electrons that can be stored and the greater the dynamic range. Normally, one thinks of increased detector element size entailing a sacrifice of spatial resolution, but this need not be so if the number of detector elements sampling the specimen cross-section remains constant.*

There is a cost associated with recording greater numbers of electrons: increased CCD readout times (for a given level of readout noise). Multiple frame averages can produce much the same result as using a CCD with larger well depth. At present, most microCT systems employ 1K × 1K (1,024 × 1,024 detector elements) or 2K × 2K CCDs with 12 bit or greater depths. (Table 4.1 gives typical formats, pixel sizes, dynamic ranges, and other characteristics of some of the CCDs used in microCT (Bonse et al., 1991; Kinney and Nichols, 1992; Bueno and Barker, 1993; Davis and Elliott, 1997).) Some use is being made of 4K-wide detector formats (i.e., in the

---
* In phosphor-optical lens-detector systems, this is achieved by using higher optical magnification.

TABLE 4.1

CCD Detector Characteristics[a]

Manufacturer	Model	Format	Pixel Size μm	Depth Bits	Readout Time s/Frame	Location
ESRF	FReLoN	2K × 2K	14	14		ESRF, ID19*
Hamamatsu	C4880-10-14A	1K × 1K	12	14	4	SPring-8, BL20B2**
	C4742-95HR	4K × 2.6K	5.9	12	0.6	"
Photometrics	Coolsnap K4	2K × 2K	7.4	12	0.33	APS, 2-BM***

[a] As wavelength sensitivity is comparable for these Si-based detectors, this information is not included.

* (2006) and (2007b) ; ** (2007a); *** (De Carlo et al., 2006) and (2008b).

reconstruction plane). Efforts to develop large detector arrays composed of a mosaic of CCD chips are also of interest, for example, Ito et al. (2007).

The characteristics of the x-ray-to-light conversion media dictate, to a large extent, what coupling scheme is optimum for a given area detector (Kinney et al., 1986; Bonse et al., 1989). Several scintillator properties are important for efficient microCT and nanoCT. First is the range of wavelengths emitted; this should match that of the peak efficiency range of the detector. Second is absorbing power of the scintillator: high-Z, high-density materials are favored for higher-energy photons. Third, the efficiency of emission is an important characteristic; weak light emission exacerbates the photon starvation encountered in tube-based systems and also can degrade counting statistics with synchrotron radiation systems. Generally speaking, microCT detector systems operate in an integration mode so that the emission persistence times are not of greatest concern. The scintillator must be defect-free, or at least possess a small number of defects, and emit relatively uniformly over areas of several square millimeters or larger. An ideal scintillator would not be damaged by the large x-ray doses accumulated over time, or at least would degrade slowly, and it would not be affected adversely by the environment (e.g., water vapor, ozone, trace hydrocarbons).

Table 4.2 lists some characteristics of scintillators used in x-ray microCT and nanoCT; thin film scintillators are a focus of current attention (Martin and Koch, 2006). Cadmium tungstate, for example, has a density of 7.9 g/cm^3, an emission peak of 475 nm and FWHM ~100 nm, and primary decay time of 14 μs. Cadmium tungstate produces 12–15 light photons/keVγ (somewhat less than NaI:Tl) (2005a) and is a good match for a typical

TABLE 4.2
Scintillator Characteristics

Material	Name	Form	Z_{eff}	Density g/cm^3	Emission max. (nm)	Light Yield Photons/keV
Bi$_4$Ge$_3$O$_{12}$	BGO		75	7.13	480	8
CdWO$_4$		Crystal	63	7.9	475	15
CsI:Tl		Columnar thin film	54.1	4.51	550	65
Gd$_2$O$_3$:Eu			61	7.1	611	19
Gd$_2$O$_2$S:Tb	P43	Powder	59.5	7.3	545	
Lu$_3$Al$_5$O$_{12}$:Ce	LAG	Crystal	61	6.73	535	20
Lu$_3$Al$_5$O$_{12}$:Eu	LAG		61	6.73	535	
Y$_3$Al$_5$O$_{12}$:Ce	YAG	Crystal	32	4.55	550	40 to 50
YAlO$_3$:Ce	YAP	Crystal				
Y$_2$O$_2$S:Eu		Powder				
Gd$_3$Ga$_5$O$_{12}$:Eu	GGG		52	7.1		
Lu$_2$O$_3$:Eu			68.8	8.4	611	20
Lu$_2$SiO$_5$:Ce	LSO		65.2	7.4	420	25
Lu$_3$Ga$_5$O$_{12}$:Eu	LGG		58.2	7.4		

Source: 2005a; Martin and Koch (2006); (2007c).

CCD (absolute quantum efficiency >0.4 between 450 and 850 nm for Kodak KAF-4301E, 2K × 2K (2005b)).

Phosphor powders on a screen or embedded in transparent media (Bonse et al., 1986; Kinney et al., 1988; Engelke et al., 1989a,b), monolithic or fiber-optic scintillator glasses (Coan et al., 2006), column-oriented polycrystalline thin films (2008a), single-crystal scintillators (Kinney and Nichols, 1992), and lithographically fabricated cellular phosphor arrays consisting of 2.5-μm spaced close-packed holes filled with plugs of phosphor (Flannery et al., 1987) have been used.* All of the phosphor "screens" except the last can be obtained in a straightforward fashion, and the reason for going to such extremes in producing a discretized micrometer-scale fluorescent screen is to prevent optical cross-talk between adjacent detector elements (Flannery and Roberge, 1987). The noise from scatter, however, remains, and little seems to be gained comparatively by the discretization (Kinney and Nichols, 1992). On the other hand, a new generation of etched column-filled scintillators is under development (Olsen

---
* Many more citations could be provided here, but those given (somewhat arbitrarily) can, at least, start the interested reader in the direction of additional papers.

et al., 2007). Single-crystal phosphors are probably the most popular for synchrotron microCT, but phosphor development continues, including materials formed through thin film processing routes (Koch et al., 1999; Martin and Koch, 2006).

As mentioned above, different x-ray-to-light converters provide different strengths. For example, a $CdWO_4$ single crystal plate 0.5-mm thick and an Y_2O_2S:Eu screen about 40-µm thick were compared (Bonse et al., 1991); the screen produced up to 15 times more light, whereas the single crystal provided considerably better spatial resolution (80 line pairs/mm at 20 percent contrast, corresponding to 6-µm resolution). Several commercial inorganic polycrystalline phosphor screens have been compared to fiber-optic glass scintillator arrays, and better performance was observed with the glass (up to 20 line pairs/mm) than the powder scintillators (Bueno and Barker, 1993). Corrections for various inhomogeneities are required, both geometrical distortions and variation in light emission.

Microchannel plates have found use in coupling powder scintillator screens to CCD cameras (Davis and Elliott, 1997). Optical lens systems have been used by many groups to provide an optical link for powder and single-crystal screens and CCD cameras. A particularly effective scheme is to use one or more low-depth-of-focus optical lens(es) combined with a single-crystal scintillator (Bonse et al., 1991; Kinney and Nichols, 1992). Low depth-of-focus restricts contributions to the image (radiograph) to light photons emitted from a narrow range of depths within the scintillator. Only a small fraction of the image-forming x-ray photons contribute to the image. On the other hand, scatter from adjacent layers of the scintillator and contributions from divergently scattered light photons from the in-focus volume are largely eliminated. Therefore, in situations of photon starvation (i.e., in tube-based systems) optical coupling of scintillator to area detector is generally not the solution of choice, whereas in photon-rich environments (synchrotron radiation sources) optical coupling is almost always adopted.

Neither output from an x-ray source nor detector response is uniform, and this must be corrected if there are not to be serious artifacts in the reconstructions. Commonly, a flat field correction is applied; that is, each radiograph is corrected on a point-by-point basis with an image recorded under the same conditions as the radiograph but with the specimen removed (i.e., a white field image). Often, a dark field correction is applied as well (an image recorded with no incident x-ray photons). Figure 4.4 shows a raw radiograph and the resulting normalized radiograph using the dark field and white field images shown. Despite the high degree of structure seen in the white field (and in the raw radiograph), essentially none of this artificial structure has propagated into the normalized radiograph. Some commercial tube-based systems record a single dark field and a single white field image at the start of the scan; this seems to suffice. Some investigators with custom-built systems forgo correcting for the

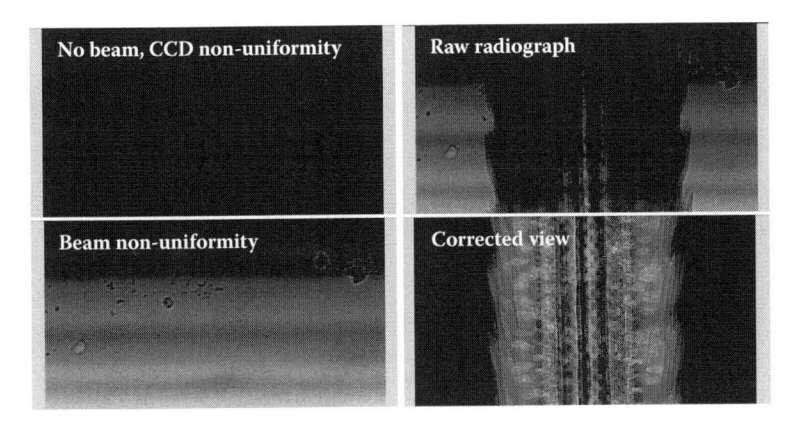

MicroCT Systems and Their Components

FIGURE 4.4
Correction for beam and detector nonuniformity. The specimen is a sea urchin spine *(Diadema setosum)*. The horizontal field of view is 1,024 pixels and the vertical 550 pixels. Data were recorded at 2-BM, APS (18 keV, 300 μm CdWO$_4$ crystal, 4X lens, and ~1.8 μm pixels) by S.R. Stock, K. Ignatiev, and F. De Carlo, 7/25/2002.

dark field variation. In December 2007 at 2-BM, APS, the standard for 2K × 2K reconstructions was to record a white field every 100 projections (rotation increment of 0.125°) and to use the average of all white fields and a single dark field recorded at the end of each dataset (to prevent thermal drift of the x-ray monochromator) for the correction.

There is a limit to the spatial resolution that can be obtained from light-emitting scintillators. This is the wavelength limit described in elementary texts for optical systems. For the materials used in x-ray microCT, this limit is about 0.3 μm, and obtaining better resolving power requires use of x-ray optics before the scintillator (asymmetric crystal beam magnifiers, Fresnel zone plate optics, etc.).

4.4 Positioning Components

Except for translation of the sample out of the beam to collect white field images, the specimen rotator is the single mechanical motion during data collection with the typical rotate-only tube-based or synchrotron microCT system. A rotator without wobble (unintended in-plane and out-of-plane translations from perfect circular paths) would be ideal, but measuring the rotator's imperfections and correcting from them improves reconstruction quality considerably (De Carlo et al., 2006; Rivers and Wang, 2006); see Section 5.2.1. Similarly, translation of the specimen into or out of the beam requires accurate repositioning. Stability over time is another requirement.

In general, one requires precision and accuracy substantially smaller than the smallest voxel size that will be reconstructed. As shown by the effect on reconstruction quality by subpixel physical shifts of the rotation axis (Section 5.1.3), imprecision of one-quarter of a voxel can degrade image quality significantly. In other words, if one were assembling a system for microCT with 1-μm voxels, precision and accuracy of one-quarter this value or smaller are certainly needed.

The rotation axis cannot wobble appreciably relative to the detector rows and columns without degrading reconstruction quality. Wobbles of even 0.03° over 2K pixels can shift projected data to an adjacent row of an area detector. For systems designed for reconstruction with 2-μm or larger voxels, high-precision mechanical bearings in the rotator are adequate, but when 1-μm voxel size or smaller is used, air bearings may be required.

In summary, high-quality mechanical components are required for microCT systems. Some postdata collection processing can be used to ameliorate the inevitable mechanical imperfections (see Chapter 5), but this does not always work well and requires a considerable investment of time for each and every specimen. It is better, therefore, to have hardware that does not require the correction steps.

4.5 Tube-Based Systems

Many investigators continue to design and build systems for microCT and nanoCT, but, for those otherwise inclined, the option of purchasing a commercial system has existed since the 1990s. This section focuses on the commercial systems, and specific purpose-built systems are described elsewhere in the book in conjunction with other topics.

Except for microCT systems explicitly employed for one type and size of specimen (e.g., rapid metrology and quality control as part of a manufacturing line for high-precision components), microCT systems must be able to accommodate a range of specimen sizes and x-ray transparencies. Generally, the more flexible a given system and the greater the range of specimen types it can accommodate, the greater the requirement for an expert operator. Having presets for resolution and x-ray energy (tube voltage in lab systems) greatly increases the efficiency of data acquisition, even for expert operators, and this is a very important consideration in most labs because of the large throughput of samples required and because a significant number of relatively inexperienced users can be expected.

Consider how specimens with different diameters can be accommodated in a fan beam or cone beam system. For simplicity, consider only the fan plane (or the central plane in a cone beam system). Figure 4.5 shows how placing a small-diameter specimen near the x-ray source and farther from the detector allows geometrical magnification to spread the

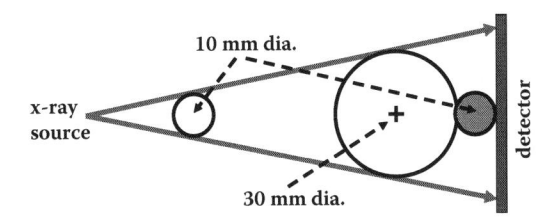

MicroCT Systems and Their Components 53

FIGURE 4.5
Sample position relative to detector and x-ray and projection onto an area detector. A 10-mm-diameter specimen near the x-ray source (white interior) and far from the detector, that is, in an optimum position for minimizing voxel size in the reconstruction. A second position for the 10-mm-diameter specimen (gray interior) near the detector and far from the source. A 30-mm-diameter specimen is shown in its optimum position for matching sample diameter and detector width.

radiograph across the available detector pixels. Placing the same small-diameter specimen near the detector would not utilize all of the available detector pixels. With the small sample near the source, the voxel size* *vox* is as small as possible and will be $vox \sim dia/N_{pix}$, where *dia* is the specimen diameter and N_{pix} is the number of detector voxels. With the small specimen away from the source, $vox \sim dia/N_{eff}$, where $N_{eff} \ll N_{pix}$. If a larger-diameter sample were to be studied, its optimum position would be closer to the detector (farther from the x-ray source) than the smaller specimen. Here, the cost of imaging a larger diameter is paid in terms of decreased geometrical magnification and greater voxel size (for a given number of detector pixels).

For a 1K detector (number of detector pixels in the plane of reconstruction), one could expect to reconstruct a 10-mm-diameter specimen with 10-µm voxels, optical performance such as penumbral blurring allowing. With a 2K detector, 5-µm voxels could be expected for the 10-mm-specimen. With a 30-mm-diameter specimen and a 2K detector, one would obtain 15-µm voxels. For comparison, microCT with synchrotron radiation and a 2K detector might involve a 2-mm-diameter specimen reconstructed with 1-µm voxels (see the following section).

Table 4.3 lists the commercial microCT systems of which the author is aware; some of a given manufacturer's models may have been inadvertently omitted. Space precludes listing all of the system characteristics that might be of interest, but reconstruction size (in voxels), reconstructed voxel size, and specimen diameter are provided. Generally, these are turnkey systems integrated with radiation shielding, generally a boxlike enclosure with interlocks, and do not occupy an inordinate amount of floor space.

---
* The discussion explicitly considers only voxel sizes in reconstructions that contain physical information. One can certainly do the reconstruction with arbitrarily small voxel size, but no information will be contained and there is a significant risk of misrepresentation of what can and cannot be seen in the data if such reconstructions are used.

TABLE 4.3

Commercial Laboratory (Absorption) MicroCT Systems with Manufacturer's Listed Voxel and Reconstruction Sizes as Well as Notes on Specimen Sizes

Manufacturer	Model (Application)	Voxel; Reconstruction Sizes	Notes
BIR (BioImaging Research)	MicroCT (specimens)	<50 µm; 1024^2	[a]
Bioscan	NanoSPECT/CT (in vivo animal)	<200 µm	[b]
Biospace	γ IMAGER-S-CT (small animal)	250 µm	[c]
Gamma Medica-Ideas	X-O (small animal)	To 43 µm; 512^3 to 2048^3	[d]
GE [e]	EXplore Vista PET/CT (small animal)	—	—
	EXplore Locus MicroCT (in vivo)	27, 45, or 90 µm isotropic	[f]
	EXplore Locus SP MicroCT (specimen)	To 8 µm isotropic	[g]
	EXplore Locus Ultra CT	—	[h]
Nittetsu Elex	Ele Scan (specimen)		[i]
	Ele Scan Mini (specimen)		[j]
Phoenix X-ray	Nanotom	To 0.5 µm	[k]
	v|tome|x| 240	To 4 µm	[l]
Scanco Medical	XtremeCT (human peripheral in vivo)	41–256 µm; 512^3 to 3072^3	[m]
	vivaCT 40 (in vivo animal)	10–72 µm isotropic; to $2048^?$	[n]
	MicroCT 80 (specimens)	10–74 µm isotropic; to 2048^2	[o]
	MicroCT 40 (specimens)	6–72 µm isotropic; to 2048^2	[p]
	MicroCT 20 (specimens)	8–34 µm isotropic; to 1024^2	[q]
Shimadzu	SMX-225CT-SV3 (specimens)	To 4096^2	[r]
Siemens [q]	Inveon Multimodality	To 15 µm	[s]
	MicroCAT	To 15 µm; to 4096^2	[t]
Skyscan	1074 (portable)	22 µm; 512^2	[u]
	1076 (in vivo)	<9 µm isotropic, <15 µm	[v]
	1078 (ultrafast in vivo)	47 µm, 94 µm; $(48 mm)^3$	[w]
	1172 (specimens)	<1 µm, 2 µm, 5 or 8 µm	[x]
	1178 (High-throughput, in vivo)	80, 160 µm; 1024^3	[y]
	2011 (nanotomography)	150, 250, 400 nm	[z]
VAMP	TomoScope 30s (rapid examination)	80 µm	[aa]
Xradia	MicroXCT	1–6 µm; 1024^2	[bb]

—continued

MicroCT Systems and Their Components

TABLE 4.3 (continued)
Commercial Laboratory (Absorption) MicroCT Systems with Manufacturer's Listed Voxel and Reconstruction Sizes as Well as Notes on Specimen Sizes

Manufacturer	Model (Application)	Voxel; Reconstruction Sizes	Notes
XRT	NanoXCT	50–70 nm; 1024^2	[cc]
	NanoXFi	<8 nm	[dd]
	X-AMIN PCX		[ee]
	XuM	<100 nm	[ff]
X-tek	Benchtop CT	5 μm	[gg]
	HMX(ST) CT	Feature detection to 1 μm	[hh]
	Venlo CT		[ii]

Source: Adapted from a table compiled and copyrighted by Steven Cool, Radiation Monitoring Devices (used with permission) and supplemented by additional entries.

Notes:
[a] Specimen diameter up to 25 mm, length up to 55 mm.
[b] 1, 2, or 4 SPECT detectors.
[c] Maximum object size 100 mm length, 90 mm diameter.
[d] Maximum object size 97 mm length, 93 mm diameter.
[e] General Electric, previously Enhanced Vision Systems.
[f] Specimen diameter up to 85 mm.
[g] Specimen diameter up to 40 mm. Cone beam system.
[h] Diameter up to 140 mm, long axis up to 100 mm/rotation.
[i] Examples of operating parameters given in user reports (Joo, Sone et al. 2003).
[j] Diameter up to 45 mm, length to 50 mm.
[k] Diameter up to 125 mm, length to 150 mm.
[l] Diameter up to 500 mm, length to 600 mm. Other variants of this industrial system available.
[m] Diameter up to 125 mm, scan length up to 150 mm.
[o] Diameters from 20 to 38 mm, scan length up to 145 mm.
[p] Diameter up to 75.8 mm, scan length up to 120 mm. Cone beam system.
[q] Diameter up to 36.9 mm, scan length up to 80 mm. Stacked (40) fan beam system.
[r] Diameter up to 17.4 mm, scan length up to 50 mm. Fan beam system.
[s] Diameter up to 140 mm.
[t] Previously CTI and Imtek.
[u] PET, SPECT, CT. Diameter to 100 mm.
[v] Cone beam. SPECT option.
[w] Diameter up to 68 mm, scan length up to 200 mm. Cone beam.
[x] Diameter up to 16 mm. Cone beam.
[y] Diameter up to 48 mm, scan length up to 140 mm. Cone beam.
[z] Diameters 20/37 mm or 35/68, depending on version.
[aa] Diameter up to 82 mm, scan length up to 210 mm.
[bb] 0.5-1 mm for maximum resolution, 11 mm maximum diameter (9 μm voxels). Cone beam.
[cc] Diameter up to 40 mm, axial length up to 37 mm. Cone beam.
[dd] Diameters 0.5 to 12 mm; 16 slices.
[ee] Phase imaging.
[ff] SEM-based instrument, phase and absorption imaging.
[gg] Field of view $(20 \mu m)^2$.
[hh] Diameter up to 50 mm.
[ii] Few details available online.

For purposes of illustration of data acquisition rates and system capabilities, consider the microCT instrument the author has in his laboratory, a Scanco MicroCT-40 system. Use of this example does not imply anything about the relative merit of this apparatus relative to the others listed in Table 4.3, and the reader should not assume that MicroCT-40 instruments newer than the author's have exactly the same characteristics. The MicroCT-40 simply is the one commercial system about which the author can speak from first-hand experience.

The Scanco MicroCT-40 system is a turnkey system employing a stacked fan beam geometry (up to 40 slices imaged simultaneously). Presets exist for the principal variables (kVp, tube current, sample diameter, reconstruction resolution), and these are listed in Table 4.4. The investigator has the choice of several different-diameter specimen holders that allow reconstruction with different voxel sizes. The smallest reconstructed voxel size (6 µm) is obtained with the 12.3-mm diameter holder and the highest-resolution scan parameters (1,000 projections with 2,048 samples). Note that this 6-µm value is not the spatial resolution!

In the Scanco MicroCT-40 system, samples are placed into the tubular holder and typically are held in place by foam packing peanuts. The internal temperature of the scanner remains more or less constant at about 25°C. Integration times for each projection can be as high as 0.3 s, and the software allows multiple frames to be collected for each projection (i.e., increasing the signal-to-noise ratio in the reconstructions). The time required to collect a high-resolution set of 40 slices with highest definition (0.3 s integration, 1,000 projections of 2,048 samples) is about 22 min. In high-resolution scans of multiple sets of 40 slices, it is interesting that the two DEC Alpha processors (dating from ~2,000) cannot reconstruct the last set of 40 slices acquired in the time required to collect the projections for the next 40 slices.

In vivo (small animal) microCT systems and nanoCT systems are relatively new commercial products. Systems are identified in Table 4.3, and discussion of nanoCT systems is postponed until Section 4.7. In vivo

TABLE 4.4

Preset Operating Parameters for the Scanco MicroCT-40 as Paradigm for Turnkey Commercial MicroCT System

Tube Settings kVP: µA	Sample Dia mm	Resolution
45: 88 or 177	12.3	250 projections, 1,024 samples, $(1,024)^2$ voxels
50: 72 or 145	16.4	500 projections, 1,024 samples, $(1,024)^2$ voxels
70: 57 or 114	20.3	1000 projections, 2,048 samples, $(2,048)^2$ voxels
	30.7	
	36.9	

microCT, like medical CT, is subject to the constraint that the animal or patient must be kept motionless and the tube and detector must be rotated. This means that small animal in vivo microCT systems are generally larger than normal microCT systems. The relatively large field of view required means that resolution is poorer than in systems such as the Scanco MicroCT-40. In addition, it is highly desirable to minimize x-ray dose in small animal in vivo microCT, but it is not as essential as in human CT. Nonetheless, in vivo microCT systems are capable of providing more than adequate spatial resolution for murine trabecular bone histomorphometry with tissue doses that are so low that they are reckoned not to interfere with the physiological processes of the imaged tissue (Kohlbrenner et al., 2001) or to affect tumor growth (Carlson et al., 2007).

4.6 Synchrotron Radiation Systems

Reviews of microCT at a given synchrotron radiation source appear periodically and typically update new capabilities (see Table 4.5). Because the components required to perform microCT are readily available, many experimental stations occasionally perform microCT in response to their users' requests. Results in the literature increasingly come from dedicated imaging/microCT beam lines, not just because they award many more shifts for microCT but also because the production facilities tailored to a small range of activities are much more efficient.

Synchrotron microCT (without lenses) is typically performed with voxel sizes between 1 and 10 µm, although routine operation with voxel sizes below 0.5 µm is possible at certain facilities, and larger voxel sizes are used upon occasion at most facilities when larger specimen diameters dictate it. Design of microCT systems is driven by the portfolio of specimen types that are expected and the features within that need to be resolved. Available resources inevitably play a role in system characteristics, and constant upgrade of capabilities is the rule at active synchrotron microCT facilities. The systems of which the author is aware are highly modular, and this allows incremental instrumental improvements.

The typical synchrotron (absorption) microCT system (that uses the parallel beam directly without focusing optics) consists of: specimen rotator, x-ray phosphor (single crystal), optical lens, and CCD detector. The available components have improved in capability and affordability over time.

Consider first the mechanical components and the physical stability that are required for high-quality reconstructions. Voxel sizes down to 1–2 µm can be achieved with (relatively) affordable positioning and optical components. Reconstructions with voxel sizes down to 0.5 µm are not uncommon (and capabilities apparently exist for voxel sizes down to 0.3 µm),

TABLE 4.5

Reviews of MicroCT at Different Synchrotron Radiation Sources[a]

Source	Reference
APS	(Rivers, Sutton et al. 1999)
	(Wang, De Carlo et al. 1999)
	(De Carlo, Albee et al. 2001)
	(Wang, De Carlo et al. 2001)
	(De Carlo and Tieman 2004)
	(Rivers, Wang et al. 2004)
	(Rivers and Wang 2006)
	(De Carlo, Xiao et al. 2006)
DESY	(Beckmann, Bonse et al. 1999)
	(Beckmann and Bonse 2000)
	(Beckmann, Lippmann et al. 2000)
	(Beckmann 2001)
	(Beckmann, Donath et al. 2004)
	(Beckmann, Vollbrandt et al. 2005)
	(Beckmann, Donath et al. 2006)
ESRF	(Weitkamp, Raven et al. 1999)
	(Schroer, Meyer et al. 2003)
	(Schroer, Cloetens et al. 2004)
	(Di Michiel, Merino et al. 2005)
	(Baruchel, Buffiere et al. 2006)
	(Martin and Koch 2006)
NSLS	(Dowd, Campbell et al. 1999)
SLS	(Stampanoni, Abela et al. 2004)
	(Stampanoni, Borchert et al. 2006)
SPring-8	(Takeuchi, Uesugi et al. 2001)
	(Takeuchi, Uesugi et al. 2002)
	(Uesugi, Tsuchiyama et al. 2003)
Other	(Lopes, Rocha et al. 2003)
	(Thurner, Wyss et al. 2004)

[a] The abbreviations for the sources are the standard ones: APS — Advanced Photon Source, DESY — Deutsches Elektronen Synchrotron, ESRF — European Synchrotron Radiation Facility, NSLS — National Synchrotron Light Source, SLS — Swiss Light Source, SPring-8 — Super Photon ring 8 GeV.

but the required stability increases system cost considerably. X-ray detector systems at most synchrotron microCT instruments consist of commercially available modules: thin single-crystal phosphors, microscope objective lenses, and CCD or other area detectors for optical wavelengths. Cadmium tungstate single-crystal phosphors precut and polished to the desired thickness are widely used and are relatively inexpensive. These crystals provide light wavelengths with acceptable efficiencies for CCD detectors. Radiation damage dictates periodic replacement of phosphor crystals (and optical lenses if they are in line with the direct beam or prisms if the optical lenses are placed off the beam axis).

Most synchrotron microCT instruments can switch between several optical lenses for different FOV and voxel sizes (Station 2BM of APS routinely uses 1.25X, 2.5X, 4X, and 5X objectives providing FOV of 5.4 mm, 2.7 mm, 1.7 mm, and 1.36 mm, respectively, when used with a 1K detector and twice these values when used with a 2K camera (De Carlo and Tieman, 2004); switches involve unscrewing one lens from the CCD camera body and manual replacement with the new lens. The lens must be refocused, and the replacement operation can be completed in a few minutes. A second example is the microCT system at ESRF ID-19. Fields of view between 40 mm and 0.57 mm are listed (reconstructed voxel sizes of 30 µm and 0.28 µm, respectively), and a rotation mount for very rapid lens changes was commissioned in 2007 (2007b). Most area detectors are $(1K)^2$ or $(2K)^2$ scientific-grade CCDs with depths of 12 bits (see for example, De Carlo et al. (2001) and Stampanoni et al. (2002b)); the specialized FReLoN detector developed at ESRF provides $(2K)^2$ elements with 14-bit depth (see Weitkamp et al. (1999)).

Significant advances in beam delivery optics include wide bandpass monochromator systems based on multilayers; these produce surprisingly uniform beams and increase throughput dramatically compared to single-crystal optics (Chu et al., 2002; Ferrie et al., 2006; 2007b). The system with which the author is familiar is based on a pair of multilayer optics with areas of different layer spacing; tuning to different energies is done by simple translation to the appropriate positions on both optical elements (Chu et al., 2002). At ID 19 of ESRF, for example, multilayer optics provide $\Delta E/E \sim 10^{-2}$ compared to $\Delta E/E \sim 10^{-4}$ for an Si 111 double-crystal monochromator (2007b), and the multilayer provides a corresponding increase in intensity.

Decreased voxel sizes in synchrotron microCT are typically achieved by increasing the magnification of the optical lens coupling the phosphor to the area detector, but there is a limit to what optical magnifications can be used. If the beam passing through the specimen is spread before the phosphor, much smaller voxel sizes can result (at the cost of decreased field of view and increased data collection times for a given brightness incident beam). Placing a perfect crystal in the beam transmitted through the specimen, orienting the crystal to diffract from a Bragg plane inclined with respect to the surface (angle of incidence less than the Bragg angle and exit angle greater than the Bragg angle), and using this diffracted beam for the reconstruction allows smaller voxel sizes for a given lens–area detector combination. This magnification is only along one direction, and use of a second orthogonally oriented crystal is required to magnify along the second direction. Asymmetric Bragg magnifiers have long been used in x-ray diffraction topography (imaging of nearly perfect crystals using diffraction contrast) (Kohra et al., 1970), and Bragg magnifiers have been used between specimen and phosphor in microCT (Sakamoto et al., 1988; Kinney et al., 1993). There has recently been renewed interest in this approach at third-generation synchrotron radiation sources (Spal, 2001; Stampanoni et al., 2001, 2002, 2003, 2005, 2006a), and some of these results are described below in the section on nanoCT. Alignment of these additional optical elements can be time consuming.

Various synchrotron microCT facilities emphasize different scientific missions, time domains, or spatial domains. The author's impressions of some of these differences follow (with apologies for the incomplete subjective nature). The reader should also realize that missions or emphases change with time (as do instrument capabilities) and that the following statements will be somewhat dated by the time this book reaches the reader's hand. At DESY the emphasis appears to be on high-energy microCT and interferometer-based phase imaging (Beckmann et al., 2000). The various facilities at ESRF appear to emphasize high spatial resolution, high temporal resolution, and phase imaging with the propagation method: for example, polychromatic radiation from a wiggler source can be used from near real-time microCT, down to 10 s per set of projections for one reconstruction (Ludwig et al., 2005). At SLS, grating-based phase imaging has received considerable emphasis. Reports from SPring-8 that have come to the author's attention are centered around high spatial resolution and on phase imaging with interferometry. At APS, GSE-CARS focuses on geological applications including measurements at high pressure (Wang et al., 2005); station 2BM (De Carlo et al., 2006) emphasizes rapid throughput (rapid reconstruction via a large dedicated computer cluster, robotic sample changer, facilities for remote access); and time-resolved microCT of evolution of fuel spray (5.1-µs temporal and 150-µm spatial resolution) has been achieved using the pulsed nature of the storage ring (Liu et al., 2004).

There is renewed interest in laminography or tomosynthesis at synchrotron radiation facilities (see Section 3.7), but it remains to be seen whether this technique becomes a mainstream option. There also are a number of commercial tube-based systems for applications such as inspection of solder bump integrity in circuit boards. Local (or region of interest) tomography also is described in Section 3.7, and this approach to high-resolution synchrotron microCT is now well accepted and can be regarded as a valuable option for certain classes of samples.

MicroCT Systems and Their Components   61

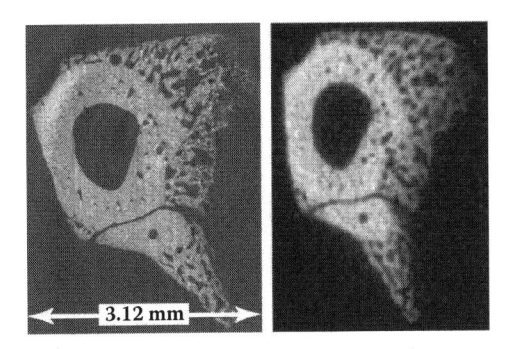

FIGURE 4.6
Synchrotron (left) and tube-based (right) microCT slices of the same sample. The murine tibia (top) and fibula (bottom) in this example of an osteoarthritis model have a considerable growth of bone outside the cortex. In these and all other slice images (unless otherwise noted), contrast is on a linear grayscale with the lighter the pixel, the higher the voxel's absorption. Synchrotron slice: 21 keV, 0.25° rotation increment, 5-μm isotropic voxels (S.R. Stock, D. Novack, K. Ignatiev, F. De Carlo. April 2004, 2-BM, APS). Lab slice: 45 kVp, 500 projections, 12 μm isotropic voxels (S.R. Stock, D. Novack, Scanco MicroCT-40).

Side-by-side comparison of synchrotron and tube-based reconstructions of the same specimen is a rarity in the literature and might be of interest, particularly to those readers new to microCT. Figure 4.6 shows matching synchrotron and tube-based microCT slices of a mouse tibia. As noted in the caption, neither slice was produced using the highest resolution of which each instrument was capable. The features in the synchrotron slice are sharper than in the lab slice, which is to be expected given the voxel sizes (5 μm vs. 12 μm, respectively).

4.7 NanoCT (Full-Field, Microscopy-Based)

Commercial nanoCT systems are listed in Table 4.3; commercial desktop systems have been described in the literature (Sasov, 2004; Tkachuk et al., 2006). Manufacturers report voxel sizes to 100 nm and perhaps smaller; concomitantly smaller specimen diameters than in microCT are required. Scanning electron microscopes possess many of the attributes required for nanoCT of small specimens, and several groups have modified SEMs for this purpose (Yoshimura et al., 2001; Mayo et al. 2003, 2006) or market SEM-based systems. SEMs produce very small diameter electron beams, that is, a very tiny x-ray source, essential for minimizing penumbral blurring and for high spatial coherence for phase imaging.

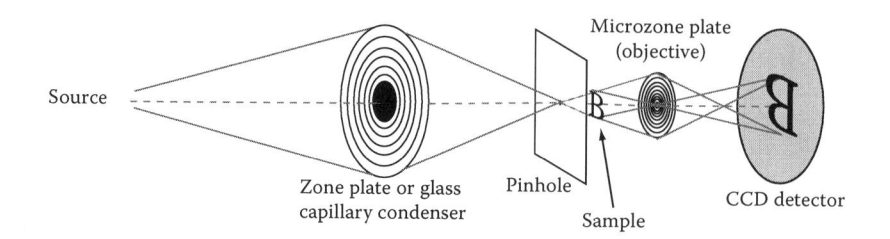

FIGURE 4.7
Lens-based systems for nanoCT employing Fresnel microzone plates. The first zone plate acts as a condenser and the second (after the specimen, schematically indicated by the letter "B") serves as the objective lens. (Adapted with permission from Withers (2007).)

Synchrotron nanoCT reconstructions have been reported using optics to provide submicrometer resolution, and this is a very rapidly developing field. Parabolic x-ray focusing lenses, asymmetric crystal magnifiers, Fresnel zone plates (Figure 4.7), or Kirkpatrick–Baez optics can be used to achieve the necessary magnification. The reader interested in more details is directed elsewhere (McNulty, 2001; Rau et al., 2001b; Schneider et al., 2001; Schroer et al., 2001a,b,c; Spal, 2001; Takeuchi et al., 2001; Uesugi et al., 2001; Stampanoni et al., 2002a,b; Schroer et al., 2003; Rau et al., 2004b; Schroer et al., 2004a; Stampanoni et al., 2005; Rau et al., 2006; Schneider et al., 2006; Stampanoni et al., 2006a; Withers, 2007).

4.8 MicroCT with Phase, Fluorescence, or Scattering Contrast

The main topic of this section is reconstruction using x-ray phase contrast. Two other topics, fluorescence microCT and microCT with x-ray scattering contrast (WAXS or SAXS), are also important but, at the level of coverage intended here, these are discussed more briefly.

4.8.1 Phase Contrast MicroCT

MicroCT using phase contrast is rapidly evolving. Covering the different methods of obtaining x-ray phase contrast radiographs and reconstructions in detail could fill an entire volume, and the coverage here is necessarily briefer.

X-rays are ever so slightly refracted when passing through solids (indices of refraction differ from one by a few parts per million), enough so that x-ray wavefronts distort when passing through regions of different electron density (see Fitzgerald (2000) for an introduction). With a suitable x-ray source (i.e., one with adequate spatial coherence), it is possible to

MicroCT Systems and Their Components

detect changes in contrast resulting from x-rays traversing volumes with different electron densities. Most frequently, phase imaging is performed at a synchrotron radiation source such as the APS, ESRF, SLS, or SPring-8; imaging can also be performed with x-ray tube sources (e.g., Pfeiffer et al. (2006)). Figure 4.8 illustrates four methods where phase effects are used to produce contrast in x-ray images.

In the propagation method (Figure 4.8a), the detector is placed much farther away from the sample than is normal for x-ray imaging (~1 m vs. ~1 cm); refracted x-rays "r" diverge and interfere with other x-rays at the detector plane, producing detectable fringes in the image at external and internal boundaries between materials with different electron densities. Here contrast is provided by differences in the second derivative of the x-ray phase (Snigerev et al., 1995; Cloetens et al., 1996, 1999a, 2000; Weitkamp et al., 2001; Ludwig et al., 2003; Weon et al., 2006). Images acquired at four or more specimen–detector separations (typically from 5 mm to 1 m or more) are required to extract the phase information (see Cloetens et al. (1999b)), and this method can be described as an analogue of the focus variation method in transmission electron microscopy (Cloetens et al., 1999a). The terms *holotomography* and *propagation method* are used in the literature.

In diffraction-enhanced imaging (DEI; Figure 4.8b), an analyzer crystal is placed in the x-ray beam after the sample; images recorded with different settings of the analyzer isolate changes in the phase angle. This method produces image contrast based on changes in the first derivative

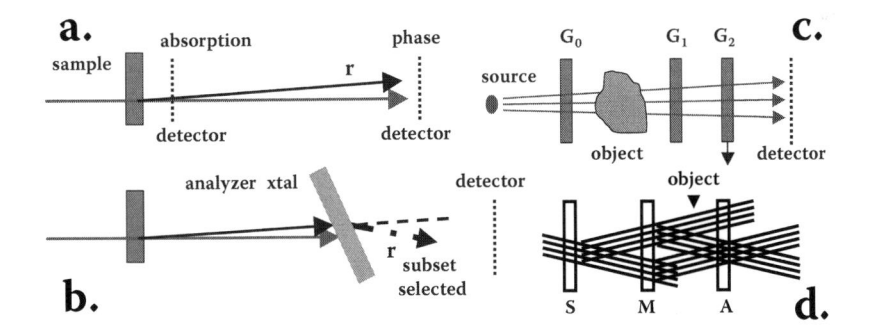

FIGURE 4.8
Methods of x-ray phase imaging: (a) propagation method. Images with the detector near the sample are dominated by absorption contrast, but placing the detector far from the specimens allows refracted x-rays "r" to interfere with transmitted x-rays and to produce edge contrast. (b) Diffraction-enhanced imaging (DEI). (c) Grating enhanced imaging. Grating G_0 is not required with synchrotron x-radiation but is needed to provide a series of small virtual sources of x-ray tube-based imaging. (d) Bonse–Hart interferometer for imaging. "S, M, and A" are crystal beam splitters, mirror, and analyzer, respectively. (Reproduced from Stock (2007).)

of the x-ray phase (Davis et al., 1995; Chapman et al., 1997). Essentially, the analyzer selects only a small angular fraction of the refracted radiation.

The grating-enhanced imaging method (Figure 4.8c), also known as Talbot interferometry, is analogous to DEI except that contrast from changes in the first derivative of the phase is provided by translation of one analyzer grating relative to a second instead of by rotation of the analyzer crystal and its periodic array (the crystal lattice); see Momose et al. (2004), Weitkamp et al. (2005), Groso et al. (2006), Momose et al. (2006), Stampanoni et al. (2006b), and Weitkamp et al. (2006). Results obtained with gratings are used below to illustrate reconstruction of pure phase contrast.

Interferometry for phase imaging is illustrated in Figure 4.8d. In the Bonse–Hart geometry, a beamsplitter "S" produces a reference beam and an imaging beam, the mirror "M" redirects the beams together, the object is placed in one of the beams exiting the mirror, and the analyzer "A" recombines the reference and object-modified beams (Momose et al., 1996; Beckmann et al., 1999; Momose et al., 1999; Bonse and Beckmann, 2001; Momose et al., 2003). An alternative is the shearing interferometer (David et al., 2002; Iwata et al., 2004), and the limited FOVs of interferometers from monolithic blocks of Si have recently been improved (Takeda et al., 2000; Yoneyama et al., 2002, 2005). Interferometers allow changes in the x-ray phase to be measured directly, not merely its derivatives.

At this point in the development of x-ray phase imaging, relatively little comparison of the different modalities has appeared in the literature. Kiss and coworkers (Kiss et al., 2003) discuss image contrast numerically for absorption versus diffraction-enhanced radiography. Pagot et al. (2005) compared radiography for phase propagation and diffraction-enhanced imaging, but it is not clear how their conclusions translate to microCT. Wernick et al. (Wernick et al., 2003, 2004; Chou et al., 2006) and Paganin et al. (2004) discuss different representations of phase imaging data, but not microCT data, and Mayo et al. (2002) examine these representations in microCT reconstructions.

Grating-based phase imaging provides an illuminating illustration of how phase microCT techniques work. Consider first the situation where no specimen is present and the spatially coherent x-ray beam passes through phase grating G_1, the lines of which show negligible absorption but substantial phase shift (Figure 4.8c). Note that grating G_0 is typically present only with imaging with an x-ray tube with large source size. Grating G_1 acts as a beamsplitter, producing the two diffracted beams used for image formation. Because the wavelength of the illuminating x-rays ($\sim 10^{-10}$ m) is much smaller than the grating period ($\sim 10^{-6}$ m), the angle between the two beams is so small that the beams overlap almost completely as they propagate away from G_1 and interfere. The interference pattern generated could be imaged directly with an x-ray detector placed an appropriate distance d_g from G_1 (see Weitkamp et al. (2005, 2006) for the relationship of d_g to x-ray wavelength λ, periodicity, and other characteristics of the grating), but lack of spatial resolution of the detector systems has led to an alternative solution, use of an absorption grating G_2 positioned d_g away from G_1. The analyzer grating G_2 acts as a transmission mask for the detector placed immediately behind it and transforms the local interference fringe position into signal intensity variation. Note that the gratings must be parallel.

Placing a specimen upstream of G_1 produces local wavefront distortions $\Phi(x,y)$ and alters the interference pattern. Phase imaging is performed by translating the analyzer grating G_2 by small increments x_g of the fringe periodicity g and recording a radiograph at each position; Figure 4.9 shows images of polystyrene spheres for different x_g (Weitkamp et al., 2005). The signal intensity $I(x,y)$ at each pixel (x,y) in the detector plane oscillates as a function of x_g, and the phases $\varphi(x,y)$ of the intensity oscillations in each pixel are related to $\Phi(x,y)$ via:

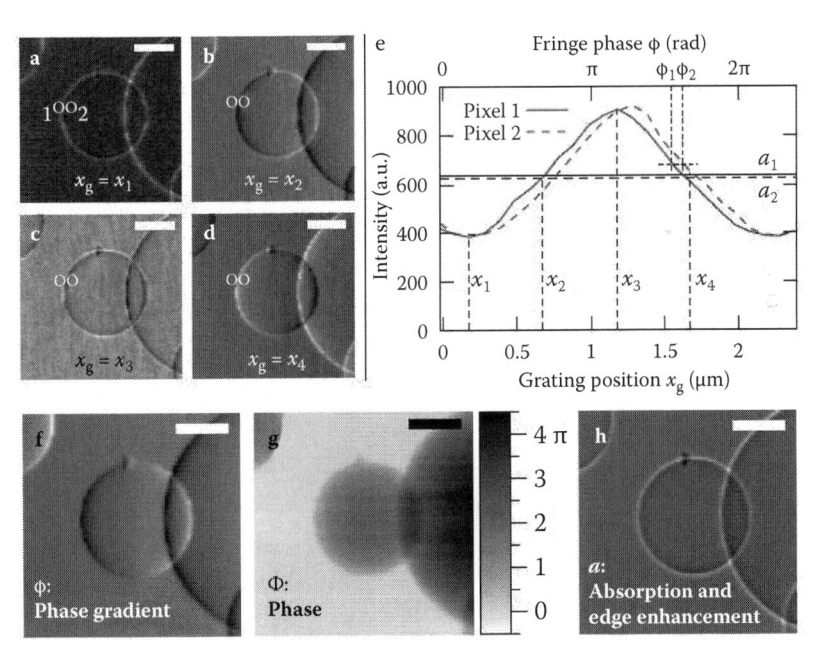

FIGURE 4.9
Principle of phase stepping (a)–(d) interferograms of polystyrene spheres (100 and 200 μm diameter), taken at the different relative positions $x_g = x_1, \ldots, x_4$ of the two interferometer gratings. (e) Intensity oscillation in the two different detector pixels $i = 1, 2$ as a function of x_g. For each pixel, the oscillation phase φ_i and the average intensity a_i over one grating period can be determined. (f) Image of the oscillation phase φ for all pixels. (g) Wavefront phase Φ retrieved from φ by integration. (h) Image of the averaged intensity a for all pixels, equivalent to a noninterferometric image. The length of the scale bar is 50 μm. (Reproduced with permission from Weitkamp et al. (2005).)

$$\varphi = (\lambda\, d_g/g_2)\, \partial\, \Phi/\partial\, x, \tag{4.1}$$

where g_2 is the period of the absorption grating (Weitkamp et al., 2005). The phase profile of the object can be retrieved from $\varphi(x,y)$ by simple one-dimensional integration (Figure 4.9g). Radiographs at as few as three positions x_g are needed to extract φ if one knows a priori that the intensity oscillation is sinusoidal, but the reconstructions in Weitkamp et al. (2005) were obtained using eight phase steps per projection. Once the set of phase radiographs is obtained at the different viewing angles, a pure phase reconstruction can be computed using the normal methods.

Synchrotron radiation is not essential for phase microCT (see Fitzgerald (2000)). The x-ray source size provided by the electron beam in an SEM provides adequate spatial coherence for phase microCT (Mayo et al., 2003), and modifying an SEM can be an effective way of studying small specimens. The fringe formation underlying the grating method described in the previous paragraph is independent of x-ray wavelength, and, provided a grating G_0 is used before the specimen (Figure 4.9c) to provide a small virtual source size (more precisely, a set of independent small sources), a relatively high-power x-ray tube and gratings can be used for phase microCT (Pfeiffer et al., 2006).

In specimens such as foams where the majority of volume is air, the total phase shift across the specimen varies relatively little, and holotomographic reconstruction can utilize the absolute values of the phase. In solid cylindrical Al–Si specimens such as that used by Cloetens et al. (~1.5 mm dia.), phase shifts will vary over 200 radians at 18 keV (Cloetens et al., 2000), and this dictates that the reconstructions employ the phase variations with respect to the phase introduced by the homogeneous matrix (i.e., the x-ray phase relative to that of the matrix).

Cloetens et al. (2000) provide a clear illustration of differences in absorption and phase enhanced tomography reconstructions produced with the propagation method. Figure 4.10 compares the same slice from an Al–Si specimen (grains of Al embedded in a matrix of very fine Al–Si eutectic) obtained under three different imaging conditions. The radiographs for the first reconstruction were absorption dominated (i.e., they were recorded with a very small specimen detector separation DS), the radiographs for the second with a single large DS (edge-enhanced interface contrast), and the radiographs at four DS were combined via the holotomography algorithm (see above) for the third. In Figure 4.10a, absorption contrast does not allow one to distinguish the Al grains and Al–Si eutectic matrix. Edge enhancement allows the two phases to be seen clearly (Figure 4.10b), but, because the Fresnel fringe intensity varies from position to position, segmentation of the grain and eutectic phases is challenging. The reconstruction with variation in refractive index decrement (Figure 4.10c) clearly shows the different metallurgical phases whose dif-

MicroCT Systems and Their Components

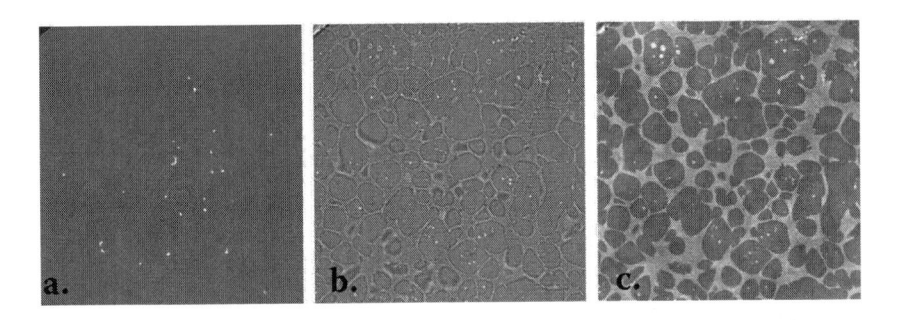

FIGURE 4.10
Matching slices of an Al–Si alloy quenched from a semi-solid state using (a) absorption contrast (sample detector separation DS = 7 mm), (b) phase (propagation) contrast (DS = 0.6 m), and (c) phase contrast (holotomography) with DS = 0.07, 0.2, 0.6, and 0.9 m. Horizontal field of view 0.9 mm, 18 keV, effective voxel size of 1.9 µm. (Reproduced from Cloetens et al. (2000) with permission of Lavoisier.)

ference in density is on the order of 0.05 g/cm^3 (Cloetens et al., 2000), and segmentation is quite straightforward.

In interferometer-based phase microCT, the spatial distributions of polystyrene (PS) and poly(methyl methacrylate) (PMMA) in a ~50 vol.% mixture were imaged (Momose et al., 2005). The polymers are immiscible (although analyses of the values of the refraction indices of both phases suggest immiscibility is not total) and form a phase-separated system. The achieved contrast resolution was, in terms of density resolution, <4 mg/cm^3, clearly beyond what is obtainable with absorption-based microCT.

Phase-based microCT has been used in a wide variety of studies, descriptions of which are folded into the various subsections of materials applications. A few examples are mentioned here in closing the subsection, including damage in composites (Cloetens et al., 1997; Buffiere et al., 1999) and structures in biological specimens (Momose et al., 1999; Wu et al., 2004). Real-time phase radiography of insect respiration produced interesting new insights (Westneat et al., 2003), and phase microCT, providing the third dimension, will undoubtedly prove very valuable.

4.8.2 Fluorescence MicroCT

Several interactions can occur between a beam of x-rays and the atoms in the specimen through which the beam passes. If the x-ray photon energy is high enough, atoms can fluoresce, emitting photons with energies characteristic of the electronic shell transition that produced the emitted photons. These characteristic x-rays have well-defined energies; an energy-sensitive detector can measure the intensity of each characteristic peak, and this intensity can be converted to the concentration of these atoms within the irradiated volume. Detection limits (atomic concentration) for x-ray fluorescence are much, much lower than for x-ray absorption, and the elemental specificity is much, much higher for the former; these advantages continue to drive development of fluorescence microCT for applications where small concentrations of elements need to be mapped.

X-ray fluorescence occurs in all directions, and the typical experimental setup positions an energy-sensitive x-ray detector to one side of the specimen in order to count the photons emitted from the irradiated volume (in the direction of the detector). Neither an area nor a ribbon-like x-ray beam appears to be practical for use in fluorescence microCT due to the confounding crossfire from different ray paths through the specimen (although receiving slit systems are being developed that may invalidate this conclusion). One consequence of sampling along only a single ray (i.e., of using a pencil beam) is that data collection rates are quite low. Quantification requires correction for absorption of the emitted characteristic x-rays along the path to the detector, and the reader is directed elsewhere for more details (Simionovici et al., 1999; Schroer et al., 2000).

An interesting option is to use a polycapillary focusing optic to localize fluorescence from a single position along the beam path. This appears to be a viable option for 3D mapping, an option that does not require sample rotation to provide a complete map of elemental distribution in a slice (Vincze et al., 2004).

Reports combining fluorescence and phase microCT in biomedical applications (organs labeled with iodine-containing contrast agents) have appeared (Takeda et al., 2004, 2006; Wu et al., 2006). Other reports of element distributions in organs include Takeda et al. (1999). Elemental maps in slices of specimens of roots have been published: K, Fe, Rb, Cl in mahogany (Schroer et al., 2001b; Simionovici et al., 2001) and K, Fe, Zn in tomato (Schroer et al., 2003). Mapping in small particles has been of interest, both for those of terrestrial origin (fly ash particles (Simionovici et al., 2001); sediment particles (Vincze et al., 2001); diatoms (Simionovici et al., 2004)) and for those of extraterrestrial origin (Si, Ca, Fe, and Cr maps in a microfragment of the Tatahouine meteriorite (Simionovici et al., 2001; Lemelle et al., 2003); S, Ca, Cr, Mn, Fe, Ni, Cu, and Zn maps in a cosmic dust particle (Schroer et al., 2004b); and Fe, Ni maps in micrometeriorites (Chukalina et al., 2003; Ignatyev et al., 2006)). Other studies include Fe nanocatalyst spatial distribution (Jones et al., 2005), trace elements in a SiC shell of a nuclear fuel particle (Naghedolfeizi et al., 2003), metal elemental maps within inclusions in diamond and quartz (Vincze et al., 2004), and light elements in biological specimens (Vincze et al., 2004). Spectroscopy related to absorption edges has been used in chemical tomographic mapping (Owens et al., 2001; Rau et al., 2001a; Schroer et al., 2003; Rau et al., 2004b).

4.8.3 Scatter MicroCT

X-rays can also be scattered by structures with electron densities differing from their surroundings. Scattering from periodic arrays of atoms reinforces intensity along certain directions in the wide-angle x-ray scattering regime, and scattering from fibrils, particles, and the like can produce peaks in scattered intensity or other characteristic scattering profiles in the small-angle regime. The larger the size or spacing of the scatterers, the smaller is the angle of the scattered intensity.

SAXS microCT is an ideal approach for studying polymer texture: absorption microCT shows no contrast, but differences in SAXS with position can be pronounced (Schroer et al., 2006; Stribeck et al., 2006). The complete SAXS pattern must be recorded for a single ray through the specimen; there would be too much overlap between patterns of adjacent rays if, for example, a ribbon beam were used. In a ~5-mm rod of warm drawn polyethylene, different layers could be resolved with SAXS microCT (see the processing subsection for more details).

Tomographic reconstructions of idealized specimens using diffracted intensity from different *hkl* (and different phases) have also been reported, for example, Stock et al. (2008). With such a tomographic approach, it may prove possible to study the spatial distribution of crystallographic texture as well as spatial distribution of absorption.

4.9 System Specification

There is certainly no one commercial or synchrotron microCT system that is "the best." It all depends on the intended application(s). For someone (new to microCT) considering the purchase of a microCT system, some reasonable steps to go through in the decision-making process and some questions to answer include the following.

1. Define the specimens of interest (present plus future applications). What is the required field of view and largest specimen dimensions (not just the diameter)? What spatial resolution and contrast sensitivity are needed? What is the needed penetrating power (i.e., the x-ray kVp range)? Very different operating parameters would be required for studying Ti samples versus studying soft tissues.

2. Define the software needed. Software for the following are or can be very important for the users: reconstruction of slices, 3D renderings of segmented data and superposition of different thresholds in renderings, numerical analysis of microstructure

(including simple 3D manual distance measurements, numerical sectioning, specimen reorientation), movies of spinning renderings or movies paging through stacks of slices, import or export of datasets, and image processing beyond simple segmentation (masking, etc.).

3. Identify the computer architecture (PC, Mac, other). How is the data backed up? What is the hard drive capacity and options for downloading data to other applications (Ethernet, DVD, flash drive)?

4. Determine how long it takes to: collect data per slice, per dataset; reconstruct slices; compute renderings; back up data.

5. Evaluate actual performance on specimens typical of the intended applications. Have the manufacturer run some specimens, consult people who already have the system, "drive" the microCT system for one or two days.

6. Determine the required levels of service support. What are previous users' experiences with a given manufacturer?

7. Ascertain how closely the cost of the identified "ideal" system matches the funds available. Can an almost-as-good system give better "bang for the buck"?

The question of whether a tube-based microCT system or synchrotron microCT is required is something each investigator must determine for himself or herself. The considerations listed for purchasing a tube-based microCT system above also apply to the synchrotron versus lab microCT decision. Synchrotron microCT can be done at several storage rings and often at more than one beam line at a single source, and deciding which instrument to use can follow the general points listed above. Unlike the author who is less than one hour by car from the APS, most potential users are fairly far from a suitable storage ring, and this affects how often and how long one can get access to the instrument. Travel costs must be factored into any decision to commit to doing synchrotron microCT, but inexpensive airfares can be obtained, given that schedules are normally formalized several months in advance.

Performing synchrotron microCT involves two steps. First is obtaining beam time, a topic discussed in this paragraph. Second is utilization of that time, something covered in the following paragraph. Obtaining access to a synchrotron microCT system almost always requires writing a fairly short, but focused, scientific proposal. Writing a successful proposal for beam time is not really very difficult, especially if one has experience and if there is a good scientific case for the work. However, first-time users would be well advised to get feedback from disinterested but experienced current users and to consult the text of others' successful proposals because what is expected in the proposal may not be evident to new users. A submitted proposal for beam time is reviewed at several levels, including by users with expertise with the technique of interest, and, if the importance and scientific case warrant, experimental time is awarded. At the APS, for example, there are currently three scheduling cycles, and the proposal deadline is currently about three months prior to the start of the cycle.

Planning experiments is very different for synchrotron "runs" compared to those done on a lab-based instrument. Access is quite limited (say once per scheduling cycle), and one is typically awarded a number of consecutive 8-hr shifts, between 3 and 21 shifts, often 9 shifts for 2-BM, APS. This means that many samples must be prepared or collected in advance and that it is difficult to make good use of the beam time if there are problems with the sample design. One does not have the luxury, when things are not going well, of taking the night off and thinking about what is going wrong (more precisely, one can do this, but the effect on productivity and on obtaining more beam time is quite undesirable). Working away from one's home laboratory also means that one cannot walk down the hall to retrieve key equipment or tools that one may unexpectedly need; many useful items can be borrowed from staff or other users at a storage ring, but locating who has the item in question can be very time consuming or, in the case of late nights or weekends, impossible. Despite the complications mentioned above, the number of samples that can be imaged per unit time (see Section 5.4) and the superior resolution and contrast sensitivity make synchrotron microCT attractive for many, but not all, investigators.

References

(2005a). CdWO$_4$ Cadmium tungstate scintillation material, Saint-Gobain Ceramics and Plastics. Retrieved 12/07 from www.detectors.saint-gobain.com.

(2005b). KAF-4301E 2084(H) x 2084(V) pixel enhanced response full-frame CCD image sensor performance specification, Eastman Kodak Co. Retrieved 1/08 from www.kodak.com/go/imagers.

(2006). FReLoN. Retrieved 1/29/08 from www.esrf.eu/UsersAndScience/Experiments/Imaging/ID22/BeamlineManual/Detectors/Ccd/Frelon.

(2007a). BL20B2 Image detectors, SPring-8. Retrieved 11/25/07 from www.spring8.or.jp/wkg/BL20B2/instrument/lang-en.

(2007b). ID19 High-resolution diffraction topography — microtomography beamline home page. Retrieved 1/24/07, from http://www.esrf.eu/UsersAndScience/Experiments/Imaging/ID19.

(2007c). Scintillation detector applications using Si diodes. Retrieved November 2007, from www.deetee.com.

(2008a). Cesium iodide scintillator films, Radiation Monitoring Devices, Inc. Retrieved 1/08 from www.rmdinc.com/products/p005.html.

(2008b). CoolSNAP K4 monochrome. Retrieved 1/29/08 from www.photomet.com/files/PDF/datasheets/k4.pdf.

Armistead, R.A. (1988). CT: Quantitative 3-D inspection. *Adv Mater Process Inc Met Prog* (Mar): 41–49.

Baruchel, J., J.Y. Buffiere, P. Cloetens, M.D. Michiel, E. Ferrie, W. Ludwig, E. Maire, and L. Salvo (2006). Advances in synchrotron radiation microtomography. *Scripta Mater* **55**: 41–46.

Beckmann, F. (2001). Microtomography using synchrotron radiation as a user experiment at beamlines BW2 and BW5 of HASYLAB at DESY. *Developments in X-ray Tomography III*. U. Bonse (Ed.). Bellingham, WA, SPIE. *SPIE Proc Vol* **4503**: 34–41.

Beckmann, F. and U. Bonse (2000). Attenuation- and phase-contrast microtomography using synchrotron radiation for the 3-dim. investigation of specimens consisting of elements with low and medium absorption. *Applications of Synchrotron Radiation Techniques to Materials Science V*. S.R. Stock, S.M. Mini, and D.L. Perry (Eds.). Warrendale,PA, MRS. *MRS Proc Vol* **590**: 265–271.

Beckmann, F., U. Bonse, and T. Biermann (1999). New developments in attenuation and phase-contrast microtomography using synchrotron radiation with low and high photon energies. *Developments in X-ray Tomography II*. U. Bonse (Ed.). Bellingham, WA, SPIE. *SPIE Proc Vol* **3772**: 179–187.

Beckmann, F., T. Donath, T. Dose, T. Lippmann, R.V. Martins, J. Metge, and A. Schreyer (2004). Microtomography using synchrotron radiation at DESY: Current status and future developments. *Developments in X-Ray Tomography IV*. U. Bonse (Ed.). Bellingham, WA, SPIE. *SPIE Proc Vol* **5535**: 1–10.

Beckmann, F., T. Donath, J. Fischer, T. Dose, T. Lippmann, L. Lottermoser, R.V. Martins, and A. Schreyer (2006). New developments for synchrotron-radiation-based microtomography at DESY. *Developments in X-Ray Tomography V*. U. Bonse (Ed.) Bellingham, WA, SPIE. *SPIE Proc Vol* **6318**: 631810–1 to –10.

Beckmann, F., T. Lippmann, and U. Bonse (2000). High-energy microtomography using synchrotron radiation. *Penetrating Radiation Systems and Applications II*. F. Doty, H. B. Barber, H. Roehrig, and E.J. Morton (Eds.). Bellingham, WA, SPIE. *SPIE Proc Vol* **4142**: 225–230.

Beckmann, F., J. Vollbrandt, T. Donath, H.W. Schmitz, and A. Schreyer (2005). Neutron and synchrotron radiation tomography: New tools for materials science at the GKSS-Research Center. *Nucl Instrum Meth A* **542**: 279–282.

Bonse, U. and F. Beckmann (2001). Multiple-beam x-ray interferometry for phase-contrast microtomography. *J Synchrotron Rad* **8**: 1–5.

Bonse, U., Q. Johnson, M. Nichols, R. Nusshardt, S. Krasnicki, and J. Kinney (1986). High resolution tomography with chemical specificity. *Nucl Instrum Meth* **A246**: 644–648.

Bonse, U., R. Nusshardt, F. Busch, R. Pahl, Q.C. Johnson, J.H. Kinney, R.A. Saroyan, and M.C. Nichols (1989). Optimization of CCD-based energy-modulated x-ray microtomography. *Rev Sci Instrum* **60**: 2478–2481.

Bonse, U., R. Nusshardt, F. Busch, R. Pahl, J.H. Kinney, Q.C. Johnson, R.A. Saroyan, and M.C. Nichols (1991). X-ray tomographic microscopy of fibre-reinforced materials. *J Mater Sci* **26**: 4076–4085.

Borodin, Y. I., E.N. Dementyev, G.N. Dragun, G.N. Kulipanov, N.A. Mezentsev, V.F. Pindyurin, M.A. Sheromov, A.N. Skrinsky, A.S. Sokolov, and V.A. Ushakov (1986). Scanning difference microscopy and microtomography using synchrotron radiation at the storage ring VEPP-4. *Nucl Instrum Meth* **A246**: 649–654.

Bowen, D.K., J.C. Elliott, S.R. Stock, and S.D. Dover (1986). X-ray microtomography with synchrotron radiation. *X-Ray Imaging II*. D.K. Bowen and L.V. Knight (Eds.). Bellingham, WA, SPIE. *SPIE Proc Vol* **691**: 94–98.

Breunig, T.M., J.C. Elliott, P. Anderson, G. Davis, S.R. Stock, A. Guvenilir, and S.D. Dover (1990). Application of x-ray microtomography to the study of SiC/Al metal matrix composite material. *New Materials and Their Applications*. D. Holland (Ed.). London, Institute of Physics. **111**: 53–60.

Breunig, T.M., J.C. Elliott, S.R. Stock, P. Anderson, G.R. Davis, and A. Guvenilir (1992). Quantitative characterization of damage in a composite material using x-ray tomographic microscopy. *X-ray Microscopy III*. A.G. Michette, G.R. Morrison, and C.J. Buckley (Eds.). New York, Springer: 465–468.

Breunig, T.M., S.R. Stock, A. Guvenilir, J.C. Elliott, P. Anderson, and G.R. Davis (1993). Damage in aligned fibre SiC/Al quantified using a laboratory x-ray tomographic microscope. *Composites* **24**: 209–213.

Bueno, C. and M.D. Barker (1993). High resolution digital radiography and 3D computed tomography. *X-ray Detector Physics and Applications II*. V.J. Orphan (Ed.). Bellingham, WA, SPIE. *SPIE Proc Vol* **2009**: 179–191.

Buffière, J.Y., E. Maire, P. Cloetens, G. Lormand, and R. Fougeres (1999). Characterization of internal damage in a MMCp using x-ray synchrotron phase contrast microtomography. *Acta Mater* **47**: 1613–1625.

Burstein, P., P.J. Bjorkholm, R.C. Chase, and F.H. Seguin (1984). The largest and smallest X-ray computed tomography systems. *Nucl Instrum Meth* **221**: 207–212.

Carlson, S.K., K.L. Classic, C.E. Bender, and S.J. Russell (2007). Small animal absorbed radiation dose from serial micro-computed tomography imaging. *Mole Imaging Biol* **9**: 78–82.

Chapman, D., W. Thomlinson, R.E. Johnson, D. Washburn, E. Pisano, N. Gmür, Z. Zhong, R. Menk, F. Arfelli, and D. Sayers (1997). Diffraction enhanced x-ray imaging. *Phys Med Biol* **42**: 2015–2025.

Chou, C.Y., M.A. Anastasio, J.G. Brankov, and M.N. Wernick (2006). A comparison of a generalized DEI method with multiple image radiography. *Developments in X-Ray Tomography V*. U. Bonse (Ed.). Bellingham, WA, SPIE. *SPIE Proc Vol* **6318**: 631819–1 to –8.

Chu, Y.S., C. Liu, D.C. Mancini, F.D. Carlo, A.T. Macrander, B. Lai, and D. Shu (2002). Performance of a double-multilayer monochromator at beamline 2-BM at the advanced photon source. *Rev Sci Instrum* **73**: 1485–1487.

Chukalina, M., A. Simionovici, L. Lemelle, C. Rau, L. Vincze, and P. Gillet (2003). X-ray fluorescence tomography for non-destructive semi-quantitative study of microobjects. *J Phys IV* **104**: 627–630.

Cloetens, P., R. Barrett, J. Baruchel, J.P. Guigay, and M. Schlenker (1996). Phase objects in synchrotron radiation hard x-ray imaging. *J Phys D* **29**: 133–146.

Cloetens, P., W. Ludwig, J. Baruchel, D.V. Dyck, J.V. Landuyt, J.P. Guigay, and M. Schlenker (1999a). Holotomography: Quantitative phase tomography with micrometer resolution using hard synchrotron radiation x-rays. *Appl Phys Lett* **75**: 2912–2914.

Cloetens, P., W. Ludwig, J.P. Guigay, J. Baruchel, M. Schlenker, and D. v. Dyck (2000). Phase contrast tomography. *X-Ray Tomography in Materials Science*. J. Baruchel, J.Y. Buffière, E. Maire, P. Merle, and G. Peix (Eds.). Paris, Hermes Science: 29–44.

Cloetens, P., M. Pateyron-Salomé, J.Y. Buffière, G. Peix, J. Baruchel, F. Peyrin, and M. Schlenker (1997). Observation of microstructure and damage in materials by phase sensitive radiography and tomography. *J Appl Phys* **81**: 5878–5886.

Cloetens, P., D. Van Dyck, J.P. Guigay, M. Schlenker, and J. Baruchel (1999b). Quantitative phase tomography by holographic reconstruction. *Developments in X-ray Tomography II*. U. Bonse (Ed.). Bellingham, WA, SPIE. *SPIE Proc Vol* **3772**: 279–290.

Coan, P., A. Peterzol, S. Fiedler, C. Ponchut, J.C. Labiche, and A. Bravin (2006). Evaluation of imaging performance of a taper optics CCD "FReLoN" camera designed for medical imaging. *J Synchrotron Rad* **13**: 260–270.

Connor, W.C., S.W. Webb, P. Spanne, and K.W. Jones (1990). Use of x-ray microscopy and synchrotron microtomography to characterize polyethylene polymerization particles. *Macromol* **23**: 4742–4747.

David, C., B. Nöhammer, H.H. Solak, and E. Ziegler (2002). Differential x-ray phase contrast imaging using a shearing interferometer. *Appl Phys Lett* **81**: 3287–3289.

Davis, G.R. (1997). Image quality in x-ray microtomography. *Developments in X-ray Tomography*. U. Bonse (Ed.). Bellingham, WA, SPIE. *SPIE Proc Vol* **3149**: 213–221.

Davis, G.R. and J.C. Elliott (1997). X-ray microtomography scanner using time-delay integration for elimination of ring artifacts in the reconstructed image. *Nucl Instrum Meth* **A394**: 157–162.

Davis, G.R. and F.S.L. Wong (1996). X-ray microtomography of bones and teeth. *Physiol Meas* **17**: 121–146.

Davis, T.J., D. Gao, T.E. Gureyev, A.W. Stevenson, and S.W. Wilkins (1995). Phase-contrast imaging of weakly absorbing materials using hard x-rays. *Nature* **373**: 595–598.

De Carlo, F. and B. Tieman (2004). High-throughput x-ray microtomography system at the advanced photon source beamline 2-BM. *Developments in X-Ray Tomography IV*. U. Bonse (Ed.). Bellingham, WA, SPIE. *SPIE Proc Vol* **5535**: 644–651.

De Carlo, F., P. Albee, Y.S. Chu, D.C. Mancini, B. Tieman, and S.Y. Wang (2001). High-throughput real-time x-ray microtomography at the Advanced Photon Source. *Developments in X-ray Tomography III*. U. Bonse (Ed.). Bellingham, WA, SPIE. *SPIE Proc Vol* **4503**: 1–13.

De Carlo, F., X. Xiao, and B. Tieman (2006). X-ray tomography system, automation and remote access at beamline 2-BM of the Advanced Photon Source. *Developments in X-Ray Tomography V*. U. Bonse (Ed.). Bellingham, WA, SPIE. *SPIE Proc Vol* **6318**: 63180K–1 to –13.

Di Michiel, M., J.M. Merino, D. Fernandez-Carreiras, T. Buslaps, V. Honkimaki, P. Falus, T. Martins, and O. Svensson (2005). Fast microtomography using high energy synchrotron radiation. *Rev Sci Instrum* **76**: 043702–1 to –7.

Dilmanian, F.A., X.Y. Wu, B. Ren, T.M. Button, L.D. Chapman, J.M. Dobbs, X. Huang, E.L. Nickoloff, E.C. Parsons, M.J. Petersen, W.C. Tomlinson, and Z. Zhong (1997). CT with monochromatic synchrotron x-rays and its potential in clinical research. *Developments in X-ray Tomography*. U. Bonse (Ed.). Bellingham, WA, SPIE. *SPIE Proc Vol* **3149**: 25–32.

Dowd, B.A., G.H. Campbell, R.B. Marr, V. Nagarkar, S. Tipnis, L. Axe, and D.P. Siddons (1999). Developments in synchrotron x-ray computed microtomography at the National Synchrotron Light Source. *Developments in X-ray Tomography II*. U. Bonse (Ed.). Bellingham, WA, SPIE. *SPIE Proc Vol* **3772**: 224–236.

Elliott, J.C. and S.D. Dover (1982). X-ray microtomography. *J Microsc* **126**: 211–213.

Elliott, J.C. and S.D. Dover (1984). Three-dimensional distribution of mineral in bone at a resolution of 15 μm determined by x-ray microtomography. *Metab Bone Dis Relat Res* **5**: 219–221.

Elliott, J.C. and S.D. Dover (1985). X-ray microscopy using computerized axial tomography. *J Microsc* **138**: 329–331.

Elliott, J.C., P. Anderson, G.R. Davis, F.S.L. Wong, and S.D. Dover (1994a). Computed tomography Part II: The practical use of a single source and detector. *J Metals* (Mar): 11–19.

Elliott, J.C., P. Anderson, X.J. Gao, F.S.L. Wong, G.R. Davis, and S.E.P. Dowker (1994b). Application of scanning microradiography and x-ray microtomography to studies of bone and teeth. *J X-ray Sci Technol* **4**: 102–117.

Engelke, K., M. Lohmann, W.R. Dix, and W. Graeff (1989a). Quantitative microtomography. *Rev Sci Instrum* **60**: 2486–2489.

Engelke, K., M. Lohmann, W.R. Dix, and W. Graeff (1989b). A system for dual energy microtomography of bones. *Nucl Instrum Meth* **A274**: 380–389.

Feldkamp, L.A. and G. Jesion (1986). 3-D x-ray computed tomography. *Rev Prog Quant NDE* **5A**: 555–566.

Feldkamp, L.A., L.C. Davis, and J.W. Kress (1984). Practical cone-beam algorithm. *J Opt Soc Am* **A1**: 612–619.

Feldkamp, L.A., D.J. Kubinski, and G. Jesion (1988). Application of high magnification to 3D x-ray computed tomography. *Rev Prog Quant NDE* **7A**: 381–388.

Ferrero, M.A., R. Sommer, P. Spanne, K.W. Jones, and C. Connor (1993). X-ray microtomography studies of nascent polyolefin particles polymerized over magnesium chloride-supported catalysts. *J Polym Sci A* **31**: 2507–2512.

Ferrie, E., J.Y. Buffiere, W. Ludwig, A. Gravouil, and L. Edwards (2006). Fatigue crack propagation: In situ visualization using x-ray microtomography and 3D simulation using the extended finite element method. *Acta Mater* **54**: 1111–1122.

Fitzgerald, R. (2000). Phase sensitive x-ray imaging. *Phys Today* **53**: 23–26.

Flannery, B.P. and W.G. Roberge (1987). Observational strategies for three-dimensional synchrotron microtomography. *J Appl Phys* **62**: 4668–4674.

Flannery, B.P., H.W. Deckman, W.G. Roberge, and K.L. D'Amico (1987). Three-dimensional x-ray microtomography. *Science* **237**: 1439–1444.

Goulet, R.W., S.A. Goldstein, M.J. Ciarelli, J.L. Kuhn, M.B. Brown, and L.A. Feldkamp (1994). The relationship between the structural and orthogonal compressive properties of trabecular bone. *J Biomech* **27**: 375–389.

Groso, A., M. Stampanoni, R. Abela, P. Schneider, S. Linga, and R. Muller (2006). Phase contrast tomography: An alternative approach. *Appl Phys Lett* **88**: 214104-1 to -3.

Hirai, T., H. Yamada, M. Sasaki, D. Hasegawa, M. Morita, Y. Oda, J. Takaku, T. Hanashima, N. Nitta, M. Takahashi, and K. Murata (2006). Refraction contrast 11x-magnified x-ray imaging of large objects by MIRRORCLE-type table-top synchrotron. *J Synchrotron Rad* **13**: 397–402.

Ignatyev, K., K. Huwig, R. Harvey, H. Ishii, J. Bradley, K. Luening, S. Brennan, and P. Pianetti (2006). XRF microCT study of space objects at SSRL. *Developments in X-Ray Tomography V*. U. Bonse (Ed.). Bellingham, WA, SPIE. *SPIE Proc Vol* **6318**: 631825-1 to -7.

Ito, K., Y. Gaponov, N. Sakabe, and Y. Amemiya (2007). A 3 x 6 arrayed CCD x-ray detector for continuous rotation method in macromolecular crystallography. *J Synchrotron Rad* **14**: 144–150.

Iwata, K., Y. Takeda, and H. Kikuta (2004). X-ray shearing interferometer and tomographic reconstruction of refractive index from its data. *Developments in X-Ray Tomography IV*. U. Bonse (Ed.). Bellingham, WA, SPIE. *SPIE Proc Vol* **5535**: 392–399.

Jones, K.W., H. Feng, A. Lanzirotti, and D. Mahajan (2005). Synchrotron x-ray microprobe and computed microtomography for characterization of nanocatalysts. *Nucl Instrum Meth B* **241**: 331–334.

Joo, Y.I., T. Sone, M. Fukunaga, S.G. Lim, and S. Onodera (2003). Effects of endurance exercise on three-dimensional trabecular bone microarchitecture in young growing rats. *Bone* **33**: 485–493.

Kak, A.C. and M. Slaney (2001). *Principles of Computerized Tomographic Imaging*. Philadelphia: SIAM (Soc. Industrial Appl. Math.).

Kinney, J.H. and M.C. Nichols (1992). X-ray tomographic microscopy (XTM) using synchrotron radiation. *Annu Rev Mater Sci* **22**: 121–152.

Kinney, J.H., U.K. Bonse, Q.C. Johnson, M.C. Nichols, R.A. Saroyan, W.N. Massey, and R. Nusshardt (1993). X-ray tomographic image magnification process, system and apparatus therefore. US Patent 5,245,648.

Kinney, J.H., Q.C. Johnson, U. Bonse, M.C. Nichols, R.A. Saroyan, R. Nusshardt, R. Pahl, and J.M. Brase (1988). Three-dimensional x-ray computed tomography in materials science. *MRS Bull* (January): 13–17.

Kinney, J.H., Q.C. Johnson, U. Bonse, R. Nusshardt, and M.C. Nichols (1986). The performance of CCD array detectors for application in high-resolution tomography. *X-ray Imaging II*. D.K. Bowen and L.V. Knight (Eds.). Bellingham, WA, SPIE. *SPIE Proc Vol* **691**: 43–50.

Kirby, B.J., J.R. Davis, J.A. Grant, and M.J. Morgan (1997). Monochromatic microtomographic imaging of osteoporotic bone. *Phys Med Biol* **42**: 1375–1385.

Kiss, M.Z., D.E. Sayers, and Z. Zhong (2003). Measurement of image contrast using diffraction enhanced imaging. *Phys Med Biol* **48**: 325–340.

Koch, A., F. Peyrin, P. Heurtier, B. Ferrand, B. Chambaz, W. Ludwig, and M. Couchaud (1999). X-ray camera for computed tomography of biological samples with micrometer resolution using $Lu_3Al_5O_{12}$ and $Y_3Al_5O_{12}$. *Medical Imaging 1999: Physics of Medical Imaging*. J.T.D. J.M. Boone, III (Ed.). Bellingham, WA, SPIE. *SPIE Proc Vol* **3659**: 170–179.

Kohlbrenner, A. et al. (2001). In vivo microtomography. Noninvasive Assessment of *Trabecular Bone Architecture and the Competence of Bone*. B.K.B.S. Majumdar (Ed.). New York: Kluwer. 213-224.

Kohra, K., H. Hashizume, and J. Yoshimura (1970). X-ray diffraction topography utilizing double-crystal arrangement of (+, +) or non-parallel (+, -) setting. *Jpn J Appl Phys* **9**: 1029–1038.

Lemelle, L., A. Simionovici, J. Susini, P. Oger, M. Chukalina, C. Rau, B. Golosio, and P. Gillet (2003). X-ray imaging techniques and exobiology. *J Phys IV* **104**: 377–380.

Liu, X., J. Liu, X. Li, S.K. Cheong, D. Shu, J. Wang, M.W. Tate, A. Ercan, D.R. Schuette, M.J. Renzi, A. Woll, and S.M. Gruner (2004). Development of ultrafast computed tomography of highly transient fuel sprays. *Developments in X-Ray Tomography IV*. U. Bonse (Ed.). Bellingham, WA, SPIE. *SPIE Proc Vol* **5535**: 21–28.

London, B., R.N. Yancey, and J.A. Smith (1990). High-resolution x-ray computed tomography of composite materials. *Mater Eval* **48**: 604–608.

Lopes, R.T., H.S. Rocha, E.F.O.D. Jesus, R.C. Barroso, L.F.D. Oliveira, M.J. Anjos, D. Braz, and S. Moreira (2003). X-ray transmission microtomography using synchrotron radiation. *Nucl Instrum Meth A* **505**: 604–607.

Ludwig, O., M. Di Michiel, L. Salvo, M. Suery, and P. Falus (2005). In-situ three-dimensional microstructural investigation of solidification of an Al–Cu alloy by ultrafast x-ray microtomography. *Metall Mater Trans A* **36**: 1515–1523.

Ludwig, W., J.Y. Buffière, S. Savelli, and P. Cloetens (2003). Study of the interaction of a short fatigue crack with grain boundaries in a cast Al alloy using x-ray microtomography. *Acta Mater* **51**: 585–598.

Martin, T. and A. Koch (2006). Recent developments in x-ray imaging with micrometer spatial resolution. *J Synchrotron Rad* **13**: 180–194.

Mayo, S., P. Miller, S.W. Wilkins, D. Gao, and T. Gureyev (2006). Laboratory-based x-ray microtomography with submicron resolution. *Developments in X-Ray Tomography V*. U. Bonse (Ed.). Bellingham, WA, SPIE. *SPIE Proc Vol* **6318**: 63181E–1 to –8.

Mayo, S.C., T.J. Davis, T.E. Gureyev, P.R. Miller, D. Paganin, A. Pogany, A.W. Stevenson, and S.W. Wilkins (2003). X-ray phase-contrast microscopy and microtomography. *Optics Express* **11**: 2289–2302.

Mayo, S.C., P.R. Miller, S.W. Wilkins, T.J. Davis, D. Gao, T.E. Gureyev, D. Paganin, D.J. Parry, A. Pogany, and A.W. Stevenson (2002). Quantitative x-ray projection microscopy: Phase contrast and multi-spectal imaging. *Science* **207**: 79–96.

McNulty, I. (2001). Current and ultimate limitations of scanning nanotomography. *X-ray Micro- and Nano-focusing: Applications and Techniques II*. I. McNulty (Ed.). Bellingham, WA, SPIE. *SPIE Proc Vol* **4499**: 23–28.

Momose, A., A. Fujii, H. Kadowaki, and H. Jinnai (2005). Three-dimensional observation of polymer blend by x-ray phase tomography. *Macromol* **38**: 7197–7200.

Momose, A., S. Kawamoto, I. Koyama, and Y. Suzuki (2004). Phase tomography using an x-ray Talbot interferometer. *Developments in X-ray Tomography IV*. U. Bonse (Ed.). Bellingham, WA, SPIE. *SPIE Proc Vol* **5535**: 352–360.

Momose, A., I. Koyama, Y. Hamaishi, H. Yoshikawa, T. Takeda, J. Wu, Y. Itai, K. Takai, K. Uesugi, and Y. Suzuki (2003). Phase-contrast microtomography using an x-ray interferometer having a 40-µm analyzer. *J Phys IV* **104**: 599–602.

Momose, A., T. Takeda, Y. Itai, and K. Hirano (1996). Phase contrast x-ray computed tomography for observing biological soft tissues. *Nature Med* **2**: 473–475.

Momose, A., T. Takeda, Y. Itai, J. Tu, and K. Hirano (1999). Recent observations with phase-contrast computed tomography. *Developments in X-ray Tomography II*. U. Bonse (Ed.). Bellingham, WA, SPIE. *SPIE Proc Vol* **3772**: 188–195.

Momose, A., W. Yashiro, M. Moritake, Y. Takeda, K. Uesugi, A. Takeuchi, Y. Suzuki, M. Tanaka, and T. Hattori (2006). Biomedical imaging by Talbot-type x-ray phase tomography. *Developments in X-Ray Tomography V*. U. Bonse (Ed.). Bellingham, WA, SPIE. *SPIE Proc Vol* **6318**: 63180T–1 to –10.

Mummery, P.M., B. Derby, P. Anderson, G.R. Davis, and J.C. Elliott (1995). X-ray microtomographic studies of metal matrix composites using laboratory x-ray sources. *J Microsc* **177**: 399–406.

Naghedolfeizi, M., J.S. Chung, R. Morris, G.E. Ice, W.B. Yun, Z. Cai, and B. Lai (2003). X-ray fluorescence microtomography study of trace elements in a SiC nuclear fuel shell. *J Nucl Mater* **312**: 146–155.

Nusshardt, R., U. Bonse, F. Busch, J.H. Kinney, R.A. Saroyan, and M.C. Nichols (1991). Microtomography: A tool for nondestructive study of materials. *Synchrotron Rad News* **4**(3): 21–23.

Olsen, U.L., X. Badel, J. Linnros, M.D. Michiel, T. Martin, S. Schmidt, and H.F. Poulsen (2007). Development of a high-efficiency high-resolution imaging detector for 30–80 keV x-rays. *Nucl Instrum Meth* **A576**: 52–55.

Owens, J.W., L.G. Butler, C. Dupard-Julien, and K. Garner (2001). Synchrotron x-ray microtomography, x-ray absorption near edge structure, extended x-ray absorption fine structure, and voxel imaging of a cobalt-zeolite-Y complex. *Mater Res Bull* **36**: 1595–1602.

Paganin, D., T.E. Gureyev, S.C. Mayo, A.W. Stevenson, Y.I. Nesterets, and S.W. Wilkins (2004). X-ray omni microscopy. *Science* **214**: 315–327.

Pagot, E., S. Fiedler, P. Cloetens, A. Bravin, P. Coan, K. Fezzaa, J. Baruchel, and J. Härtwig (2005). Quantitative comparison between two phase contrast techniques: Diffraction enhanced imaging and phase propagation imaging. *Phys Med Biol* **50**: 709–724.

Pfeiffer, F., T. Weitkamp, O. Bunk, and C. David (2006). Phase retrieval and differential phase contrast imaging with low brilliance x-ray sources. *Nature Phys* **2**: 258–261.

Rau, C., V. Crecea, C.P. Richter, K.M. Peterson, P.R. Jemian, U. Neuhäusler, G. Schneider, X. Yu, P.V. Braun, T.C. Chiang, and I.K. Robinson (2006). A hard x-ray KB-FZP microscope for tomography with sub-100 nm resolution. *Developments in X-Ray Tomography V*. U. Bonse (Ed.). Bellingham, WA, SPIE. *SPIE Proc Vol* **6318**: 63181G–1 to –6.

Rau, C., K.M. Peterson, P.R. Jemian, T. Terry, M.T. Harris, S. Vogt, C.P. Richter, U. Neuhäusler, G. Schneider, and I.K. Robinson (2004a). The evolution of hard x-ray tomography from the micrometer to the namometer length scale. *Developments in X-ray Tomography IV*. U. Bonse (Ed.). Bellingham, WA, SPIE. *SPIE Proc Vol* **5535**: 709–714.

Rau, C., A. Somogyi, and A. Simionovici (2004b). Tomography with chemical speciation. *Developments in X-Ray Tomography IV*. U. Bonse (Ed.). Bellingham, WA, SPIE. *SPIE Proc Vol* **5535**: 29–35.

Rau, C., A. Somogyi, A. Bytchkov, and A.S. Simionovici (2001a). XANES microimaging and tomography. *Developments in X-ray Tomography III*. U. Bonse (Ed.). Bellingham, WA, SPIE. *SPIE Proc Vol* **4503**: 249–255.

Rau, C., T. Weitkamp, A.A. Snigirev, C.G. Schroer, B. Benner, J. Tümmler, T.F. Günzler, M. Kuhlmann, B. Lengeler, C.E.K. III, K.M. Döbrich, D. Michels, and A. Michels (2001b). Tomography with high resolution. *Developments in X-ray Tomography III*. U. Bonse (Ed.). Bellingham, WA, SPIE. *SPIE Proc Vol* **4503**: 14–22.

Rivers, M.L., S.R. Sutton, and P. Eng (1999). Geoscience applications of x-ray computed microtomography. *Developments in X-ray Tomography II*. U. Bonse (Ed.). Bellingham, WA, SPIE. *SPIE Proc Vol* **3772**: 78–86.

Rivers, M.L. and Y. Wang (2006). Recent developments in microtomography at GeoSoilEnviroCARS. *Developments in X-Ray Tomography V*. U. Bonse (Ed.). Bellingham, WA, SPIE. *SPIE Proc Vol* **6318**: 63180J–1 to –15.

Rivers, M.L., Y. Wang, and T. Uchida (2004). Microtomography at GeoSoilEnviroCARS. *Developments in X-Ray Tomography IV*. U. Bonse (Ed.). Bellingham, WA, SPIE. *SPIE Proc Vol* **5535**: 783–791.

Sakamoto, K., Y. Suzuki, T. Hirano, and K. Usami (1988). Improvement of the spatial resolution of monochromatic x-ray CT using synchrotron radiation. *Jpn J Appl Phys* **27**: 127–130.

Sasov, A. (2004). X-ray nanotomography. *Developments in X-Ray Tomography IV*. U. Bonse (Ed.). Bellingham, WA, SPIE. *SPIE Proc Vol* **5535**: 201–211.

Sasov, A.Y. (1987a). Microtomography: I. Methods and equipment. *J. Microsc* **147**: 169–178.

Sasov, A.Y. (1987b). Microtomography: II. Examples of applications. *J. Microsc* **147**: 179–192.

Sasov, A.Y. (1989). X-ray microtomography. *Radiation Methods*. New York, Plenum: 315–321.

Schneider, G., C. Knöchel, S. Vogt, D. Wei, and E.H. Anderson (2001). Nanotomography of labeled cryogenic cells. *Developments in X-ray Tomography III*. U. Bonse (Ed.). Bellingham, WA, SPIE. *SPIE Proc Vol* **4503**: 156–165.

Schneider, P., R. Voide, M. Stuaber, M. Stampanoni, L.R. Donahue, P. Wyss, U. Sennhauser, and R. Müller (2006). Assessment of murine bone ultrastructure using synchrotron light: Towards nanocomputed tomography. *Developments in X-Ray Tomography V*. U. Bonse (Ed.). Bellingham, WA, SPIE. *SPIE Proc Vol* **6318**: 63180C–1 to –9.

Schroer, C.G., B. Benner, T.F. Gunzler, M. Kuhlmann, B. Lengeler, C. Rau, T. Weitkamp, A. Snigirev, and I. Snigireva (2001a). Magnified hard x-ray microtomography: Toward tomography with sub-micron resolution. *Developments in X-ray Tomography III*. U. Bonse (Ed.). Bellingham, WA, SPIE. *SPIE Proc Vol* **4503**: 23–33.

Schroer, C.G., B. Benner, T.F. Günzler, M. Kuhlman, B. Lengeler, W.H. Schröder, A.J. Kuhn, A. Simionovici, A. Snigirev, and I. Snigireva (2003). High resolution element mapping inside biological samples using fluorescence microtomography. *J Phys IV* **104**: 353.

Schroer, C.G., B. Benner, T.F. Gunzler, M. Kuhlmann, B. Lengeler, W.H. Schroder, A.J. Kuhn, A.S. Simionovici, A. Snigirev, and I. Snigireva (2001b). High resolution element mapping inside biological samples using fluorescence microtomography. *Developments in X-ray Tomography III*. U. Bonse (Ed.). Bellingham, WA, SPIE. *SPIE Proc Vol* **4503**: 230–239.

Schroer, C.G., P. Cloetens, M. Rivers, A. Snigirev, A. Takeuchi, and W. Yun (2004a). High-resolution 3D imaging microscopy using hard x-rays. *MRS Bull* **29**: 154–156 and 157–165.

Schroer, C.G., T.F. Gunzler, M. Kuhlmann, O. Kurapova, S. Feste, M. Schweitzer, B. Lengeler, W.H. Schroder, M. Drakopoulos, A. Somogyi, A.S. Simionovici, A. Snigirev, and I. Snigireva (2004b). Fluorescence microtomography using nanofocusing refractive x-ray lenses. *Developments in X-Ray Tomography IV*. U. Bonse (Ed.). Bellingham, WA, SPIE. *SPIE Proc Vol* **5535**: 162–168.

Schroer, C.G., M. Kuhlmann, T.F. Gunzler, B. Benner, O. Kurapova, J. Patormmel, B. Lengeler, S.V. Roth, R. Gehrke, A. Snigirev, I. Snigireva, N. Stribeck, A. Almendarez-Camarillo, and F. Beckmann (2006). Full-field and scanning microtomography based on parabolic refractive x-ray lenses. *Developments in X-Ray Tomography V*. U. Bonse (Ed.). Bellingham, WA, SPIE. *SPIE Proc Vol* **6318**: 63181H–1 to –9.

Schroer, C.G., M. Kuhlmann, T.F. Gunzler, B. Lengeler, M. Richwin, B. Griesebock, D. Lutzenkirchen-Hecht, R. Frahm, E. Ziegler, A. Mashayekhi, D.R. Haeffner, J.D. Grunwaldt, and A. Baiker (2003). Mapping the chemical states of an element inside a sample using tomographic x-ray absorption spectroscopy. *Appl Phys Lett* **82**: 3360–3362.

Schroer, C.G., B. Lengeler, B. Benner, T.F. Gunzler, M. Kuhlmann, A.S. Simionovici, S. Bohic, M. Drakopoulos, A. Snigirev, I. Snigireva, and W.H. Schroder (2001c). Microbeam production using compound refractive lenses: Beam characterization and applications. *X-Ray Micro- and Nano-Focusing: Applications and Techniques II*. I. McNulty (Ed.). Bellingham, WA, SPIE. *SPIE Proc Vol* **4499**: 52–63.

Schroer, C.G., J. Meyer, M. Kuhlmann, B. Benner, T.F. Günzler, B. Lengeler, C. Rau, T. Weitkamp, A. Snigirev, and I. Snigireva (2003). Nanotomography based on hard x-ray microscopy with refractive lenses. *J Phys IV* **104**: 271.

Schroer, C.G., J. Tuemmler, T. F. Guenzler, B. Lengeler, W.H. Schroeder, A.J. Kuhn, A.S. Simionovici, A. Snigirev, and I. Snigireva (2000). Fluorescence microtomography: External mapping of elements inside biological samples. *Penetrating Radiation Systems and Applications II*. F.P. Doty, H.B. Barber, H. Roehrig, and E.J. Morton (Eds.). Bellingham, WA, SPIE. *SPIE Proc Vol* **4142**: 287–296.

Seguin, F.H., P. Burstein, P.J. Bjorkholm, F. Homburger, and R.A. Adams (1985). X-ray computed tomography with 50-μm resolution. *Appl Optics* **24**: 4117–4123.

Simionovici, A., M. Chukalina, M. Drakopoulos, I. Snigireva, A. Snigirev, C. Schroer, B. Lengeler, K. Janssens, and F. Adams (1999). X-ray fluorescence microtomography: Experiment and reconstruction. *Developments in X-Ray Tomography II*. U. Bonse (Ed.). Bellingham, WA, SPIE. *SPIE Proc Vol* **3772**: 304–310.

Simionovici, A.S., M. Chukalina, B. Vekemans, L. Lemelle, P. Gillet, C.G. Schroer, B. Lengeler, W.H. Schröder, and T. Jeffries (2001). New results in x-ray computed fluoresecence tomography. *Developments in X-ray Tomography III*. U. Bonse (Ed.). Bellingham, WA, SPIE. *SPIE Proc Vol* **4503**: 222–229.

Simionovici, A.S., B. Golosio, M.V. Chukalina, A. Somogyi, and L. Lemelle (2004). Seven years of x-ray fluorescence computed microtomography. *Developments in X-Ray Tomography IV*. U. Bonse (Ed.). Bellingham, WA, SPIE. *SPIE Proc Vol* **5535**: 232–242.

Snigerev, A., I. Snigereva, V. Kohn, S. Kuznetsov, and I. Schelokov (1995). On the possibilities of x-ray phase contrast microimaging by coherent high-energy synchrotron radiation. *Rev Sci Instrum* **66**: 5486–5492.

Spal, R. D. (2001). Submicrometer resolution hard x-ray holography with the asymmetric Bragg diffraction microscope. *Phys Rev Lett* **86**: 3044–3047.

Spanne, P. and M.L. Rivers (1987). Computerized microtomography using synchrotron radiation from the NSLS. *Nucl Instrum Meth* **B24/25**: 1063–1067.

Stampanoni, M., R. Abela, G. Borchert, and B.D. Patterson (2004). New developments in synchrotron-based microtomography. *Developments in X-Ray Tomography IV*. U. Bonse (Ed.). Bellingham, WA, SPIE. *SPIE Proc Vol* **5535**: 169–181.

Stampanoni, M., G. Borchert, and R. Abela (2003). Two-dimensional asymmetrical Bragg diffraction for submicrometer computer tomography. *Crystals, Multilayers, and Other Synchrotron Optics*. T. Ishikawa, A.T. Macrander, and J.L. Wood (Eds.). Bellingham, WA, SPIE. *SPIE Proc Vol* **5195**: 54–62.

Stampanoni, M., G. Borchert, and R. Abela (2005). Towards nanotomography with asymmetrically cut crystals. *Nucl Instrum Meth A* **551**: 119–124.

Stampanoni, M., G. Borchert, and R. Abela (2006a). Progress in microtomography with the Bragg magnifier at SLS. *Rad Phys Chem* **75**: 1956–1961.

Stampanoni, M., G. Borchert, R. Abela, and P. Ruegsegger (2002a). Bragg magnifier: A detector for submicrometer x-ray computer tomography. *J Appl Phys* **92**: 7630–7635.

Stampanoni, M., G. Borchert, P. Wyss, R. Abela, B. Patterson, S. Hunt, D. Vermeulen, and P. Rüegsegger (2002b). High resolution x-ray detector for synchrotron-based microtomography. *Nucl Instrum Meth A* **491**: 291–301.

Stampanoni, M., A. Groso, A. Isenegger, G. Mikuljan, Q. Chen, A. Bertrand, S. Henein, R. Betemps, U. Frommherz, P. Bohler, D. Meister, M. Lange, and R. Abela (2006b). Trends in synchrotron-based tomographic imaging: The SLS experience. *Developments in X-Ray Tomography V*. U. Bonse (Ed.). Bellingham, WA, SPIE. *SPIE Proc Vol* **6318**: 63180M–1 to –14.

Stampanoni, M., P. Wyss, G.L. Borchert, D. Vermeulen, and P. Rüegsegger (2001). X-ray tomographic microscope at the Swiss Light Source. *Developments in X-ray Tomography III*. U. Bonse (Ed.). Bellingham, WA, SPIE. *SPIE Proc Vol* **4503**: 42–53.

Stock, S.R. (1999). Microtomography of materials. *Int Mater Rev* **44**: 141–164.

Stock, S.R. (2007). X-ray phase microradiography and x-ray absorption microcomputed tomography, compared in studies of biominerals. *Handbook of Biomineralization*. P. Behrens and E. Bäuerlein (Eds.). Weinheim, Wiley-VCH. **2**: 389–400.

Stock, S.R., F. De Carlo, and J.D. Almer (2008). High energy x-ray scattering tomography applied to bone. *J Struct Biol* **161**: 144–150.

Stock, S.R., L.L. Dollar, G.B. Freeman, W.J. Ready, L.J. Turbini, J.C. Elliott, P. Anderson, and G.R. Davis (1994). Characterization of conductive anodic filament (CAF) by x-ray microtomography and by seial sectioning. *Electronic Packaging Materials Science VII*. P. Børgesen, K.F. Jansen, and R.A. Pollak (Eds.). Pittsburgh, *Mater Res Soc*. **323**: 65–69.

Stock, S.R., A. Guvenilir, T.L. Starr, J.C. Elliott, P. Anderson, S.D. Dover, and D.K. Bowen (1989). Microtomography of silicon nitride/silicon carbide composites. *Ceram Trans* **5**: 161–170.

Stribeck, N., A.A. Camarilla, U. Nochel, C. Schroer, M. Kuhlmann, S.V. Roth, R. Gehrke, and R.K. Bayer (2006). Volume-resolved nanostructure survey of a polymer part by means of SAXS microtomography. *Macromol Chem Phys* **207**: 1139–1149.

Suzuki, Y., K. Usami, K. Sakamoto, H. Kozaka, T. Hirano, H. Shiono, and H. Kohno (1988). X-ray computerized tomography using monochromated synchrotron radiation. *Japan J Appl Phys* **27**: L461–L464.

Takeda, T., A. Momose, Q. Yu, J. Wu, K. Hirano, and Y. Itai (2000). Phase contrast x-ray imaging with a large monolithic x-ray interferometer. *J Synchrotron Rad* **7**: 280–282.

Takeda, T., J. Wu, T.T. Lwin, A. Yoneyama, Y. Hirai, K. Hyodo, N. Sunaguchi, T. Yuasa, M. Minami, K. Kose, and T. Akatsuka (2006). Progress in biomedical application of phase contrast x-ray imaging and fluorescent x-ray CT. *Developments in X-Ray Tomography V*. U. Bonse (Ed.). Bellingham, WA, SPIE. *SPIE Proc Vol* **6318**: 63180W-1 to –12.

Takeda, T., J. Wu, A. Yoneyama, Y. Tsuchiya, T.T. Lwin, Y. Hirai, T. Kuroe, T. Yuasa, K. Hyodo, F.A. Dilmanian, and T. Akatsuka (2004). SR biomedical imaging with phase contrast and fluoresecent x-ray CT. *Developments in X-Ray Tomography IV*. U. Bonse (Ed.). Bellingham, WA, SPIE. *SPIE Proc Vol* **5535**: 380–391.

Takeda, T., Q. Yu, T. Yashiro, T. Yuasa, Y. Itai, and T. Akatsuka (1999). Human thyroid specimen imaging by fluorescent x-ray computed tomography with synchrotron radiation. *Developments in X-ray Tomography II*. U. Bonse (Ed.). Bellingham, WA, SPIE. *SPIE Proc Vol* **3772**: 258–267.

Takeuchi, A., K. Uesugi, Y. Suzuki, and S. Aoki (2001). Hard x-ray microtomography using x-ray imaging optics. *Jpn J Appl Phys* **40**: 1499–1503.

Takeuchi, A., K. Uesugi, H. Takano, and Y. Suzuki (2002). Submicrometer-resolution three-dimensional imaging with hard x-ray imaging microtomography. *Rev Sci Instrum* **73**: 4246–4249.

Thurner, P.J., P. Wyss, R. Voide, M. Stauber, B. Müller, M. Stampanoni, J.A. Hubbell, R. Müller, and U. Sennhauser (2004). Functional microimaging of soft and hard tissue using synchrotron light. *Developments in X-ray Tomography IV*. U. Bonse (Ed.). Bellingham, WA, SPIE. *SPIE Proc Vol* **5535**: 112–128.

Tkachuk, A., M. Feser, H. Cui, F. Duewer, H. Chang, and W. Yun (2006). High resolution x-ray tomography using laboratory x-ray sources. *Developments in X-Ray Tomography V*. U. Bonse (Ed.). Bellingham, WA, SPIE. *SPIE Proc Vol* **6318**: 63181D–1 to –8.

Uesugi, K., Y. Suzuki, N. Yagi, A. Tsuchiyama, and T. Nakano (2001). Development of submicrometer resolution x-ray CT system at SPring-8. *Developments in X-Ray Tomography III*. U. Bonse (Ed.). Bellingham, WA, SPIE. *SPIE Proc Vol* **4503**: 291–298.

Uesugi, K., A. Tsuchiyama, H. Yasuda, M. Nakamura, T. Nakano, Y. Suzuki, and N. Yagi (2003). Micro-tomographic imaging for material sciences at BL47XU in SPring-8. *J Phys IV* **104**: 45–48.

Vincze, L., B. Vekemans, I. Szaloki, F.E. Brenker, G. Falkenberg, K. Rickers, K. Aerts, R.V. Grieken, and F. Adams (2004). X-ray fluorescence microtomography and polycapillary based confocal imaging using synchrotron radiation. *Developments in X-Ray Tomography IV*. U. Bonse (Ed.). Bellingham, WA, SPIE. *SPIE Proc Vol* **5535**: 220–231.

Vincze, L., B. Vekemans, I. Szaloki, K. Janssens, R.V. Grieken, H. Feng, K.W. Jones, and F. Adams (2001). High resolution X-ray fluorescence micro-tomography on single sediment particles. *Developments in X-ray Tomography III*. U. Bonse (Ed.). Bellingham, WA, SPIE. *SPIE Proc Vol* **4503**: 240–248.

Wang, Y., F. De Carlo, I. Foster, J. Insley, C. Kesselman, P. Lane, G. von Laszewsk, D. Mancini, I. McNulty, M.H. Su, and B. Tieman (1999). Quasi-realtime x-ray microtomography system at the Advanced Photon Source. *Developments in X-ray Tomography II*. U. Bonse (Ed.). Bellingham, WA, SPIE. *SPIE Proc Vol* **3772**: 318–327.

Wang, Y., F. De Carlo, D.C. Mancini, I. McNulty, B. Tieman, J. Bresnahan, I. Foster, J. Insley, P. Lane, G. Von Laszewski, C. Kesselman, M.H. Su, and M. Thiebaux (2001). A high-throughput x-ray microtomography system at the Advanced Photon Source. *Rev Sci Instrum* **72**: 2062–2068.

Wang, Y., T. Uchida, F. Westferro, M.L. Rivers, N. Nishiyama, J. Gebhardt, C.E. Lesher, and S.R. Sutton (2005). High-pressure x-ray tomography microscope. *Rev Sci Instrum* **76**: 073709.

Weitkamp, T., C. David, C. Kottler, O. Bunk, and F. Pfeiffer (2006). Tomography with grating interferometers at low-brilliance sources. *Developments in X-Ray Tomography V*. U. Bonse (Ed.). Bellingham, WA, SPIE. *SPIE Proc Vol* **6318**: 63180S–1 to –10.

Weitkamp, T., A. Diaz, C. David, F. Pfeiffer, M. Stampanoni, P. Cloetens, and E. Ziegler (2005). X-ray phase imaging with a grating interferometer. *Optics Express* **13**: 6296–6304.

Weitkamp, T., C. Rau, A.A. Snigirev, B. Benner, T.F. Gunzler, M. Kuhlmann, and C.G. Schroer (2001). In-line phase contrast in synchrotron-radiation microradiography and tomography. *Developments in X-ray Tomography III*. U. Bonse (Ed.). Bellingham, WA, SPIE. *SPIE Proc Vol* **4503**: 92–102.

Weitkamp, T., C. Raven, and A. Snigirev (1999). Imaging and microtomography facility at the ESRF beamline ID 22. *Developments in X-Ray Tomography II*. U. Bonse (Ed.). Bellingham, WA, SPIE. *SPIE Proc Vol* **3772**: 311–317.

Weon, B.M., J.H. Je, Y. Hwu, and G. Margaritondo (2006). Phase contrast X-ray imaging. *Int J Nanotechnol* **3**: 280–297.

Wernick, M.N., J.G. Brankov, D. Chapman, Y. Yang, C. Muehleman, Z. Zhong, and M.A. Anastasio (2004). A preliminary study of multiple image computed tomography. *Developments in X-ray Tomography IV*. U. Bonse (Ed.). Bellingham, WA, SPIE. *SPIE Proc Vol* **5535**: 369–379.

Wernick, M.N., O. Wirjadi, D. Chapman, Z. Zhong, N.P. Galatsanos, Y. Yang, J.G. Brankov, O. Oltulu, M.A. Anastasio, and C. Muehleman (2003). Multiple image radiography. *Phys Med Biol* **48**: 3875–3895.

Westneat, M.W. et al. (2003). Tracheal respiration in insects visualized with synchrotron x-ray imaging. *Science* **299**: 558-560.

Withers, P.J. (2007). X-ray nanotomography. *Mater Today* **10**: 26–34.

Wu, J., T. Takeda, T.T. Lwin, I. Koyama, A. Momose, A. Fujii, Y. Hamaishi, T. Kuroe, T. Yuasa, Y. Suzuki, and T. Akatsuka (2004). Microphase contrast x-ray computed tomography for basic biomedical study at SPring-8. *Developments in X-Ray Tomography IV*. U. Bonse (Ed.). Bellingham, WA, SPIE. *SPIE Proc Vol* **5535**: 740–747.

Wu, J., T. Takeda, T.T. Lwin, N. Sunaguchi, T. Fukami, T. Yuasa, M. Minami, and T. Akatsuka (2006). Fusion imaging of fluorescent and phase contrast x-ray computed tomography using synchrotron radiation in medical biology. *Developments in X-Ray Tomography V*. U. Bonse (Ed.). Bellingham, WA, SPIE. *SPIE Proc Vol* **6318**: 631828–1 to –8.

Yoneyama, A., A. Momose, I. Koyama, E. Seya, T. Takeda, Y. Itai, K. Hirano, and K. Hyodo (2002). Large area phase contrast x-ray imaging using a two crystal x-ray interferometer. *J Synchrotron Rad* **9**: 277–281.

Yoneyama, A., T. Takeda, Y. Tsuchiya, J. Wu, T.T. Lwin, K. Hyodo, and Y. Hirai (2005). High energy phase contrast x-ray imaging using a two-crystal x-ray interferometer. *J Synchrotron Rad* **12**: 534–536.

Yoshimura, H., C. Miyata, C. Kuzuryu, A. Hori, T. Obi, and N. Ohyama (2001). X-ray computed tomography using projection x-ray microscope. *Developments in X-Ray Tomography III*. U. Bonse (Ed.). Bellingham, WA, SPIE. *SPIE Proc Vol* **4503**: 166–171.

5
MicroCT in Practice

The preceding chapter discussed microCT systems and the component subsystems from which they are assembled. Systems contain various nonidealities, and their effects on (and amelioration in) reconstructions are the subject of this chapter. Section 5.1 describes reconstruction artifacts, and Section 5.2 discusses the precision and accuracy of reconstructions. Sections 5.3 through 5.5 cover techniques for contrast enhancement, data acquisition challenges, and speculations for future apparatus development, respectively.

5.1 Reconstruction Artifacts

Reconstruction software must cope with various nonidealities intrinsic to the experimental apparatus and to the x-ray sources and return the highest fidelity reconstructions practical. An ideal microCT or nanoCT apparatus would have positioning component errors that are always much smaller than the smallest information-containing voxel size specified in the system design. This ideal system would employ a bright, highly stable x-ray source amenable to flat field correction. It is best to collect the highest quality data possible, but software can ameliorate the effects of instrument nonidealities and from less-than-optimum sampling dictated by experimental requirements.

Sasov used data from a narrow cone beam system and compared slices reconstructed with a fan beam algorithm, a cone beam algorithm, and a spiral scan algorithm (Sasov, 2001). The quality of slices at the ends of the stack was compared to that at the center, and this study should be considered by those interested in different strategies for rapid data collection.

5.1.1 Motion Artifacts

Small motions of the specimen during collection of the projections for a reconstruction must be avoided. Figure 5.1 of a living sea urchin in seawater shows an extreme example of the effect of motion. Despite wedging the live sea urchin tightly in place, it moved significantly, smearing the image beyond recognition (compare with Figure 5.1b, a slice of an intact,

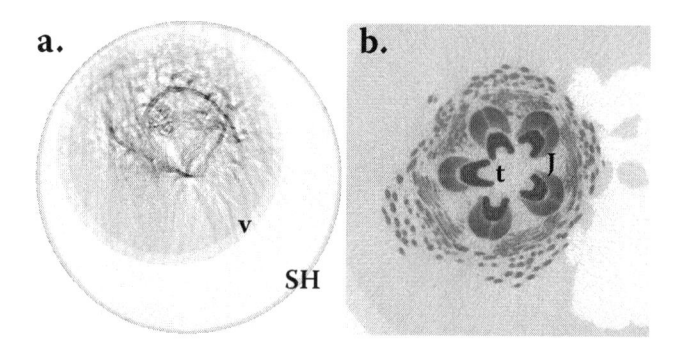

FIGURE 5.1
Motion artifacts illustrated. Here, the darker the pixel, the more absorbing the corresponding voxel is. (a) Live sea urchin (*Lytechinus variegatus*) in seawater showing motion of the animal but not of the vial "v" containing the seawater and scanner sample holder "SH." 1,024 × 1,024 voxel field of view, 37-μm isotropic voxels, 250 projections of 1,024 samples, 0.3 s integration per projection, 45 kVp, 88 μA. Scanco MicroCT-40. (b) Fixed sea urchin (*Eucidaris tribuloides*) tooth "t," jaw structure "J." 512 × 512 voxel field of view, 37-μm isotropic voxels, 500 projections of 1,024 samples, 0.3 s integration per projection, 70 kVp, 57 μA. Scanco MicroCT-40.

fixed sea urchin). Note the images of the specimen holder "SH" and the vial "v" are sharp in Figure 5.1a. Motion artifacts can also occur inside a specimen, particularly for soft tissues which can relax (shift positions) under the influence of gravity or slight amounts of drying.

5.1.2 Ring Artifacts

Figure 5.2 shows ring artifacts in a synchrotron microCT slice, rings that exist even after white-field normalization has been performed. Figure 5.2b shows the result of a robust white field normalization procedure: most of the rings are eliminated. Unfortunately, ring artifacts do not merely detract from the appearance of an image, but they also can interfere with accurate segmentation and quantification of the amount of phases present and their geometrical properties. Correction with a median filter often does not work particularly well, and use of a purpose-built filter taking advantage of the concentric nature of the rings seems to work well to minimize the rings (Bentz et al., 2000). Ring reduction using a 21-point smooth to the average of all rows of the sinogram (2D plot showing transmitted intensity in the radiograph, for one slice, that is, one row of the radiograph, as a function of rotation angle; see Figure 3.11) offers substantial improvement over uncorrected reconstructions (Rivers and Wang, 2006): the actual high-frequency content of the slices does not appear to be affected, only the rings. Sinogram correction algorithms for ring reduction have also been investigated for lab microCT data (Ketcham, 2006).

MicroCT in Practice

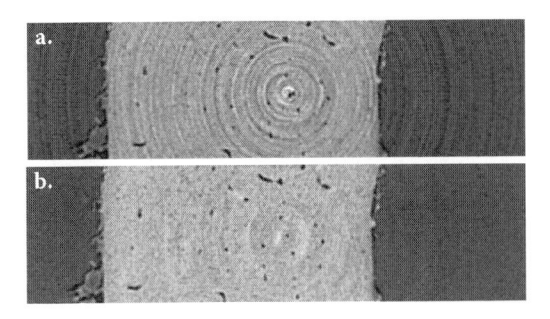

FIGURE 5.2
Section of a (1K)² reconstructed slice of a small section of bone cut from a rabbit femur showing ring artifacts. (a) Ring artifacts persisting after normal white field correction. (b) Elimination of most of the ring artifacts after a more robust white field correction. The lighter the pixel, the greater the absorption of the corresponding voxel is. 21 keV, 0.25° rotation increment, 5-μm isotropic voxels (S. R. Stock, K. Ignatiev, N. M. Rajamannan, and F. De Carlo. March 2003, 2-BM, APS).

These and other artifacts in lab microCT (Davis and Elliott, 2006) and in synchrotron microCT (Vidal et al., 2005) have been recently discussed.

5.1.3 Reconstruction Center Errors

Accurate reconstruction requires that the center of rotation be defined very precisely, to within a fraction of a voxel of that in the intended reconstruction. A recentering algorithm (Brunetti and De Carlo, 2004) that seems to work quite well is based on knowledge of the form of artifacts from centering errors: tails of (apparent) mass extending from features such as the corners of the specimen or voids or pores within the specimen (Figure 5.3). These tails increase the numbers of voxels with nonzero values, and iterating through different trial centers for a representative slice of the volume allows one to select the "best" center for reconstructing the rest of the volume, even in the presence of beam fluctuations, noise, and low contrast. Donath et al. (2006) developed metrics for optimization of center-of-rotation corrections. The center of rotation can also be refined by eye, but this is impractical when more than a few specimens are being imaged. As Figure 5.4 shows, even an inaccuracy in the center of rotation less than one voxel can appreciably degrade the quality of a reconstruction (Ignatiev et al., 2007).

5.1.4 Mechanical Imperfections Including Rotation Stage Wobble

In current generations of microCT systems (employing ribbon or area beams), the only required specimen motion is rotation (translation along the rotation axis in order to enlarge the scanned volume does not affect the quality of reconstructions). Specimen axis wobble and rotation axis

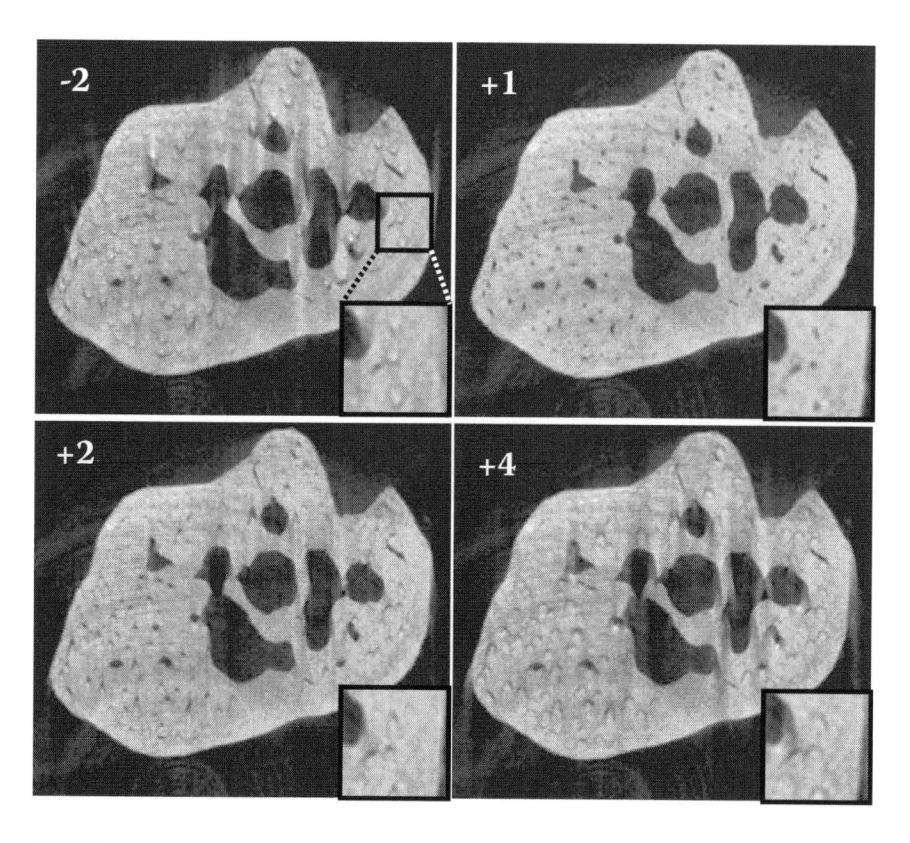

FIGURE 5.3
Illustration of errors introduced by incorrect reconstruction centers. Sections of the same slice reconstructed with translation of the center by the number of voxels indicated in the upper left portion of the image. The correct center (and clearest reconstruction) is with the +1 voxel center. Inset in the lower right of each image is a 2X enlargement of the area shown by the box in the upper left image. Mouse tibia in phosphate buffered saline, and the vertical field of view is 280 voxels. The lighter the pixel, the greater the absorption of the corresponding voxel is. 17 keV, 0.25° rotation increment, 5-μm isotropic voxels (S. R. Stock, N. M. Rajamannan, X. Xiao, and F. De Carlo. December 2005, 2-BM, APS).

misalignment can be significant sources of error in reconstructions. Reconstruction software typically uses the pixels of a row of the detector as the input for a single slice's reconstruction. Tilt of the rotation axis from perpendicular to detector rows brings material from adjacent slices into and out of the beam for specific ranges of angles; this degrades the fidelity of the reconstruction. Such tilts are best avoided by very careful alignment (note tilts of even 0.03° over 2K pixels can shift projected data to an adjacent row), but this can be corrected by postcollection rotation of the projection (and resampling of the pixels) to align the rows precisely perpendicular to the actual rotation axis. Automatic routines for this geometric correction are described elsewhere (Weitkamp and Bleuet, 2004).

MicroCT in Practice

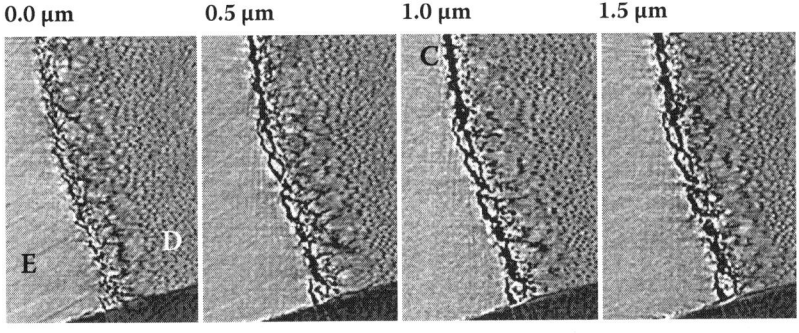

FIGURE 5.4
Effect of subvoxel changes in center of reconstruction illustrated using a section of a slice of the bovine dentinoenamel junction (DEJ). The reconstructed voxel size was slightly smaller than 2 μm, and the physical shifts of the rotation center are given above each image: the lighter the pixel, the more absorbing the voxel is. A crack "C" runs vertically between the enamel "E" and dentin "D." The dentin tubules, open cylinders surrounded by hypermineralized dentin and running nearly perpendicular to the plane of the slice and about the same diameter as the voxel size, are clearest in the reconstruction with the 1.0-μm shift. The horizontal field of view in each image is 220 voxels. 26 keV, 0.125° rotation increment, $(2K)^2$ reconstruction (K. Ignatiev, S. R. Stock, F. De Carlo. October 2007, 2-BM, APS).

5.1.5 Undersampling

Figure 5.5 shows how angular undersampling can deleteriously affect reconstruction quality. The specimen extends from the lower right corner of the field of view, and air is seen to the left and above the specimen. For this $(1K)^2$ reconstruction, angular sampling is adequate with 0.25° between projections (Figure 5.5a), and image quality decreases significantly with increasing angular step size (Figure 5.5b–d) and is particularly evident in the air. Even with 1° between projections (Figure 5.5b), it has become difficult to resolve the internal structure seen in Figure 5.5a.

5.1.6 Beam Hardening

As mentioned in Section 2.6, use of polychromatic radiation produces an effect called beam hardening: the average photon energy of the beam penetrating the sample increases with increasing sample thickness because the lower-energy photons are absorbed at a much higher rate than the higher-energy photons. Thus, Equations (3.1) through (3.3) are no longer strictly valid as written. Beam hardening combined with scattering leads to cupping in reconstructed slices of a uniform object, that is, a radial gradient in the linear attenuation coefficient with abnormally low values at the interior and high values at the periphery. One must always check for this effect when analyzing data from lab microCT systems; only rarely must it be considered with synchrotron systems because these normally

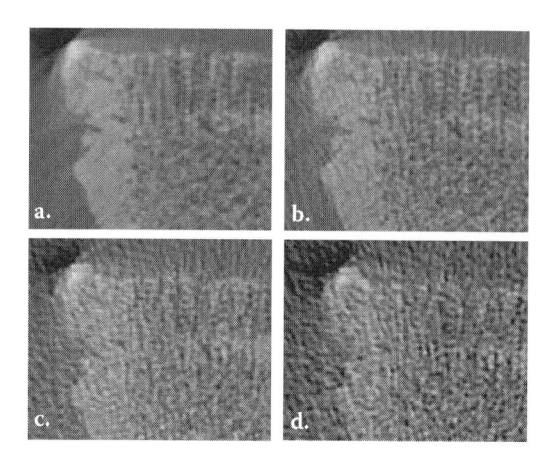

FIGURE 5.5
Illustration of the effect of angular sampling on reconstruction quality. In these small portions of the (1K)^2 slice, the lighter the pixel is, the more absorbing the corresponding voxel. (a) 0.25° increment between projections. (b) 1° increment. (c) 2° increment. (d) 4° increment. The same set of projections was used for each reconstruction, that is, all projections for (a), every fourth for (b), and so on. Test plate of sea urchin *Lytechinus variegatus* (i.e., a section of the calcite mineralized ellipsoidal endoskeleton enclosing and protecting the urchin's critical organs). The horizontal field of view in each image is 150 voxels. Voxel size equaled 5 μm and energy was 21 keV (S. R. Stock, K. I. Ignatiev, and F. De Carlo, 2-BM, APS, June 2003).

employ monochromatic radiation. Figure 5.6 shows a portion of a slice through an aluminum specimen; the graph shows lower values of the linear attenuation coefficient in the interior compared to the outer portions of the specimen. Dual energy techniques, however, offer promise of correcting for the effect of collecting data using a range of wavelengths (Engler and Friedman, 1990).

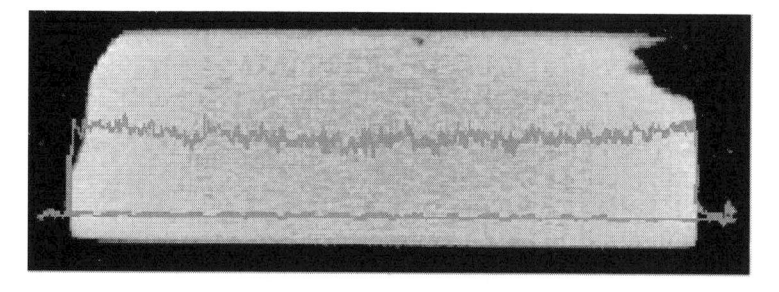

FIGURE 5.6
Beam hardening seen in an aluminum (AA 2090) sample imaged at 70 kVp. The line graph shows the variation of the linear attenuation coefficient as a function of position across the 7.45-mm wide specimen. Linear attenuation coefficients vary from 3 cm^{-1}, and significant cupping is seen in the middle of the specimen. Black represents the lowest levels of attenuation. Imaging with a Scanco MicroCT-40 system: 500 projections each with 0.3 s integration time and 512 samples, reconstruction with 12 μm isotropic voxels (S. R. Stock, K. Ignatiev).

5.1.7 Streak Artifacts

Nonphysical streaks can radiate from high-absorption objects within a lower attenuation matrix (such star artifacts are described in the medical CT literature, for example, originating from metal implants in bone). Figure 5.7a shows streaks from wires of a small strain gauge attached to a rat tibia. The wires are essentially opaque and affect the values of the linear attenuation coefficient elsewhere in the specimen. Streaks can often be seen emanating from and parallel to long flat specimen faces (Figure 5.7b), and, if one has the option of designing specimen geometry, such features should be avoided. Streak artifacts generally are attributed to the effect of scatter. One wonders, however, whether the streak originates from the large, very sharp jump in attenuation from inside to outside the specimen when the projection direction is parallel to the long specimen face; the finite sampling frequencies present in the data may be insufficient to track the jump adequately.

5.1.8 Phase Contrast Artifacts

Synchrotron microCT data from sources such as APS, ESRF, SLS, and SPring-8 seem to invariably have a strong component of phase contrast in the reconstructions. The "hot edges" seen in radiographs along the projected boundary between materials with different electron density

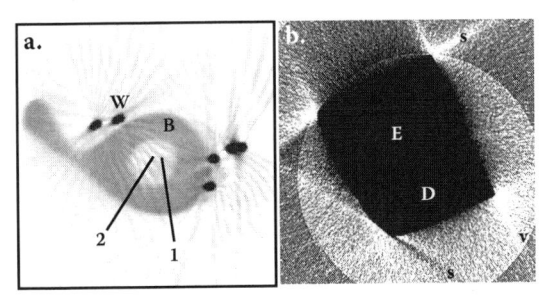

FIGURE 5.7
Streak artifacts: in both images, the darker the pixel is, the more absorbing the voxel, and the slices were recorded with a Scanco MicroCT-40 system. (a) Streaks produced by high absorption features (strain gauge wires "W"). The linear attenuation coefficients for the identified features are: W 8 cm^{-1}, bone B 2.9 cm^{-1}, high absorption streak 1 1.1 cm^{-1}, and low absorption streak 2 (adjacent to 1) 0.3 cm^{-1}. 70 kVp, 114 µA, 250 projections each with 0.3 s integration and 1,024 samples, reconstruction with 36-µm isotropic voxels, horizontal field of view of 9.2 mm. (b) Streaks "s" emanating from flat specimen surfaces into the surrounding air. The sample is in a small vial "v" and is a section of a bovine incisor containing enamel "E" and dentin "D." The field of view is 3.01 mm, and the histogram has been equalized to show the streaks more clearly. 45 kVp, 177 µA, 1,000 projections each with 0.3 s integration and 2,048 samples, reconstruction with 12-µm isotropic voxels. (From the dataset reported in Vieira et al. (2006).)

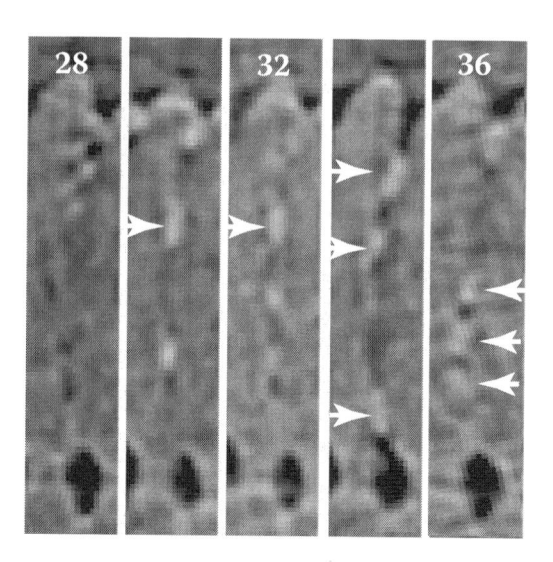

FIGURE 5.8
Anomalous contrast from phase effects seen in channels between calcite carinar process plates in a tooth of the sea urchin *Lytechinus variegatus*. Sections of slices 28, 30, 32, 34, and 36 are shown from left to right, respectively. The white features identified with arrowheads have values of the linear attenuation coefficient three times greater than is possible for calcite. 14 keV, $(1K)^2$ reconstruction, ~1.3 μm isotropic voxels (Stock et al., 2003c).

produce edge sharpening in the reconstructions; that is, this contrast can be thought of as a physical analogue of the sharpening filters used in reconstruction software. Sometimes these phase effects can lead to anomalously large values of the linear attenuation coefficient in positions within a specimen that are not easily recognizable as being near surfaces. In one study (Stock et al., 2003c), such unexpected contrast (Figure 5.8) was not recognized until later (Stock et al., 2003a). This effect may or may not be responsible for contrast interpreted as solute segregation (e.g., high local Zn concentration in an Al engineering alloy (Ohgaki et al., 2006)) or as hypermineralization bands in bone (Stock, 2008). Considerable care must be taken, therefore, in the interpretation of voxel values in synchrotron microCT. In situations such as those mentioned above, viewing a movie paging through a stack of slices can be very helpful.

5.2 Performance: Precision and Accuracy

The accuracy of microCT reconstructions has been considered in many papers. Many direct comparisons (i.e., microCT slices matched with

MicroCT in Practice

physical sections) have established beyond doubt that microCT provides accurate reconstructions of specimen volumes; furthermore, microCT's limitations are also well established. (That is not to say that microCT data cannot be, and has not been, misinterpreted or the technique misused or misapplied.) Also important is understanding what can be resolved (spatial resolution) and what can be detected (sensitivity limits). Knowledge of the dependence of spatial resolution and contrast sensitivity on experimental parameters as well as the intrinsic microstructural characteristics is central to proper interpretation of microCT data. Use of a physical phantom for calibrating 3D microCT systems is important (Perilli et al., 2006; Du et al., 2007) and is an often overlooked necessity for high-accuracy, high-precision work.

5.2.1 Correction for Nonidealities

Section 5.1 discussed not only various artifacts that can be found in CT reconstructions but also several methods for correcting for nonidealities in microCT data collection. Section 5.1.2 mentioned ring correction by knowledge-based smoothing of sinograms. Section 5.1.3 covered how correct rotation centers can be obtained for reconstructions, and Section 5.1.4 mentioned geometrical correction for mechanical errors in microCT apparatus.

5.2.2 Partial Volume Effects

Partial volume effects can significantly bias results obtained with microCT or any volumetric imaging technique, particularly if the choice of voxel size is ill-considered relative to the dimensions of the object of interest. Consider the two tubular specimens in Figure 5.9 that have the same outer

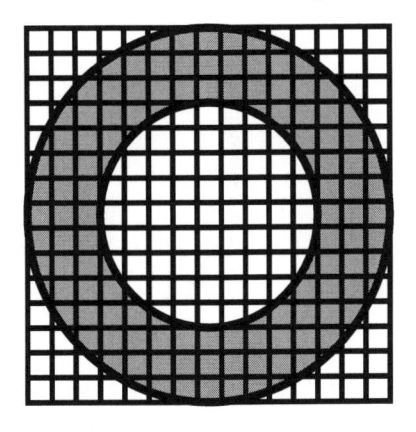

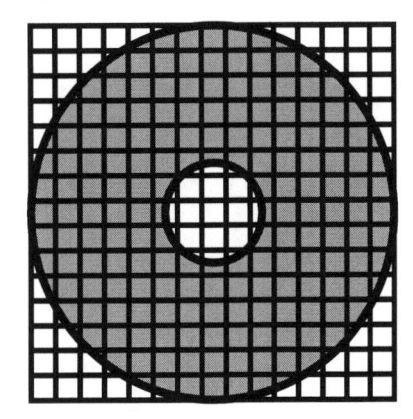

FIGURE 5.9
Two tubular specimens with the same outer diameter but different wall thicknesses, sampled with identical voxel sizes.

diameter (taken as 1.5 mm) but differing wall thicknesses (0.3 mm, left, and 0.55 mm, right). Both are sampled with the same-size voxels (0.1 mm × 0.1 mm in the plane of the reconstruction), and, for simplicity, everything normal to the plane is ignored. For the left-hand specimen, 55 percent of the voxels are partial, but the right-hand specimen contains only 10.5 percent partial voxels. In a calculation partitioning the slice into solid and empty space using a specific fraction of occupancy, conclusions about the left-hand specimen will be much more sensitive to the selected threshold than the sample shown at the right.

The above example is not an academic exercise. Such considerations apply in studies comparing the response to treatment of mice from genetic strains with very different bone phenotypes but with comparable long bone outer diameters, particularly if the imaging is done with voxel dimensions a substantial fraction of the characteristic dimensions of the bone. Furthermore, the clarity of actual scan datasets deteriorated rapidly as voxel size increased, but datasets artificially created from data collected at much higher resolution maintained their clarity to much higher voxel sizes (Cooper et al., 2007).

5.2.3 Detection Limits for High-Contrast Features

In many cases, a microCT system may be required to detect the presence of small features smaller than the tiniest possible voxel size. If the small features have high enough contrast (e.g., high Z precipitates in a low Z matrix or cracks in a homogenous solid) and are reasonably well dispersed, it is certainly possible to detect these features when they are smaller than the voxel size. Here, cracks in a solid are used as a paradigm for feature detection based on differences in absorption contrast.

Before discussing detection limits established using metal matrix composites, consider the requirements for reliably detecting cracks with very small openings, say 0.5 µm. Assuming such a crack extends across the entire voxel and that the signal-to-noise ratio in the reconstruction is adequate for detection of one-quarter voxel of empty space, projections of the object must be recorded with 2-µm pixels in order for the sample to be reconstructed with the required voxel size for crack detection. If a 2K × 2K detector array were used, sample diameters up to 4 mm could be studied successfully. It is important to emphasize that detecting a 0.5-µm wide crack in a 4-mm thick specimen requires sensitivity to sample thickness changes of $\sim 10^{-4}$. Cracks consist of spatially correlated voxels of low absorption, so this helps in detecting such narrow cracks. On the other hand, cracks often follow the interface between phases (i.e., reinforcement and matrix in composites), and this hinders detection.

Crack detectability has been studied in Al/SiC monofilament (uniaxial fiber alignment, ~140-µm fiber diameter with ~30-µm C core) composites using a thin wedge pushed into a sample parallel to the fiber axes (Breunig et al., 1992,1993). Two orientations of the wedge were examined, perpendicular to the eight plies of the composite and 45° to the plies. A pencil beam system with a 10-μm diameter collimator, a Pd filter, and an energy-sensitive detector set for AgK_α radiation allowed slices perpendicular to the monofilament axes to be reconstructed with 12.5-μm isotropic voxels. Crack opening displacement in the plane of the plies was measured as a function of distance from the tip of the wedge by direct measurement of the opening or by comparing the separation of fibers on either side of the crack with that of the same monofilaments in the first slice beyond the tip of the crack.

Results of the two measurement techniques agreed within 2 μm (within about 15 percent of one voxel). The wedge at 45° to the plies was inserted 2.1 mm into the ~1.5 mm × ~1.5 mm sample, which was not far enough to cause significant fiber fracture, but it did cause significant fiber displacement, fiber–matrix debonding, and ductile rupture of the Al matrix. The wedge perpendicular to the plies caused substantial fiber fracture. Cracks open as little as 1 to 2 μm, that is, 10 percent of a voxel's width, could be detected. Opening as a function of distance from the wedge tip was very different for the two geometries and reflected whether fiber fracture had occurred. With the 45° wedge, opening decreased fairly uniformly and quite rapidly until about 700 μm from the tip of the wedge, whereas with the 90° wedge, opening was roughly constant and much larger (than that in the 45° sample), between 100 and 1,000 μm from the wedge tip after which the crack became invisible within 100 μm (Stock, 1999).

The relatively well-behaved gradients in the crack opening produced in the studies cited above allowed detailed assessment of how well the crack opening can be quantified for a given level of noise in the reconstructed images. Besides controlling the amount of opening allowed, the fibers provided a built-in fiducial for measuring the crack opening: the change in fiber–fiber separations. In other words, the difference in two monofilaments' separation across the crack at a particular position and that far from the crack tip should give an accurate measure of the opening at the position in question. Most samples in which the crack opening needs to be measured do not have these fiducials, so crack opening must be quantified by summing the openings in adjacent, partially open voxels. It is difficult to trust the robustness of such a procedure without at least once checking its results against those of a fiducial-based opening measurement; thus, the results of the wedge studies offer important guidance not only for how small an increment of crack opening can be quantified in monolithic samples but also for detection limits (in terms of partial voxels) for other high-contrast features (Stock, 1999).

The results of Breunig and coworkers cited above have been confirmed in more recent synchrotron microCT studies of cracked specimens. Sometimes higher sensitivity can be obtained. At synchrotrons such as ESRF and APS, it requires special effort to suppress phase contrast, but its presence substantially improves crack visibility beyond that possible with pure absorption contrast. Details are summarized elsewhere (Stock, 2008).

5.2.4 Geometry*

Assessing the accuracy of microCT reconstructions requires comparison of reconstructions with the results of another independent technique on the same specimen. Lab microCT agreed with physical measurements of tooth dimensions in one study (Kim et al., 2007). Lab microCT versus microMRI (magnetic resonance imaging) provided one comparison (Borah et al., 2001), and synchrotron microCT versus scanning acoustic microscopy of osteonal bone furnished a second example verifying microCT's accuracy (Dalstra et al., 2004). MicroCT quantification of porosity in a bone cement was found to be much more repeatable and robust with respect to segmentation thresholds than either radiography or optical microscopy (Cox et al., 2006). Bone formation in polymeric scaffolds was evaluated by proton magnetic resonance microscopy and microCT (Washburn et al., 2004).

Physical sectioning compared to microCT slices has been the subject of still other studies confirming validity of reconstructions: confocal optical microscopy of thin serial sections contrasted with microCT of lung specimens (Kriete et al., 2001), histology versus microCT of cortical bone (Wacheter et al., 2001), and histomorphometry versus microCT of biopsies of cancellous bone (Cortet et al., 2004; Chappard et al., 2005).

Calcein labeling is a standard method used to show where new bone is formed, and, in a longitudinal in vivo study of rat tibiae, positions of calcein labeling matched positions where microCT showed new bone had formed (Waarsing et al., 2004). Finite element modeling of microCT-derived bone structures has been validated by testing of rapid prototyping models produced from the microCT data (Su et al., 2007).

MicroCT determination of cortical porosity and of mineral levels showed good agreement with the results of axial ultrasound velocity measurements in the human radius (Bossy et al., 2004). MicroCT versus radiography of the same portions of human femora showed good agreement for measures of cortical porosity (Cooper et al., 2004). Over a significant range of strains ($-1 < \varepsilon < -1.7$), Martin et al. found excellent agreement between the volume fraction of cavities measured with synchrotron microCT and macroscopic measurements of density (Martin et al., 2000). The distribution of particle sizes in pumice clasts showed good agreement between synchrotron microCT and the classic crushing + sieving + winnowing method (Gualda and Rivers, 2006). Yarn dimensions and spacings in 3D textiles and their variability were identical in measurements performed

---
* This section appeared in Stock, 2008.

with lab microCT, surface scanning, and optical microscopy of cross-sections (Desplentere et al., 2005).

Sheppard et al. (2006) found good agreement between simulations of mercury invasion capillary pressure based on 3D microCT pore quantification in four varied specimens and actual measurements performed on the same specimens, with the main differences being attributable to microporosity below the resolution limit in the reconstructions. In a specimen of packed monodisperse beads (3.0-mm nominal diameter), the algorithms of Sheppard et al. (2006) determined a mean particle diameter of 2.98 mm (full width at half maximum ~0.06 mm, slightly less than one 63.4-µm voxel) in the reconstruction; in a second specimen of unconsolidated sandstone, the distribution of particle diameters from lab microCT agreed with results of laser light scattering

Accuracy and precision of lung tumor volume measurements were determined from respiratory gated in vivo microCT of a mouse model (Cody et al., 2005). Lung tumor volumes were both reproducible (2 percent operator variability) and accurate (6 percent average error), and tumor number assessed at necropsy correlated significantly with microCT. Relatively poor contrast between soft tissue types (tumors, blood vessels) was typical of absorption microCT, but the authors employed careful differentiation procedures. Spatial resolution was somewhat limited in both microCT (91-µm isotropic voxels) and in optical inspection (0.5-mm detection limit for tumors). Despite these limitations, this study is convincing, in no small part because of the thorough account provided.

Davis (1999) discussed expected image quality and accuracy in data obtained with a lab cone beam microCT system; different specimen geometries were examined for different cone beam angles. Correction for beam hardening in microCT quantification and the effect of beam hardening on resolution have been discussed (Van De Casteele et al., 2004). Determination of actual versus nominal resolving power was described elsewhere (Seifert and Flynn, 2002).

One investigation of the reproducibility of microCT data collection found the results stable with respect to replication, and displacement of the 6-mm-long ROI by up 4 mm along the axis of the trabecular cored specimen produced little change in microscopic parameters (Nägele et al., 2004). Olurin et al. (2002) examined the dependence of morphometric indices for closed-cell Al foam and for two thicknesses of Al foil as a function of scan parameters in a lab microCT system and found characteristics such as volume fraction and mean feature thickness did not depend appreciably on voxel size for their specimens and scanner.

Examination of potential differences in reconstructions produced by different microCT instruments has been the subject of other studies. For example, results of second-generation versus third-generation synchrotron microCT as well as absorption versus phase microCT have been compared for cat claws (bone as well as tough cornified tissue; Ham et al., 2006). Comparisons of lab and synchrotron microCT are particularly informative for applications such as bone or tooth around metal implants where there is a large difference in absorptivity: one such study examined bone surrounding Ti-implants (Bernhardt et al., 2004) and a second characterized bone around dental implants (Cattaneo et al., 2004).

Gureyev and coworkers have compared what they term quasi-local tomography with absorption and with phase contrast and found that accurate quantitative data could be obtained through quite small fractions of the specimen volume (Gureyev et al., 2007). A study comparing errors (and their spatial distribution) between reconstructions with local tomography, corrected local tomography, and lower-resolution microCT scans is also of interest (Xiao et al., 2007).

Repeated imaging of the same trabecular bone specimen with three different systems (voxel sizes between 14 µm and 2 µm) showed that the larger size provided adequate parameterization of the trabecular structure (Peyrin et al., 1998), a result to be expected because trabeculae are typically ~100-µm thick, that is, on the order of seven voxels along the minimum dimension. The effect of different scanning and reconstruction voxel sizes on trabecular bone parameters was examined for one instrument, and, for the extreme case (voxel size of 110 µm versus mean trabecular thickness of 120 µm), differences in specific surface area (i.e., per unit volume of bone) were as large as 100 percent (Kim et al., 2004). This study suggests that morphometry studies performed on low-resolution pQCT systems should be evaluated very carefully before being accepted (see Schmidt et al. (2003) for a discussion of circumstances where pQCT is accurate; Busignies et al. (2006) for a demonstration that actual low-resolution data is substantially less clear than equivalent resolution data artificially generated from high-resolution datasets; and MacNeil and Boyd (2007) for a comparison of high-resolution pQCT, available for longitudinal studies of patients, and microCT).

In a study of liquid foam with synchrotron microCT, variation of segmentation threshold by ±3 units (on a 256-level grayscale) altered the volume fraction of liquid phase on the order of ±2 percent (Lambert et al., 2005). Lab microCT data were collected on trabecular bone biopsies (6-, 23-, and 230-week-old porcine vertebrae). The three datasets were investigated systematically using a range of segmentation levels that an observer might select; the segmented scans were converted into FEM, and a variation of 0.5 percent in threshold produced a 5 percent difference in bone volume fraction and a 9 percent difference in maximal stiffness for the most sensitive data (6-week sample with lowest bone volume fraction; Hara et al., 2002).

5.2.5 Linear Attenuation Coefficients

In many applications, reconstructed slices are presented in a 0–255 contrast scale and analyzed with simple segmentation. High-definition studies require better contrast sensitivity, for example, where one is looking for spatial variation of mineralization levels in bone. One current area of interest in high-contrast sensitivity microCT centers on healthy and impaired bone, and discussion in this section focuses on mineralized tissue. Before turning to mineralized tissue, however, consider an inorganic material system, aluminum–silicon carbide, whose phases differ little in linear attenuation coefficient.

Figure 5.10 shows a portion of a slice of an Al/SiC composite sample imaged at 21 keV with 1.4-μm isotropic voxels. In addition to processing-related pores "p" within the Al matrix, the carbon cores at the center of each fiber are very clear, and two concentric zones of differing contrast are seen in most of the SiC fibers (inset image at upper left of Figure 5.10). Spatial resolution is high enough to also allow visualization of the thin carbon layer coating the SiC. Within (40 voxel)^2 areas of the Al matrix away from pores or high-absorption impurities "i" (e.g., the small white box in the figure), the mean linear attenuation coefficient ± one standard deviation was about 6.4 ± 1.7 cm^{-1}. Within the 20 × 40 voxel area of the outer zone of the SiC fiber (this was the largest rectangular area that would not overlap with the inner zone of SiC), the value averaged about 6.9 ± 1.5 cm^{-1}. These values are similar (after conversion to the same x-ray energy) to those observed earlier in a companion specimen (Kinney et al., 1990). The 7–8 percent difference in the linear attenuation coefficient provides

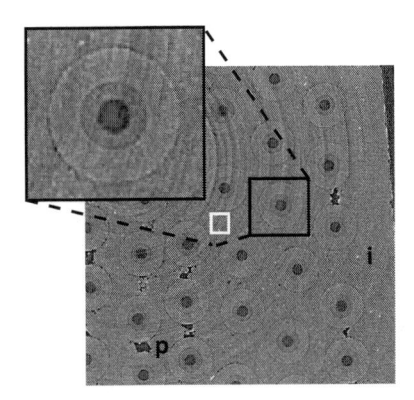

FIGURE 5.10
Slice of an Al/SiC composite showing high absorption impurities "i" and processing-related porosity "p" in the Al matrix. An enlargement (3X) of a single SiC fiber overlaps the upper left of the section of the slice. The small white box shows one area used for determining the mean linear attenuation coefficient. 20 keV, 1.4 μm isotropic voxels, 0.25° rotation increment, 0.72-mm horizontal field of view (S. R. Stock, X. Xiao, and F. De Carlo, 2-BM, APS, March, 2006).

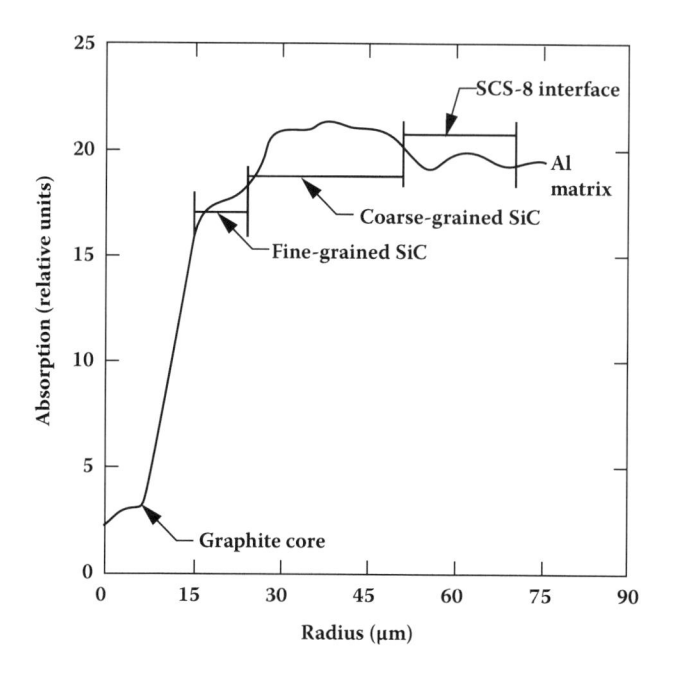

FIGURE 5.11
Radial variation of linear attenuation coefficient in SiC monofilament in an Al/SiC composite. The differences reflect the different microstructures present (Kinney et al. (1990); Breunig (1992); © Materials Research Society).

enough contrast for the SiC fiber visibility, despite the variability in mean linear attenuation of ~25 percent. Although intrinsic material variability and ring artifacts contribute, the standard deviation values are larger than expected mainly because along the longest specimen direction, transmissivity was a bit greater than 5 percent in the 12-bit radiographs.

Figure 5.11 shows the radial variation of the linear attenuation coefficient for a SiC monofilament fiber in a matrix of Al and identifies the different microstructural zones of the fiber (Kinney et al., 1990). Radial averaging over many slices was needed to overcome the noise in the data and to reveal the regions of very slightly different composition consistent with reported variations of stoichiometry (Nutt and Wawner, 1985; Lerch et al., 1988). Often, changing the x-ray energy can improve contrast, but in the SiC and Al the linear attenuation coefficients track each other closely for the x-ray energies that can be used for microCT.

Several studies of mineralized tissue have interpreted specific values of the linear attenuation coefficient. These include a comparison of quantitative backscattering imaging in the SEM versus microCT of mineralized tissue (Mechanic et al., 1990; Fearne et al., 1994; Elliott et al., 1997). Scanning acoustic microscopic maps of specimens have been compared to microCT data (Raum et al., 2006a,b). Several quantitative studies have appeared on demineralization of tooth enamel (Elliott et al., 1998; Dowker et al., 2004; Delbem et al., 2006; Vieira et al., 2006). Others have examined linear attenuation coefficients quantitatively in bone (Nuzzo et al., 2002; Borah et al., 2005).

Interpretation of linear attenuation coefficient μ is limited by irregular microstructural gradients, by partial volume effects, by counting statistics, and by other sources of noise. In analyzing the mean value <μ> for a given phase or region, an area typically is defined and the number average for the voxels and its standard deviation σ are computed. Increasing the number of photons sampling the specimen can improve the variance in μ, at least from this source. According to Equation (3.8), increasing the number of counts transmitted through a specimen by a factor of four will decrease σ (due to counting statistics) by a factor of two. Contributions to the total variance from other sources are, of course, not affected.

Figure 5.12 presents a practical example of the effect of counting statistics on feature visibility in bone that has undergone some remodeling (i.e., replacement of pre-existing bone by the action of osteoclasts and osteoblasts; see the figure caption for further details). The reconstructed slice in Figure 5.12b was collected with a single frame of the CCD and the matching area in Figure 5.12a with a 32-frame average. The remodeled osteons in Figure 5.12a appear more clearly against the surrounding matrix of older interosteonal bone than in Figure 5.12b. Figure 5.12c shows the histogram of inter- and intraosteon areas, that is, the overall histogram. Figure 5.12d superimposes histograms of an interosteon area and an intraosteon area for the 32-frame averaged data, and Figure 5.12e superimposes the same plots for the single-frame data. The histograms are more clearly separated in Figure 5.12d than in Figure 5.12e, and this explains the qualitatively better visibility.

Consideration of experimental values of <μ> and σ for different areas of the slices reveals quantitatively the extent of improvement. For the slice with the longer counting time (32-frame average), <μ> ±σ was 11.5 ± 0.7 cm^{-1} for the interosteon area and 10.2 ± 0.9 cm^{-1} for intraosteon area. For the shorter counting time (single frame), the values were 11.3 ± 0.8 cm^{-1} for the interosteon area and 10.4 ± 0.8 cm^{-1} for the intraosteon area. With longer counting times, the standard deviations do not improve, certainly not by a factor of anything near $\sqrt{32}$, and the variance in this sample is not dominated by counting statistics but is probably determined by subvoxel structural variability and other effects.

In data from tube-based microCT systems, beam hardening limits interpretation of gradients of density, particularly when the denser portions of the sample are at its periphery. This was the situation in a study of model pharmaceutical tablets, and careful correction for beam hardening appears to have allowed valid quantification of density gradients (Busignies et al., 2006).

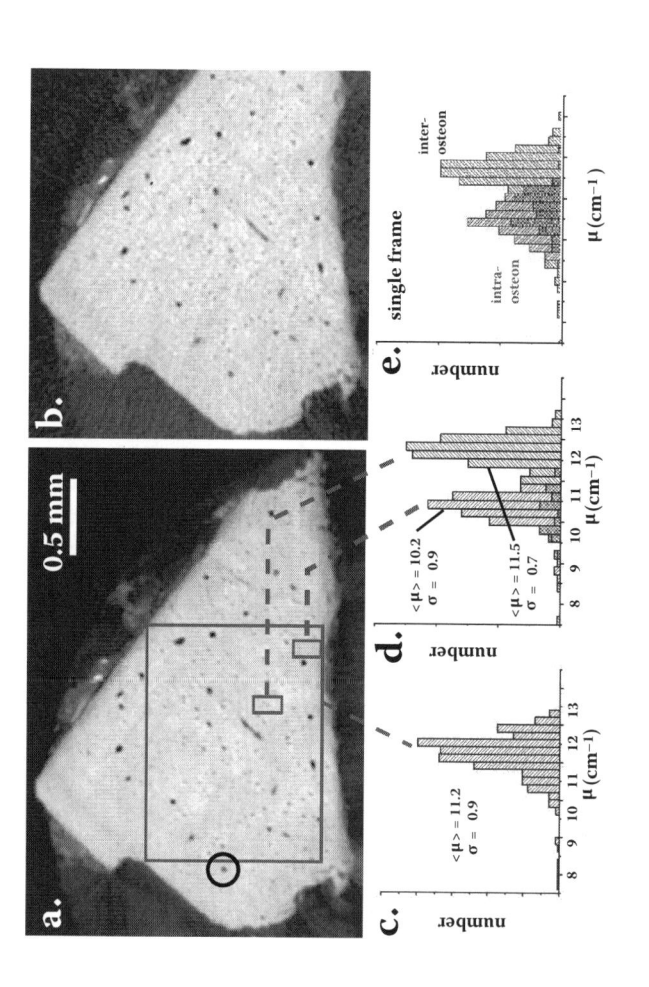

FIGURE 5.12 (SEE COLOR INSERT FOLLOWING PAGE 144.)
Effect of counting time on contrast sensitivity in a bone specimen (canine 2–3 yr fibula) imaged with (a) 32-frame average and (b) single-frame average. The images are of the same slice and at the same magnification, but (b) covers a slightly smaller horizontal field of view. The mineral level within remodeled osteons (e.g., that inside the black circle in (a) centered on the low absorption Haversian canal) is lower than that in the older bone between the osteons and is more clearly visible in (a) than in (b). (c) Histogram of linear attenuation coefficient values μ (mean <μ> and standard deviation σ) for the large boxed area in (a). (d) Histograms of the two small areas indicated in (a) (intraosteonal and interosteonal bone) superimposed. (e) Histograms of the areas in (b) matching those in (d) showing the lower-contrast sensitivity in the single-frame data compared to the frame-averaged data. 21 keV, 5 μm isotropic voxels, 0.25° rotation increment, 0.90-mm horizontal field of view (S. R. Stock and F. De Carlo, 2-BM, APS, November, 2004).

5.3 Contrast Enhancement

High-contrast penetrant liquids can be used to enhance visibility of features such as cracks. Figure 5.13 shows matching slices of the keel of a tooth of the sea urchin *Arbacia punctulata* before and after infiltration with a polytungstate solution that has a much higher linear attenuation coefficient than the calcite of the tooth. The keel consists of an array of single crystal prisms running nearly normal to the plane of the slice. At the stage of mineralization shown, the prisms have not yet been cemented together, and the polytungstate has filled the gaps between prisms that were originally filled with soft tissue and fluid. The polytungstate delineated boundaries between prisms, white in Figure 5.13b, are much clearer than before infiltration.

High-atomic-number gases such as Xe can also be used to enhance visibility of internal structures. Respiration of Xe, for example, has been

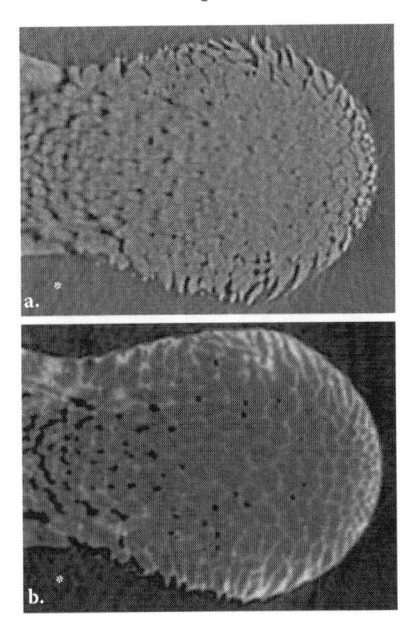

FIGURE 5.13
Contrast enhancement through use of a high-absorption penetrant. Keel of sea urchin tooth (*Arbacia punctulata*) (a) before and (b) after infiltration with a solution of 60 percent polytungstate (see Stock et al. (2003b) for details) for 3.9×10^3 s. In (b), the solution has filled many of the gaps between prisms in the keel (white polygonal network because the solution is much more absorbing than the calcite mineral) and greatly enhanced the visibility of the individual prisms. Note that the orientation of the two slices could only be closely matched for a small area (just above the asterisk). 21 keV, 1.8-μm isotropic voxels, 0.25° rotation increment, 0.90-mm horizontal field of view (S. R. Stock, K. I. Ignatiev, and F. De Carlo, 2-BM, APS, November, 2003).

used to map areas of active gas transport in lungs (Bayat et al., 2001, 2006). Respired Xe enters and persists within the fatty tissue of the brain and has been used to visualize the soft tissue of the brain.

Absorption edge difference imaging can increase sensitivity to small concentrations of the element of interest and is absolutely straightforward at most synchrotron imaging beam lines. Applications include transport in low-porosity materials (Altman et al., 2005a) and in sands (Wildenschild et al., 2002); mapping of flame retardants (Br, Sb) in polymers (Butler et al., 2001); mapping Cs adsorption on iron oxide–hydroxide particles (Altman et al., 2005b); and mapping new bone formation through administration of Pb or Sr labels (Kinney and Ryaby, 2001). Tetrachloroethane with 8 vol.% iodobenzene was used in model studies of organic, water-immiscible phase distribution in porous water-filled materials (Schnaar and Brusseau, 2005, 2006). Multienergy data collection and reconstruction algorithms have also received attention for materials where there are no convenient absorption edges (Ham et al., 2004).

Sensitivity limits to contrast agents have been investigated numerically. Sensitivity to a fixed concentration of KI in water was clearly much better in a coarse sand (mean particle diameter d_{50} = 0.58 mm in a 6-mm-diameter sample) than in a fine sand (d_{50} = 0.17 mm in a 1.5-mm-diameter sample) because the larger photon flux in the former produces a much higher signal-to-noise ratio (Wildenschild et al., 2002); this example is particularly compelling because the specimens are self-similar; that is, the relative sizes of pore and particle do not vary.

Altman et al. (2005a) examined how well porosity could be quantified in various geological specimens; they employed fluid containing high-contrast ions (Cs or I) and compared reconstructions obtained above and below the ion's absorption edge. This difference imaging approach maximizes sensitivity for the atomic species in question, and Altman et al. (2005a) established detection limits and uncertainties using a series of solution concentrations in model, large-diameter specimens and examined the effect of pore diameter on sensitivity to Cs in absorption edge difference imaging.

5.4 Data Acquisition Challenges

The large amounts of data collected with lab microCT systems (potentially running 24 hrs of every day) and with synchrotron microCT (see below) are very challenging to handle. Although the specifics vary from system to system, considering a single example (from the author's recent experience at the synchrotron microCT facility at station 2-BM of APS) illustrates the main points (Stock, 2008). Of course, by the time the reader encounters

MicroCT in Practice

this section, the specifics will be somewhat dated but still illustrate the challenges facing investigators. Note also that the data described below were undersampled with respect to number of projections for samples spanning the 2K detector elements, and the situation actually would be worse than presented.

In November 2006, nine eight-hour shifts were assigned, and the principal goal was to image a large number of sea urchin spines for a detailed comparison of design variations within one phylogenic family. As such, maximizing throughput was essential, and this dictated that the 2-BM specimen placement robot (De Carlo et al., 2006) would be used. One shift was lost to overnight problems with the robot, but otherwise the robot performed smoothly. A total of 91 specimen volumes were imaged during the eight shifts (1.3 specimens/hr); with the present hardware and operating system, one reckons that the maximum is 3 specimens/hr (1,200 sec/specimen). It is instructive, therefore, to examine the durations of various portions of the data collection cycle and their effect on throughput.

The November 2006 run was done with a 2K × 2K detector, and views were collected every 0.25° (significant angular undersampling outside the central region of 1K voxel diameter). Typical acquisition times were 0.2 sec/view or less using the DMM (double multilayer monochromator), for a total of ~150 sec of actual image acquisition. The other 87 percent of the time was occupied mainly by sample motions and handshaking between various hardware components. Removal and placement of specimens occupied no more than a couple of minutes per specimen. Loading the specimen holders onto the robot sample tray (up to 24 specimens per tray) required about 10 min (the specimens were placed on the holders while the previous tray was being collected), so this was a negligible delay when spread over multiple specimens and multiple trays.

The difference between specimens actually imaged (91) and the expected throughput (192 specimens) arises from sources other than those listed above. Some "wasted" time is inevitable between finishing one tray and starting the next, but this is a very minor component. In the absence of other effects, writing to an old disk array (late 1990s technology) was about twice as slow as writing to a newer disk array (2006 parallelized and scalable array); however, data for most of the trays described above were written with the newer array. The ~50-percent decrease in data collection rate is probably produced by nonoptimum tuning of the various network and hardware components (De Carlo, 2006).

Reconstructing the datasets at the tomography facility at station 2-BM of APS is very rapid compared to other aspects of data handling, and most users can expect to leave the facility at the end of their shifts with a significant fraction of their data reconstructed but probably not in hand. Currently, data are in HDF-5 format; each stack of 2K slices from a single specimen amounts to 20–25 GByte, depending on the amount of dynamic compression possible, and writing these data to DVDs is no longer practical. During the November 2006 run, transferring 1 TByte of data (40–50 specimens) from the data analysis cluster (Linux) to an external USB-2 hard drive attached to a PC running Windows required 40 hr. When the drives were attached to a Linux machine and formatted as ext2, the transfer took 15 hr (De Carlo, 2006).

Data collection rates will continue to increase. One hopes that the overall infrastructure (network, disks, etc.) keeps pace, or at least does not lag farther behind. Of course, the biggest bottleneck of all is the lag between bringing the data home and actually doing something with the data (even just paging through slices).

5.5 Speculations

Several trends for future micro- and nanoCT are clear from recently published studies. Consideration of what constitutes an "adequate" study is discussed in the next paragraph. The remainder of the section speculates on directions of future studies. Data representations in the literature have become increasingly sophisticated, and these are discussed in Chapter 6 after some of the quantification methodologies are introduced.

As the number of microCT publications has increased, expectations for the quality and depth of analysis in published studies have risen. What was a strong PhD thesis in the early to mid-1990s became, by the late 1990s (in this author's opinion as standards vary from discipline to discipline, institution to institution, or country to country), only an adequate MS thesis. The same is true of papers in archival journals. Although long explications of analysis methods and of the principles of microCT data collection remain appropriate for theses, very little of this should appear in journal papers, given previous coverage in the literature. One now expects not only interpretation of geometry defined via single thresholds of structures but also (brief but rigorous) consideration of numerical values of linear attenuation coefficients. If binary segmentation is used for numerical analysis, short but detailed examination of the effect of threshold choice should be incorporated; more complicated segmentation routines require presentation of more details.

Future studies will incorporate more elaborate loading apparatus and environmental chambers (furnaces, cooling stages, high-pressure chambers) and more elaborate and better calibrated monitoring of experimental conditions. Peripherals purpose-built by manufacturers for their commercial microCT systems are already appearing and will undoubtedly appear in future publications. Emphasis will surely continue on repeated observations of the same specimen: a wider range of in situ and in vivo studies with increased data acquisition rates. More studies will appear on very

MicroCT in Practice

fast phenomena using gating to freeze movement such as found in sprays; such gating is quite involved, however, so the number of such studies will remain relatively small.

Studies looking at evolution in the structure of individual specimens should emphasize incorporation of proper boundary conditions, which, in practice, means larger volumes of material surrounding the volume of interest. Incorporation of microstructure directly into finite element or other numerical models will be an area that will continue to grow. Reconstructions using 2K × 2K detectors are now standard, and introduction of 4K detector widths (in the plane of reconstruction) will be a direct approach for preserving spatial resolution while examining large-diameter samples. One should not forget that increasing image definition by a factor of two requires an equal increase in the number of projections; decreasing the voxel size by a factor of two requires the collection of 2^4 more photons to obtain the same signal-to-noise ratio as in the original reconstruction. More frequent use of local tomography also is expected, although specimen size will be limited if there is significant absorption: contrast will decrease due to the extra absorption of the material outside the region of interest, and the effect of noise, accompanying the excess, non-image-forming decrease in the number of transmitted photons, will degrade the reconstruction.

MicroCT will be applied more often as part of studies integrating it with other scales of testing and analysis or with other techniques such as x-ray microbeam diffraction mapping. Such multimode studies are described in the following chapters on applications, and one should not forget that methods other than those employing x-rays can be used. One expects more studies will center on key specimens linking the microscale (samples with optimum dimensions for contrast sensitivity) with the macroscopic scale of more normal engineering specimens. Some intermediate-sized specimens may also need to be studied to complete the linkage between different structural scales. Although such studies are not as novel as they were a few years ago, the earlier demonstrations may actually make it easier to organize the resources required for more detailed, multiscale research programs.

More nanoCT and phase micro- and nanoCT studies will appear in the future, especially as more commercial nanoCT systems are installed. One expects in the near future to see commercial phase microCT imaging systems using the grating method or perhaps the analyzer crystal (diffraction-enhanced imaging) method. The propagation method will probably not be used because of the large but precise translations needed. Stability issues may be a critical determinant of whether it proves practical to produce commercial phase microCT systems.

Crystal optics of any sort can be finicky, and implementing a robust system of collecting lab-based diffraction-enhanced phase radiographs will probably require considerable additional hardware (e.g., feedback circuitry) to guard against optics drift. Collecting views at three or more positions of the analyzer crystal will increase data collection times, therefore, by at least a factor of three compared to normal absorption microCT acquisition (and this does not include the effect of decreased flux from the optics, i.e., from wavelength rejection due to beam monochromatization). Grating-based phase microCT systems should be relatively stable, the micrometer-sized translation of the second analyzer grating should not be too difficult to implement with current piezoelectric translators, and all wavelengths passing through the specimen contribute to image formation. Whether the translation of gratings in a commercial system can be controlled robustly enough for day-in and day-out data collection by users or staff without extensive experience remains to be proven.

In conclusion, the author sees a period of consolidation in the area of materials microCT characterization. Instruments are widespread, and new applications will certainly appear, but truly novel developments or applications will be few as investigators concentrate on exploiting areas pioneered in the last decade. Of course, the thing about most profound innovations is that they seem to come out of nowhere, so it will be interesting to see whether the next few years bring surprises in the area of nano- and microCT.

References

Altman, S.J., W.J. Peplinski, and M.L. Rivers (2005a). Evaluation of synchrotron x-ray computerized microtomography for the visualization of transport processes in low porosity materials. *J Contam Hydrol* **78**: 167–183.

Altman, S.J., M.L. Rivers, M.D. Reno, R.T. Cygan, and A.A. Mclain (2005b). Characterization of adsorption sites on aggregate soil samples using synchrotron x-ray computerized microtomography. *Env Sci Technol* **39**: 2679–2685.

Bayat, S., G. Le Duc, L. Porra, G. Berruyer, C. Nemoz, S. Monfraix, S. Fiedler, W. Thomlinson, P. Suortti, C.G. Standertskjold-Nordenstam, and A.R.A. Sovijarvi (2001). Quantitative functional lung imaging with synchrotron radiation using inhaled xenon as contrast agent. *Phys Med Biol* **46**: 3287–3299.

Bayat, S., L. Porra, H. Suhonen, C. Nemoz, P. Suortti, and A.R.A. Sovijärvi (2006). Differences in the time course of proximal and distal airway response to inhaled histamine studied by synchrotron radiation CT. *J Appl Physiol* **100**: 1964–1973.

Bentz, D.P., D.A. Quenard, H.M. Kunzel, J. Baruchel, F. Peyrin, N.S. Martys, and E.J. Garboczi (2000). Microstructure and transport properties of porous building materials. II: Three-dimensional x-ray tomographic studies. *Mater Struct* **33**: 147–153.

Bernhardt, R., D. Scharnweber, B. Müller, P. Thurnier, H. Schliephake, P. Wyss, F. Beckmann, J. Goebbels, and H. Worch (2004). Comparison of microfocus and synchrotron x-ray tomography for the analysis of osteointegration around Ti6AlV4 implants. *Euro Cells Mater* **7**: 42–51.

Borah, B., G.J. Gross, T.E. Dufresne, T.S. Smith, M.D. Cockman, P.A. Chmielewski, M.W. Lundy, J.R. Hartke, and E.W. Sod (2001). Three-dimensional microimaging (MRµI and µCT), finite element modeling, and rapid prototyping provide unique insights into bone architecture in osteoporosis. *Anat Rec* **265**: 101–110.

Borah, B., E.L. Ritman, T.E. Dufresne, S.M. Jorgensen, S. Liu, J. Sacha, R.J. Phipps, and R.T. Turner (2005). The effect of risedronate on bone mineralization as measured by microcomputed tomography with synchrotron radiation: Correlation to histomorphometric indices of turnover. *Bone* **37**: 1–9.

Bossy, E., M. Talmant, F. Peyrin, L. Akrout, P. Cloetens, and P. Laugier (2004). An in vitro study of the ultrasonic axial transmission technique at the radius: 1 MHz velocity measurement are sensitive to both mineralization and intracortical porosity. *J Bone Miner Res* **19**: 1548–1556.

Breunig, T.M. (1992). Nondestructive evaluation of damage in SiC/Al metal/matrix composite using x-ray tomographic microscopy. Atlanta, Georgia Institute of Technology.

Breunig, T.M., J.C. Elliott, S.R. Stock, P. Anderson, G.R. Davis, and A. Guvenilir (1992). Quantitative characterization of damage in a composite material using x-ray tomographic microscopy. *X-ray Microscopy III*. A.G. Michette, G.R. Morrison, and C.J. Buckley (Eds.). New York, Springer: 465–468.

Breunig, T.M., S.R. Stock, A. Guvenilir, J.C. Elliott, P. Anderson, and G.R. Davis (1993). Damage in aligned fibre SiC/Al quantified using a laboratory x-ray tomographic microscope. *Composites* **24**: 209–213.

Brunetti, A. and F. De Carlo (2004). A robust procedure for determination of center of rotation in tomography. *Developments in X-Ray Tomography IV*. U. Bonse (Ed.). Bellingham, WA, SPIE. *SPIE Proc Vol* **5535**: 652–659.

Busignies, V., B. Leclerc, P. Porion, P. Evesque, G. Couarraze, and P. Tchoreloff (2006). Quantitative measurements of localized density variations in cylindrical tablets using x-ray microtomography. *Euro J Pharm Biopharm* **64**: 38–50.

Butler, L.G., K. Ham, H. Jin, and R.L. Kurtz (2001). Tomography at the Louisiana State University CAMD synchrotron: Application to polymer blends. *Developments in X-ray Tomography III*. U. Bonse (Ed.). Bellingham, WA, SPIE. *SPIE Proc Vol* **4503**: 54–61.

Cattaneo, P.M., M. Dalstra, F. Beckmann, T. Donath, and B. Melsen (2004). Comparison of conventional and synchrotron-radiation-based microtomography of bone around dental implants. *Developments in X-ray Tomography IV*. U. Bonse (Ed.). Bellingham, WA, SPIE. *SPIE Proc Vol* **5535**: 757–764.

Chappard, D., N. Retailleau-Gaborit, E. Legrand, M. Baslé, and M. Audran (2005). Comparison insight bone measurements by histomorphometry and µCT. *J Bone Miner Res* **20**: 1177–1184.

Cody, D.D., C.L. Nelson, W.M. Bradley, M. Wislez, D. Juroske, R.E. Price, X. Zhou, B.N. Bekele, and J.M. Kurie (2005). Murine lung tumor measurement using respiratory-gated microcomputed tomography. *Invest Radiol* **40**: 263–269.

Cooper, D., A. Turinsky, C. Sensen, and B. Hallgrimsson (2007). Effect of voxel size on 3D microCT analysis of cortical bone porosity. *Calcif Tiss Int* **80**: 211–219.

Cooper, D.M.L., J.R. Matyas, M.A. Katzenberg, and B. Hallgrimsson (2004). Comparison of microcomputed tomographic and microradiographic measurements of cortical bone porosity. *Calcif Tiss Int* **74**: 437–447.

Cortet, B., D. Chappard, N. Boutry, P. Dubois, A. Cotton, and X. Marchandise (2004). Relationship between computed tomographic image analysis and histomorphometry for microarchitectural characterization of human calcaneus. *Calcif Tiss Int* **75**: 23–31.

Cox, B.D., R.K. Wilcox, M.C. Levesley, and R.M. Hall (2006). Assessment of a three-dimensional measurement technique for the porosity evaluation of PMMA bone cement. *J Mater Sci Mater Med* **17**: 553–557.

Dalstra, M., E. Karaj, F. Beckmann, T. Andersen, and P.M. Cattaneo (2004). Osteonal mineralization patterns in cortical bone studied by synchrotron-radiation-based computed microtomography and scanning acoustic microscopy. *Developments in X-ray Tomography IV*. U. Bonse (Ed.). Bellingham, WA, SPIE. *SPIE Proc Vol* **5535**: 143–151.

Davis, G.R. (1999). Image quality and accuracy in x-ray microtomography. *Developments in X-ray Tomography II*. U. Bonse (Ed.). Bellingham, WA, SPIE. *SPIE Proc Vol* **3772**: 147–155.

Davis, G.R. and J.C. Elliott (2006). Artifacts in x-ray microtomography of materials. *Mater Sci Technol* **22**: 1011–1018.

De Carlo, F. (2006). Personal communication.

De Carlo, F., X. Xiao, and B. Tieman (2006). X-ray tomography system, automation and remote access at beamline 2-BM of the Advanced Photon Source. *Developments in X-Ray Tomography V*. U. Bonse (Ed.). Bellingham, WA, SPIE. *SPIE Proc Vol* **6318**: 63180K–1 to –13.

Delbem, A.C.B., A.E.M. Vieira, K.T. Sassaki, M.L. Cannon, S.R. Stock, X. Xiao, and F. De Carlo (2006). Quantitative analysis of mineral content in enamel using synchrotron microtomography and microhardness analysis. *Developments in X-Ray Tomography V*. U. Bonse (Ed). Bellingham, WA, SPIE. *SPIE Proc Vol* **6318**: 631824–1 to –5.

Desplentere, F., S.V. Lomov, D.L. Woerdeman, I. Verpoest, M. Wevers, and A. Bogdanovich (2005). Micro-CT characterization of variability in 3D textile architecture. *Compos Sci Technol* **65**: 1920–1930.

Donath, T., F. Beckmann, and A. Schreyer (2006). Image metrics for the automated alignment of microtomography data. *Developments in X-Ray Tomography V*. U. Bonse (Ed.). Bellingham, WA, SPIE. *SPIE Proc Vol* **6318**: 631818–1 to –9.

Dowker, S.E.P., J.C. Elliott, G.R. Davis, R.M. Wilson, and P. Cloetens (2004). Synchrotron x-ray microtomographic investigation of mineral concentrations at micrometer scale in sound and carious enamel. *Caries Res* **38**: 514–522.

Du, L.Y., J. Umoh, H.N. Nikolov, S.I. Pollmann, T.Y. Lee, and D.W. Holdsworth (2007). A quality assurance phantom for the performance evaluation of volumetric micro-CT systems. *Phys Med Biol* **52**: 7087–7108.

Elliott, J.C., P. Anderson, G.R. Davis, F.S.L. Wong, S.E.P. Dowker, and N. Kozul (1997). Microtomography in medicine and related fields. *Developments in X-ray Tomography*. U. Bonse (Ed.). Bellingham, WA, SPIE. *SPIE Proc Vol* **3149**: 2–12.

Elliott, J.C., F.S.L. Wong, P. Anderson, G.R. Davis, and S.E.P. Dowker (1998). Determination of mineral concentration in dental enamel from x-ray attenuation measurements. *Conn Tiss Res* **38**: 61–72.

Engler, P. and W.D. Friedman (1990). Review of dual-energy computed tomography techniques. *Mater Eval* **48**: 623–629.

Fearne, J.M., J.C. Elliott, F.S. Wong, G.R. Davis, A. Boyde, and S.J. Jones (1994). Deciduous enamel defects in low-birth-weight children: Correlated X-ray microtomographic and backscattered electron imaging study of hypoplasia and hypomineralization. *Anat Embryol* **189**: 375–381.

Gualda, G.A.R. and M. Rivers (2006). Quantitative 3D petrography using x-ray tomography: Application to Bishop Tuff pumice clasts. *J Volcanol Geothermal Res* **154**: 46–62.

Gureyev, T.E., Y.I. Nesterets, and S.C. Mayo (2007). Quantitative quasi-local tomography using absorption and phase contrast. *Opt Comm* **280**: 39–48.

Ham, K., H.A. Barnett, T. Ogunbakin, D.G. Homberger, H.H. Bragulla, K.L.M. II, S. Willson, and L.G. Butler (2006). Imaging tissue structures: Assessment of absorption and phase contrast x-ray tomography imaging at 2nd and 3rd generation synchrotrons. *Developments in X-Ray Tomography V*. U. Bonse (Ed.). Bellingham, WA, SPIE. *SPIE Proc Vol* **6318**: 631822-1 to -10.

Ham, K., C.S. Willson, M.L. Rivers, R.L. Kurtz, and L.G. Butler (2004). Algorithms for three-dimensional chemical analysis with multi-energy tomographic data. *Developments in X-Ray Tomography IV*. U. Bonse (Ed.). Bellingham, WA, SPIE. *SPIE Proc Vol* **5535**: 286–292.

Hara, T., E. Tanck, J. Homminga, and R. Huiskes (2002). The influence of micro-computed tomography threshold variations on the assessment of structural and mechanical trabecular bone properties. *Bone* **31**: 107–109.

Ignatiev, K., S.R. Stock, and F. De Carlo (2007). Personal communication.

Ketcham, R.A. (2006). New algorithms for ring artifact removal. *Developments in X-Ray Tomography V*. U. Bonse (Ed.). Bellingham, WA, SPIE. *SPIE Proc Vol* **6318**: 631800-1 to -7.

Kim, D.G., G.T. Christopherson, X.N. Dong, D.P. Fyhrie, and Y.N. Yeni (2004). The effect of microcomputed tomography scanning and reconstruction voxel size on the accuracy of stereological measurements in human cancellous bone. *Bone* **35**: 1375–1382.

Kim, I., K.S. Paik, and S.P. Lee (2007). Quantitative evaluation of the accuracy of micro-computed tomography in tooth measurement. *Clin Anat* **20**: 27–34.

Kinney, J.H. and J.T. Ryaby (2001). Resonant markers for noninvasive, three-dimensional dynamic bone histomorphometry with x-ray microtomography. *Rev Sci Instrum* **72**: 1921–1923.

Kinney, J.H., S.R. Stock, M.C. Nichols, U. Bonse, T.M. Breunig, R.A. Saroyan, R. Nusshardt, Q.C. Johnson, F. Busch, and S.D. Antolovich (1990). Nondestructive investigation of damage in composites using x-ray tomographic microscopy. *J Mater Res* **5**: 1123–1129.

Kriete, A., A. Breithecker, and W. Rau (2001). 3D imaging of lung tissue by confocal microscopy and micro-CT. *Three-Dimensional and Multidimensional Microscopy: Image Acquisition and Processing VIII*. J.A. Conchello, C.J. Cogswell, and T. Wilson (Eds.). Bellingham, WA, SPIE. *SPIE Proc Vol* **4261**: 40–47.

Lambert, J., I. Cantat, R. Delannay, A. Renault, F. Graner, J.A. Glazier, I. Veretennikov, and P. Cloetens (2005). Extraction of relevant physical parameters from 3D images of foams obtained by x-ray tomography. *Colloids Surf A* **263**: 295–302.

Lerch, B.A., D.R. Hull, and T.A. Leonhardt (1988). As-received microstructures of a SiC/Ti-15-3 composite, NASA Lewis. TM-100938.

MacNeil, J.A. and S.K. Boyd (2007). Accuracy of high-resolution peripheral quantitative computed tomography for measurement of bone quality. *Med Eng Phys* **29**: 1096–1105.

Martin, C.F., C. Josserond, L. Salvo, J.J. Blandin, P. Cloetens, and E. Boller (2000). Characterization by x-ray micro-tomography of cavity coalescence during superplastic deformation. *Scripta Mater* **42**: 375–381.

Mechanic, G.L., S.B. Arnaud, A. Boyde, T.G. Bromage, P. Buckendahl, J.C. Elliott, E.P. Katz, and G.N. Durnova (1990). Regional distribution of mineral and matrix in the femurs of rats flown on Cosmos 1887 biosatellite. *FASEB J* **4**: 34–40.

Nägele, V.K., H Vogt, T.M Link, R. Müller, E.M Lochmüller, and F. Eckstein (2004). Technical considerations for microstructural analysis of human trabecular bone from specimens excised from various skeletal sites. *Calcif Tiss Int* **75**: 15–22.

Nutt, S.R. and F.E. Wawner (1985). Silicon carbon filaments: Microstructure. *J Mater Sci* **20**: 1953–1960.

Nuzzo, S., M. H. Lafage-Proust, E. Martin-Badosa, G. Boivin, T. Thomas, C. Alexandre, and F. Peyrin (2002). Synchrotron radiation microtomography allows the analysis of three-dimensional microarchtiecture and degree of mineralization of human iliac crest biopsy specimens: Effects of etidronate treatment. *J Bone Miner Res* **17**: 1372–1382.

Ohgaki, T., H. Toda, M. Kobayashi, K. Uesugi, M. Niinom, T. Akahori, T. Kobayashi, K. Makii and Y. Aruga (2006). In situ observations of compressive behaviour of aluminium foams by local tomography using high-resolution tomography. *Phil Mag* **86**: 4417–4438.

Olurin, O.B., M. Arnold, C. Körner, and R.F. Singer (2002). The investigation of morphometric parameters of aluminium foams using micro-computed tomography. *Mater Sci Eng A* **328**: 334–343.

Perilli, E., F. Baruffaldi, M.C. Bisi, L. Cristofolini, and A. Cappello (2006). A physical phantom for the calibration of three-dimensional X-ray microtomography examination. *J Microsc* **222**: 124–134.

Peyrin, F., M. Salome, P. Cloetens, A.M. Laval-Jeantet, E. Ritman, and P. Rüegsegger (1998). MicroCT examinations of trabecular bone samples at different resolutions: 14, 7 and 2 micron level. *Technol Health Care* **6**: 391–401.

Raum, K., R.O. Cleveland, F. Peyrin, and P. Laugier (2006a). Derivation of elastic stiffness from site-matched mineral density and acoustic impedance maps. *Phys Med Biol* **512**: 747–758.

Raum, K., I. Leguerney, F. Chandelier, M. Talmant, A. Saied, F. Peyrin, and P. Laugier (2006b). Site-matched assessment of structural and tissue properties of cortical bone using scanning acoustic microscopy and synchrotron radiation microCT. *Phys Med Biol* **51**: 733–746.

Rivers, M.L. and Y. Wang (2006). Recent developments in microtomography at GeoSoilEnviroCARS. *Developments in X-Ray Tomography V*. U. Bonse (Ed.). Bellingham, WA, SPIE. *SPIE Proc Vol* **6318**: 63180J–1 to –15.

Sasov, A. (2001). Comparison of fan-beam, cone-beam and spiral scan reconstruction in x-ray microCT. *Developments in X-ray Tomography III*. U. Bonse (Ed.). Bellingham, WA, SPIE. *SPIE Proc Vol* **4503**: 124–131.

Schmidt, C., M. Priemel, T. Kohler, A. Weusten, R. Müller, M. Amling, and F. Eckstein (2003). Precision and accuracy of peripheral quantitative computed tomography (pQCT) in the mouse skeleton compared with histology and microcomputed tomography (μCT). *J Bone Miner Res* **18**: 1486–1496.

Schnaar, G. and M.L. Brusseau (2005). Pore-scale characterization of organic immiscible-liquid morphology in natural porous media using synchrotron x-ray microtomography. *Env Sci Technol* **39**: 8403–8410.

Schnaar, G. and M.L. Brusseau (2006). Characterizing pore-scale dissolution of organic immiscible liquid in natural porous media using synchrotron x-ray microtomography. *Env Sci Technol* **40**: 6622–6629.

Seifert, A. and M.J. Flynn (2002). Resolving power of 3D x-ray microtomography systems. *Medical Imaging 2002: Physics of Medical Imaging*. M.J.Y.L.E. Antonuk (Ed.). Bellingham, WA, SPIE. *SPIE Proc Vol* **4682**: 407–413.

Sheppard, A.P., C.H. Arns, A. Sakellariou, T.J. Senden, R.M. Sok, H. Averdunk, M. Saadatfar, A. Limaye, and M.A. Knackstedt (2006). Quantitative properties of complex porous materials calculated from x-ray μCT images. *Developments in X-ray Tomography V*. U. Bonse (Ed.). Bellingham, WA, SPIE. *SPIE Proc Vol* **6318**: 631811–1 to –15.

Stock, S.R. (1999). Microtomography of materials. *Int Mater Rev* **44**: 141–164.

Stock, S.R. (2008). Recent advances in x-ray microtomography applied to materials. *Int Mater Rev* **58**: 129–181.

Stock, S.R., J. Barss, T. Dahl, A. Veis, J.D. Almer, and F. De Carlo (2003c). Synchrotron x-ray studies of the keel of the short-spined sea urchin *Lytechinus variegatus*: Absorption microtomography (microCT) and small beam diffraction mapping. *Calcif Tiss Int* **72**: 555–566.

Stock, S.R., K.I. Ignatiev, T. Dahl, A. Veis, and F.D. Carlo (2003a). Three-dimensional microarchitecture of the plates (primary, secondary and carinar process) in the developing tooth of *Lytechinus variegatus* revealed by synchrotron x-ray absorption microtomography (microCT). *J Struct Biol* **144**: 282–300.

Stock, S.R., S. Nagaraja, J. Barss, T. Dahl, and A. Veis (2003b). X-Ray microCT study of pyramids of the sea urchin *Lytechinus variegatus*. *J Struct Biol* **141**: 9–21.

Su, R., G.M. Campbell, and S.K. Boyd (2007). Establishment of an architecture-specific experimental validation approach for finite element modeling of bone by rapid prototyping and high resolution computed tomography. *Med Eng Phys* **29**: 480–490.

Van De Casteele, E., D.V. Dyck, J. Sijbers, and E. Raman (2004). The effect of beam hardening on resolution in X-ray microtomography. *Progress in Biomedical Optics and Imaging — Medical Imaging 2004: Imaging Processing*. J.M. Fitzpatrick and M. Sonka (Eds.). Bellingham, WA, SPIE. *SPIE Proc Vol* **5370** III: 2089–2096.

Vidal, F.P., J.M. Letang, G. Peix, and P. Cloetens (2005). Investigation of artifact sources in synchrotron microtomography via virtual X-ray imaging. *Nucl Instrum Meth B* **234**: 333–348.

Vieira, A.E.M., A.C.B. Delbem, K.T. Sassaki, M.L. Cannon, and S.R. Stock (2006). Quantitative analysis of mineral content in enamel using laboratory microtomography and microhardness analysis. *Developments in X-Ray Tomography V*. U. Bonse (Ed.). Bellingham, WA, SPIE. *SPIE Proc Vol* **6318**: 631823–1 to –5.

Waarsing, J.H., J.S. Day, J.C. v. d. Linden, A.G. Ederveen, C. Spanjers, N.D. Clerck, A. Sasov, J.A.N. Verhaar, and H. Weinans (2004). Detecting and tracking local changes in the tibiae of individual rats: A novel method to analyse longitudinal in vivo microCT data. *Bone* **34**: 163–169.

Wacheter, N.J., P. Augat, G.D. Krischak, M. Mentzel, L. Kinzl, and L. Claes (2001). Prediction of cortical bone porosity in vitro by microcomputed tomography. *Calcif Tiss Int* **68**: 38–42.

Washburn, N.R., M. Weir, P. Anderson, and K. Potter (2004). Bone formation in polymeric scaffolds evaluated by proton magnetic resonance microscopy and x-ray microtomography. *J Biomed Mater Res Pt A* **69**: 738–747.

Weitkamp, T. and P. Bleuet (2004). Automatic geometric calibration for x-ray microtomography based on Fourier and Radon analysis. *Developments in X-Ray Tomography IV*. U. Bonse (Ed.). Bellingham, WA, SPIE. *SPIE Proc Vol* **5535**: 623–627.

Wildenschild, D., J.W. Hopmans, C.M.P. Vaz, M.L. Rivers, D. Rickard, and B.S.B. Christensen (2002). Using x-ray tomography in hydrology: Systems, resolutions and limitations. *J Hydrol* **267**: 285–297.

Xiao, X., F.D., Carlo, and S. Stock (2007). Practical error estimation in zoom-in and truncated tomography reconstructions. *Rev Sci Instrum* **78**: 063705-1–7.

6

Experimental Design, Data Analysis, Visualization

This chapter has three main subjects and serves as an introduction for the approaches used in the chapters on applications. The first topic is design of microCT characterization experiments; although most of the factors involved are considered automatically by experienced tomographers and the presentation of these details will undoubtedly bore those individuals, certain aspects will perhaps not occur to newcomers and are, therefore, worth describing. The second is data analysis strategies, that is, how relevant numerical quantities might be extracted for different classes of specimens. The third is data visualization, that is, the presentation of numerical data in a form allowing 2D, 3D, 4D, and 5D interpretation of sample sets. Data representation in 2D and 3D is familiar to most readers; 4D representations include temporal observation of a 3D structure or superposition of a numerical quantity (say magnitude of a crack opening) onto a 3D structure (the crack plane); 5D representations might combine the latter representation (measured quantity folded into the 3D structure) with temporal evolution.

6.1 Experiment Design

Before delving into the details of microCT experimental design, posing several questions will help to focus attention on the big picture and on the choices to be made. The first two questions concern the imaging conditions and are not independent of specimen considerations that follow.

- What spatial resolution is needed for the application? Specifically, what spatial resolution (voxel size) is optimum, what might be adequate, and what might be marginal?
- What contrast sensitivity is required for the application? This might depend on whether and how the data is to be segmented (is simple binary segmentation sufficient, i.e., selection of a threshold voxel value and division of all voxels into two classes, one the material of interest and the other "empty" space) or on whether the voxel values are to be interpreted numerically.

One way of describing the combination of different spatial and contrast resolutions is to use a high/medium/low definition to describe the imaging characteristics. High definition refers to imaging with high spatial resolution and high contrast sensitivity. In this context, though, high spatial resolution refers not to absolute voxel dimensions but, rather, to the number of voxels per unit specimen diameter. Medium definition might refer to high spatial definition combined with moderate contrast sensitivity or the reverse, moderate spatial definition combined with high contrast sensitivity. Questions relating to specimen specifics include:

- What is the specimen absorptivity (and what are the available x-ray energies)?
- What specimen constraints exist (minimum representative specimen diameter, minimum specimen volume; what is the required specimen aspect ratio, i.e., length to width of the cross-section; what is the acceptable length of time out of freezer or culture chamber)?
- How much hard drive space will be required? How long will it take to acquire data? How slow (fast) is reconstruction? Once analysis routines have been perfected, how long will it take to analyze the required number of specimens?

Questions of the number of specimens required for statistical significance are the province of the specific experiments and are not covered here.

Before turning to the question of extracting numerical data or informative 3D representations from microCT data, the reader may find that it is useful to review the design and constraints outlined below for one recent study. This study involved the time course of Portland cement degradation by sulfate ions, a significant durability issue and potential safety concern in certain environments.

Sulfate ions, in particular, from the external environment, can produce deleterious phase transformations in Portland cement, a major construction material. One phase that forms is gypsum, and it is associated with loss of adhesion and strength. A second phase involved in the attack is ettringite; it is associated with expansion and cracking. Sulfate attack of Portland cement remains incompletely understood, and the microCT study aimed at improving this situation by combining 3D measures of damage, including cracking, with 3D x-ray diffraction mapping of the spatial distribution of crystalline reaction products (Stock et al., 2002; Naik, 2003; Jupe et al., 2004; Naik et al., 2004; Wilkinson et al., 2004; Naik et al., 2006). Exposure in these accelerated tests was interrupted at different points over one year, and the noninvasive sampling allowed the same specimen to be interrogated repeatedly. Some results are presented in Chapter 10.

Several constraints limited data collection for the sulfate attack study. First, a tube-based microCT system was to be used, and the sulfate damage study had to compete for access with several other projects. Nonetheless, numerous variables (and hence specimens) and four exposure times (up to 52 weeks) were to be studied, including two cement types, multiple sulfate ion concentrations, two cations, and two water-to-cement ratios. Furthermore, the specimens had to be large enough that edge effects would not dominate the observations and that a significant volume in the specimen interior would remain unaffected by the damage. On the other hand, the sample size needed to be small enough that there was adequate x-ray transmissivity at the maximum tube voltage (70 kVp) and that the voxel dimensions were small enough to capture important damage processes.

A cylindrical specimen geometry was selected because it was optimum for microCT and because rotation about the sample axis during x-ray diffraction pattern collection would sample more grains at a constant distance from the surface. A 12-mm diameter and 40-mm length were selected based on the sample transmissivity, on the dimensions of inexpensive (disposable) plastic tubes serving as molds for casting the cement, and on the desire to study effects of sample "corners" as well as to study a length of uniform cylinder away from the cylinder's ends (16 mm of length was one sample diameter or more from the ends). The available time for examining each specimen after each increment of sulfate exposure was about four hours, and the investigators selected the following for each specimen as compromises between this temporal constraint, sensitivity to small cracks, and coverage of a significant volume:

- 37-μm voxels for $(1K)^2$ reconstruction,
- Maximum integration time per projection (0.35 s), and
- 390 slices covering 19.5 mm of specimen length (one slice every ~50 μm).

Although details of damage at the 5–10-μm level were lost, the microCT voxel size was significantly smaller than the sampling volume for diffraction, an important consideration for the combined interpretation of two types of data.

6.2 Data Analysis

This section introduces some principles of data analysis and some of the methods that have been used to analyze microCT data. The intent is neither to go into the depth that one would expect of an image analysis text nor to indicate which method(s) is (are) best for given application types. This latter consideration is taken up along with the applications in the following chapters. Before describing the numerical approaches, it is useful to consider the quantities that must be measured and the type and amount of sampling required to return a reliable assessment of these quantities.

Numerical measurements of microstructure are typically used to test hypothetical relationships between macroscopic mechanical properties and the volume fraction of a particular phase, the amount and distribution of fluid or gas transport, and the channel dimensions and network characteristics, treatment with a given drug and prevention of degradation of trabecular architecture in osteoporosis models, and so on. Microstructural quantities include volume fraction of a phase V_V (in bone quantification software bundled with commercial software systems and in some literature, this is written BV/TV, the ratio of bone volume to total volume), surface area per unit volume of the phase of interest S_V, mean volumes of cells, mean diameter of particles d_{50}, mean thickness <Th> of structural elements (or diameter of channels), distribution of Th valves, distribution of particle sizes, structural connectivity, and structure model index (SMI valves indicate whether the objects in a specimen tend to be rodlike, platelike, or spherical; see Chapter 8). Measures of anisotropy of microstructural features have also received attention, but these tend to be context-specific and are covered only in conjunction with the associated application.

One should also remain aware of the fact that analogous quantities may be measured in different subject areas and called by different names. Quite elaborate analysis routines that one does not normally encounter may be available in the literature, and that is one reason that the applications chapters tend to be organized by structure type rather than traditional discipline.

MicroCT sample volumes may be as little as the thickness of a single slice (the volume, of course, would be the thickness multiplied by the area of the slices). As illustrated in Figure 6.1a, a single slice may sometimes give an accurate measure of the microstructure (here the slice is perpendicular to the axes of six circular and parallel rods). The black disks, viewed in perspective, show the intersection of the slice (dashed lines on the cube surfaces) with the rods. A single slice, however, may not represent the actual structure accurately, as Figure 6.1b shows. The slice plane labeled "*i*" misses the rods entirely and the plane "*ii*" intersects two rods. If the specimen has an isotropic microstructure, one slice may give an accurate value of the volume fraction (a basic stereology result for random structures shows that the area fraction of a phase A_A equals V_V), but, absent the certainty of a truly isotropic microstructure, it is better to sample a volume. Furthermore, measurement of quantities such as thicknesses of structures or diameters of particles are almost always

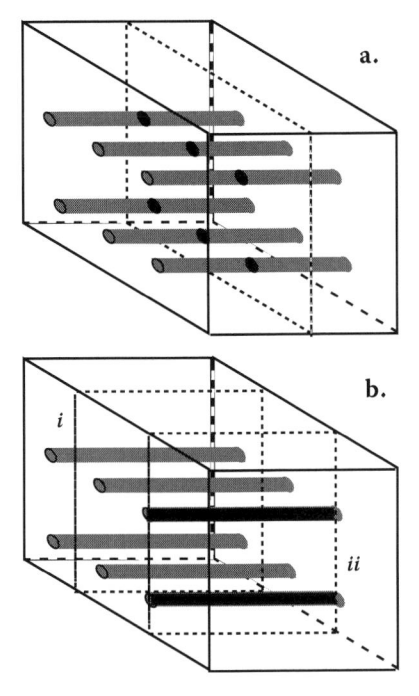

Experimental Design, Data Analysis, Visualization

FIGURE 6.1
Illustration of sampling by a single plane. (a) The single plane perpendicular to the axes of the rods captures their cross-sections (black disks), volume fraction, and spacing accurately, but this is quite fortuitous. (b) In a second sampling geometry, plane "*i*" misses the rods entirely and "*ii*" intercepts two rods (black rectangles).

biased unless volumetric sampling and analysis are used (the situation in Figure 6.1a is quite fortunate, and this sampling is rarely attained in practice without 3D datasets).

6.2.1 Segmentation

In many circumstances, the specimen consists of several discrete phases with distinct absorption or phase contrasts. These individual phases occupy one or more regions within the field of view. Often, only one of these phases is of interest for computational purposes, and one segments the image (volume) into the phase of interest and everything else. Typically, the segmentation process replaces voxel values of the phase of interest by a binary value of "1" (in physical terms, these voxels are treated as solid) and the values of all other voxels with "0" (empty space). Segmentation is often used synonymously with thresholding, and visualizations of volumes employing multiple thresholds and partially transparent solids can be very effective (see Figure 6.14 below).

Binary segmentation and selection of a reasonable threshold separating the phase of interest from all other phases is perhaps the most popular approach. Specimens containing high-contrast phases — for example, bone where the bone/marrow contrast is about 10/1 for a lab microCT system operating with an effective x-ray energy of ~25 keV (Rüegsegger et al., 1996) — have histograms with peaks separated by a valley; these materials can be segmented by choosing a threshold within the valley. Such a situation is pictured in Figure 6.2 for a slice of murine trabecular bone. As has been shown by comparing results on the same specimen reconstructed with different voxel sizes, the specific threshold needed to produce the same volume fraction (a valid constraint, given that the data are from the same specimen) will necessarily vary with voxel size (Hangartner, 2007). Likewise, valid comparison between high-porosity specimens embedded in plastic and similar unembedded samples (e.g., trabecular bone) requires use of different thresholds (Perilli et al., 2007).

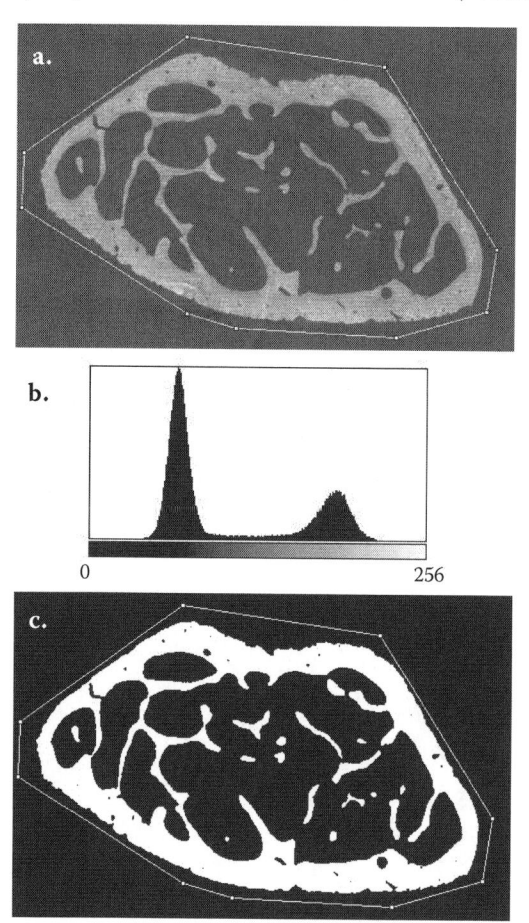

FIGURE 6.2
Binary segmentation applied to a slice of murine trabecular bone. (a) Portion of a synchrotron microCT slice of the murine femur in ethanol, ~2.8-µm isotropic voxels, 512-voxel horizontal field of view, 1K × 1K reconstruction, 17 keV (S. R. Stock, N. M. Rajamannan, F. De Carlo, 2-BM APS, June 2005). (b) Histogram of the polygonal area (>10^5 voxels) within the thin lines in (a). (c) The voxels with grayscale values 128 and above are shown as white; all other voxels are black.

Confounding effects that necessitate less routine segmentation include large populations of partial voxels resulting, for example, from structures with minimum dimensions on the order of the voxel size (see Figures 5.9 and 6.3). In addition, situations exist where the material itself contains a spread of linear attenuation coefficients; that is, the material lacks well-defined, well-separated peaks in the histogram. This is the situation shown in Figure 5.12, and segmentation of remodeled osteons from older bone is problematic at best. Robb and coworkers discuss the range of segmentation approaches used for microCT (Rajagopalan et al., 2005).

In the absence of a well-defined valley between histogram peaks, one frequent segmentation approach is assignment by inspection: the operator examines a typical slice or slices and selects the threshold which best (to his or her eye) preserves the important fine-scale features of both the solid and the surrounding empty space. Given that this is a highly subjective process, considerable effort has been devoted to assessing the robustness of conclusions derived from small shifts in threshold; the general consensus is that absolute numbers (V_V, Th, etc.) will change somewhat with changing threshold, but, as long as the features being quantified have minimum dimensions greater than perhaps four voxels, comparisons between specimens will be valid (see Section 5.2 on microCT accuracy for more details). The author uses this trial-and-error method but with the following precaution: the threshold is chosen based on evaluation of a preliminary subset of specimens, a subset explicitly excluded from the actual statistical comparison of different treatment groups (Stock et al., 2004b).

It is not always possible for a single threshold value to adequately define the phase of interest over the entire specimen. This occurs, for example, when significant beam hardening is present. An alternative to a single, global threshold value is use of the gradient in grayscale value to define the boundary between phases (Rajon et al., 2006). Adaptive or dynamic thresholding, which explicitly accounts for varying background, is another approach (Ramaswamy et al., 2004; Brunke et al., 2005; Feeney et al., 2006; Burghardt et al., 2007). Yet another approach is to use two thresholds $T_A < T_B$, with T_A including phase A (e.g., empty space) and none of phase B (e.g., solid) and T_B that includes phase B and none of A; those voxels with values between the thresholds are assigned via indicator kriging to phase A or phase B in such a way as to maximize smoothness of the surface between the phases (Oh and Lindquist, 1999; Mendoza et al., 2007).

In structures where the solid phase is particularly thin and difficult to resolve reliably, as it often is for walls between cells, the cell boundaries can be determined by a distance transformation plus watershed algorithm (Gonzalez et al., 2004; see the following subsections), as, for example, in a study of evolution of a liquid foam described in Chapter 8. In specimen systems where partial volumes and low contrast dominate, selective iterative thresholding may be useful: here an initial threshold is guessed and the results of this binarization are numerically compared to those of different iterations (Montanini, 2005). Segmentation by "snake" or active contour models for boundary detection can also be effective (Sheppard et al., 2006): an initial contour is deformed toward the boundary to be detected so as to minimize a functional designed to have its local minimum at the boundary (Caselles et al., 1997). Spowage et al. (2006) described the steps required to identify surface porosity in foams. Lindquist (2001) explicitly compared three quite different thresholding methods, and the interested reader can take this paper as a starting point for learning more about thresholding.

6.2.2 Distance Transform Method

The distance transform method is a powerful, albeit simple, approach for analyzing structures; here it is defined in terms of the most popular application in microCT, determining an accurate mean "trabecular" thickness or distribution of thicknesses for bone or related structures (Hildebrand and Rüegsegger, 1997; Hildebrand et al., 1999). Analysis proceeds by calculating the metric distance of each solid voxel to the nearest solid-(empty) space surface; that is, this distance is the radius of a sphere centered on this voxel and fitting inside the structure. Redundant (smaller) spheres are eliminated, producing a set of centers of maximal spheres filling the structure completely (Figure 6.3). Thicknesses for each portion of the structure are twice the radii, and this allows maps of local thickness value to be produced in 3D renderings as well as the mean thickness or distribution of thicknesses. Mean spacing <Sp> between structural elements is calculated with the same method by simply switching the background and object voxels.

6.2.3 Watershed Segmentation

In addition to its use in determining thicknesses and the like, the distance transform can be used for segmentation employing the watershed transformation (Gonzalez et al., 2004), a methodology based on the idea of catchment basins in geography. A watershed is the ridge dividing areas drained by different rivers or reservoirs, and the approach is to transform the image into a form whose catchment basins are the regions we want to identify. An example of such regions is the interior of cells in foams with the "ridge-lines" being the cell walls. The negative of distance transform yields the ridge lines dividing the regions of the structure, and this can be very effective when a significant fraction of cell walls cannot be detected (Lambert et al., 2005).

6.2.4 Other Methods

One of the shortcomings of voxel-value-based thresholding is that voxels are selected primarily on the basis of their value and not on their

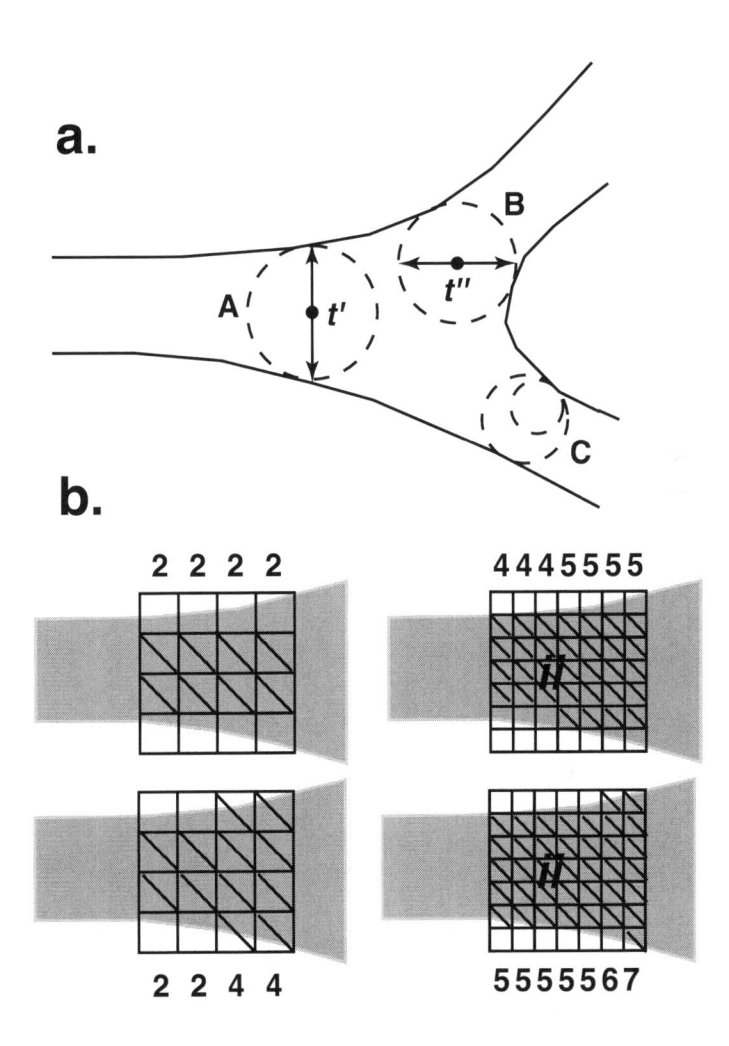

Experimental Design, Data Analysis, Visualization 123

FIGURE 6.3
Two-dimensional illustration of distance transformation determination of plate or strut thickness, a process actually applied in 3D. (a) The dashed circles show that local thicknesses are t' at A and t" at B. At C, the largest sphere defines the local thickness. (b) Effect of threshold level and voxel size on the thickness determined. The gray areas represent the same section of "bone" and the large square represents the region of interest (ROI) for analysis. The left column shows the ROI sampled by 4 × 4 voxels and the right column shows the ROI sampled with (smaller) 7 × 7 voxels. Diagonal slashes show voxels considered to be "bone": in the upper row the voxels must be totally occupied, whereas in the lower row, voxels more than one-half occupied are considered solid. Above and below each column of voxels is the number of "bone" voxels in that column. (Reprinted from Stock et al. (2004a) with permission.)

location. Boundary-following algorithms and related techniques (such as the watershed function) emphasize connected volumes; the boundaries of open channels in SiC/SiC cloth-based composites, for example, were determined and the channel widths between upper and lower boundaries were rapidly computed as a function of position (Lee, 1993; Lee et al., 1998). A related type of analysis is measurement of crack opening by programs that follow the crack from position to position and measure the width as a second step; this works especially well if the crack is first measured under the maximum applied load so that it produces the greatest amount of contrast (Guvenilir et al., 1997; Guvenilir and Stock, 1998; Guvenilir et al., 1999).

The difficulty with this approach is that considerable operator intervention is required to fill local gaps (snakes or membranes may be used) or to identify connections that are not obvious locally, that is, structures that may connect after quite some distance out of plane. Percolation operators may help in this regard, that is, allowing virtual particles to bounce around the open structure (random walk) while recording all positions the particle reached. Cluster labeling within segmented data, for example, in transport studies of porous substances (Nakashima et al., 2004), is related to this last approach.

Rajagopalan et al. (2005) expressed microCT images in terms of their 2D discrete Fourier transform and produced segmentations based on the phase component of the transform. This approach was motivated by the success of image matching methods based on phase images, for example, cross-correlation metrics.

Simultaneous solid-phase and void-phase burn is very useful for analysis of particle characteristics. Starting at an interface, a value is assigned to each voxel that equals the number of voxels it is away from the interface. The local maximum of burn number is a particle center, and all voxels previously identified as solid are assigned to one or another particle, with additional steps required to accurately partition contacting grains. Particle volume, surface area, orientation, aspect ratio, and contact statistics flow directly from the assigned particles (Thompson et al., 2006).

Iterative opening and closing in 3D (n erosions* followed by n dilations with increasing n until no volume remains in the image; note that n erosions remove all structures smaller than $2n$ voxels) is the basis of another method (granulometry) of rapidly computing the distribution of wall

---

* Erosion refers to the operation on a segmented image where the outermost voxel is removed from all around the boundary of the object in question. Dilation refers to the opposite operation, addition of a voxel to all boundary positions. In a single (one-voxel) erosion of a segmented image, all isolated single voxels are removed (in fact, all structures with largest dimension three voxels or less are eliminated). Dilation of the eroded structure restores the structure except where it was completely removed. A moment's reflection reveals that dilation followed by erosion serves to fill in small voids in the segmented objects; see Figure 6.4.

thicknesses in a cellular material (Elmoutaouakkil et al., 2003); the derivative of the remaining volume with respect to structuring element size n gives the (size) distribution of thicknesses (Brunke et al., 2005). Figure 6.4 illustrates the effect of two erosions followed by two dilations and the effect of two dilations followed by two erosions using the slice shown in Figure 6.2.

Skeletonization refers to a representation of an object by its centerline, that is, by reducing the object to a one-voxel-wide string or branched string extending to the extreme ends of the original object. This representation of data simplifies analysis of complex arrays of objects, particularly those that are channel- or fiberlike. Skeletonization analysis was used, for example, to

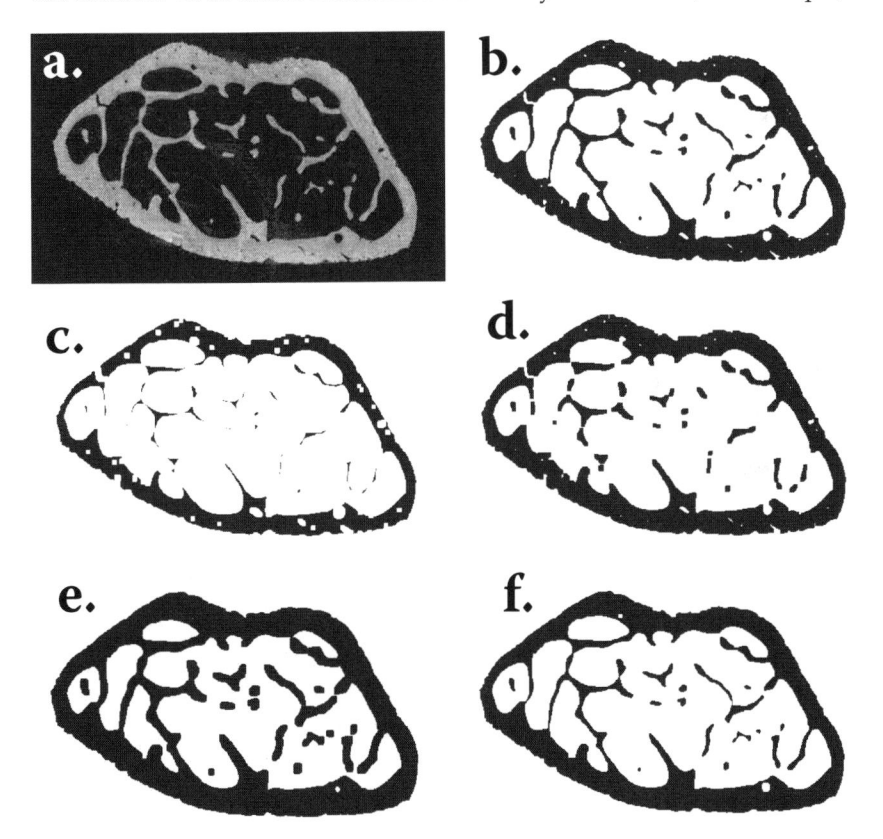

FIGURE 6.4
Illustration of erosion + dilation and dilation + erosion for the slice shown in Figure 6.2. (a) Grayscale image in 8-bit format (0 to 255 contrast levels). (b) Image thresholded at 128. (c) Result of two erosions applied to (b). (d) Result of two dilations following the two erosions. Some small thin structures were removed (lower right). (e) Result of two dilations applied to (b). (f) Result of two erosions following two dilations. Comparison of (d) and (f) shows significant differences: for example, the fine porosity in the bone cortex persists in (d) but is eliminated in (f).

study two bonded stainless steel fiber assemblies and to show that the distribution of fiber segment lengths between the two specimens differed, as did the distribution of fiber orientations (Tan et al., 2006). Such orientation data are often shown on stereographic projections; see Cullity and Stock (2001) or other materials texts for an introduction to this type of plot.

In biology there are many vessel systems (arteries and veins; kidney microvasculature and glomeruli; lung bronchi, bronchioles, and alveoli; cortical bone Haversian and Volkmann canals) with branching that can be likened to trees. Hundreds of branches are present in a typical organ, and automated analysis routines are essential, if an adequate number of replicates are to be analyzed. This last requirement is particularly important in biological studies where interindividual variability is very large. It is important to note the treelike nature of vessel systems and branches because this differs from the situation pertaining to cellular materials and condition analysis algorithms. After extraction of the vessel tree, quantitative data can be computed, and numerical analysis of the tree characteristics can be performed. One approach, partition into mother/child/sibling relationships, also known as generational analysis, has been shown to be effective for six or more hierarchical levels (Wan et al., 2002); this focuses analysis of functionality and dimensional changes onto equivalent portions of the network. Wan et al. (2002) studied the coronary arterial tree of a rat and focused quantification on arterial lumen cross-sectional area, interbranch segment length, branch surface area at equivalent generation, and interbranch and intrabranch levels. Use of the self-similarity of the arterial trees can improve analysis efficiency (Johnson et al., 1999).

Structural connectivity is important for the mechanical integrity of the array of struts and plates (trabeculae) in cancellous bone and also for pores' function in gas and fluid transport in plants. Significant errors can result if a volume is divided for computational purposes, and this is a particularly important consideration for high-definition synchrotron microCT images of trabecular bones (datasets currently comprising 8 GB or more voxels). Labeling of connected objects on both sides of a common face of two subvolumes and comparing the common border is one promising approach (Apostol and Peyrin, 2007). Porosity connectivity in apples was also determined to depend highly on the voxel size of the reconstructed images, and the representative elemental volumes (REV, range of volumes over which a valid statistical average can be computed) were determined for this type of fruit (Mendoza et al., 2007). The REV approach was also applied to paper materials studied with synchrotron microCT (Rolland Du Roscoat et al., 2007). A related method that the authors termed the "mean window technique" was used to determine the properties of a particulate-reinforced metal composite (Borbély et al., 2006). Levitz (2007) suggests determining chord distribution functions, linear graphs of retraction (related to skeletonization), and correlated Gaussian fields as useful techniques for analyzing the properties of different porous materials.

Experimental Design, Data Analysis, Visualization 127

MicroCT datasets can be used as the basis for incorporating specimen-specific microarchitecture or other microstructure into finite element models. This has been an active area in bone research (discussed in Chapter 8), and research employing this methodology continues (Fu et al., 2006; Gong et al., 2007; Su et al., 2007), including approaches where local mineral levels are used to define local elastic constants to the individual matrix elements of the modeled solid (Mulder et al., 2007). Kim and coworkers found that prediction of apparent mechanical properties and structural properties agreed well with experiment, regardless of which of the three thresholding methods they used (Kim et al., 2007).

In certain circumstances, interpretation of the structures within the volume as an array of idealized solids (spheres, plates, ellipsoids, etc.) can be helpful. It is important, however, to match the idealized form as closely as possible with the actual object, say, by keeping the volume identical and by minimizing the number of outlier voxels. For example, study of different structural metal foam types revealed differences in cell anisotropy (distribution of cell volumes and of aspect ratios quantified as equivalent ellipsoids; stereographic projections showing distribution of cell axes vs. orientation) that correlated with altered mechanical properties (Benaouli et al., 2000, 2005).

6.2.5 Image Texture

Images of any type often contain regions with different textures, and, without going into detail, analogy provides a simple illustration of what is meant by texture. Texture refers to the structure contained within a region, and most readers would have little difficulty visualizing smooth, rough, or periodic textures of surfaces (see Figure 11.19 of Gonzalez et al. (2004)). MicroCT datasets can be analyzed by using the texture of different regions, but this approach is largely unexplored.

One recent study did, however, apply texture analysis to 3D microCT datasets of five porous specimens (Jones et al., 2007). The specimens were mineral carbon forms from different geographical locations with similar topological structure that differed mainly in textural quality. Robust measures of structural texture were extracted from the grayscale images in the form of a set of 96 texture features that constituted the texture vector for a particular sample. The texture vector was then related to the texture space defined by prior measurements on "known" training specimens; that is, probabilities could be computed for the likelihood that a given specimen belonged to each population. One expects that, with further development, this approach will be valuable not only for classification of complex specimens but also for rapid (automated?) identification of key structural differences. Texture features at CT-level resolutions were correlated (by binning microCT data on many cancellous bone specimens) with the underlying microarchitecture in another study (Showalter et al., 2006).

6.2.6 Interpretation of Voxel Values

The values of the linear attenuation coefficient can be used to analyze changes in microstructure or to quantify volume fraction of unresolved solid in two-phase materials. Stock and coworkers used lab microCT to image a demipyramid of the sea urchin *Lytechinus variegatus* at a resolution too coarse for the trabecular structure of stereom to be resolved (Stock et al., 2003). Because this ossicle consisted of only two phases (high Mg calcite and air), the 3D distribution of linear attenuation coefficient values was interpreted in terms of partial volumes of calcite. A similar approach (analysis of volume fraction in partial voxels) for quantifying crack opening as a function of applied stress and 3D position is described in Chapter 11. Interpretation of voxel values as levels of mineral in bone and importation of these 3D maps as variable Young's moduli in finite element models of trabecular bone is another interesting use of this type of data (Mulder et al., 2007).

6.2.7 Tracking Evolving Structures

There are quite a number of circumstances when prior and altered microstructures must be compared, directly or indirectly, in the same specimen. Using the specimen as its own control is an incredibly powerful advantage, given the considerable a priori knowledge available. On the other hand, it can be incredibly frustrating to accomplish in practice because of difficulties aligning structures and sorting out actual changes from misor reorientations. Indirect comparisons are much simpler and include measurements of volume fraction of a phase (bone volume fraction), mean particle sizes (mean trabecular thickness), distributions of particle sizes, and so on. There are other analysis options avoiding point-by-point registration, an example of which is covered after registration is considered.

Automatic tracking of the displacement of small, readily identifiable features is one method of quantifying deformation fields. In one study, thousands of micropores within the solid were tracked automatically for each increment of deformation, and a microstructure gauge formalism was used to calculate maps of the 3D strain tensor components (Toda et al., 2006a,b). A particle tracking approach has also been used to map displacement fields during mechanical processing (Nielsen et al., 2003, 2004; McDonald et al., 2006; Zettler et al., 2006). Particles or micropores can also be used as fiducials to measure changes in crack opening as a function of applied stress (Breunig et al., 1992, 1993; Toda et al., 2003, 2004). Tracking the motion of specific features also seems to be a viable technique for strain quantification in trabecular bone (Verhulp et al., 2004; Liu and Morgan, 2007).

Alignment of altered structures is an active area in biomedical research. Displacements in CT images of lungs can be followed through the formalism of image warping (Fan and Chen, 1999). Strains in trabeculae from their displacement during loading are a second area of research covered in more detail in Chapter 8. Determination of longitudinal changes in trabeculae by direct comparison of the reconstructed volumes was an ambitious approach employed in an in vivo lab microCT comparison of OVX and control rats (Waarsing et al., 2004). In this study, image registration (translation and rotation between volumetric datasets) was performed to maximize mutual information. The registration algorithm worked well for the controls, for which changes were very small, but the large changes in the OVX animals required a modified approach based on registration of a relatively small number of large invariant structures. For in vivo studies, a special imaging chamber was devised to allow microCT and microPET intermodal registration (Chow et al., 2006). Intramodal methodologies for registering longitudinal microCT datasets have also received attention (Boyd et al., 2006). Also of interest, although it does not concern biomedical subjects, is the volumetric digital image correlation method applied to triaxial compression of rock in order to calculate the 3D strain fields (Lenoir et al., 2007).

Registration issues can be avoided to some extent by extraction of the quantity of interest and mapping this quantity in the 3D or 4D (three spatial dimensions plus time) space occupied by the specimen. An example of this is measurement of the crack opening as a function of position and its mapping on the 3D crack surface (i.e., the crack's center position) as a fourth dimension, say, as color. Although this sounds quite mysterious at this juncture, actually going through the steps in the analysis in the context of data representations in the following section, one hopes, illustrates what is meant.

6.3 Data Representation

MicroCT datasets can be used in many ways to explain changes in specimens and to reveal differences between groups of similarly treated specimens. Such illustrations can be as simple as qualitative comparisons of comparable slices or as complex as three-, four- or five-dimensional representations of volumetric data of evolving specimens.

The simplest representation of microCT data is simple presentation of 2D sections through volumetric data. A number of slice images have been presented above. Often, movies paging through the stack of slices can be very effective in establishing an overall impression of the structure or in identifying a particular subset of the volume from detailed interrogation. In some cases, of course, only isolated slices or sets of widely spaced slices are available. In this case, analysis options are limited. Provided the sampling adequately represents the structure, then V_V and the variance of V_V across the specimen can be determined with some confidence. Unless there is a priori knowledge about the structure and its orientation with respect to the tomographic rotation axis, apparent thicknesses and other measurements are liable to differ substantially and in an unpredictable way from the actual values.

In volumetric microCT, one can section the volume numerically along more than one plane. Representation of anisotropy of structure by combining micrographs from three orthogonal planes on the visible faces of a cube, for example, has been common for many years in metallurgy. Sectioning through three orthogonal planes and showing all three planes from an isometric viewpoint is also very effective for microCT data. Figure 6.5 shows three orthogonal planes through the reconstructed volume of several different types of sea urchin spines (Stock et al., 2006). As described in the figure caption, some of the structure is difficult to appreciate in single sections. Sectioning through stacks of slices along cylindrical surfaces and unwrapping the surface into a flat 2D map (circle cuts) is another way of presenting data simply (Lee, 1993).

Another simple method of displaying structure is to apply a threshold to a dataset and to view the resulting 3D rendering of the voxels more absorbing than the threshold. An example of this sort is used in Figure 6.6 to show the trabecular structure in a mouse femur. Note that part of the cortex and trabecular bone has been numerically removed to reveal the center. It is always useful to remember that renderings showing the low absorption voxels of an object can be as effective as those showing the high absorption voxels. Figure 6.7 shows the low absorption voxels within a small volume of a rabbit femur; this dataset was noisy and the threshold used truncated some of the channels, making them appear discontinuous in places, but paging through the slices reveals these breaks are not, in fact, present.

Thresholded 3D renderings, however, are not always superior to sets of three orthogonal grayscale sections for visualizing structures. Figure 6.8 compares the two representations of the stereom of a sea urchin demipyramid, a structure with nearly 50 vol.% empty space. In the author's opinion, Figure 6.8d, the 3D rendering with a threshold yielding about 50 vol.% solid, is a much poorer visualization of the structure than Figure 6.8c, the three orthogonal grayscale section representation. It is very difficult to make sense out of very crowded renderings unless one does significant editing or color coding of specific structures such as used in some recent studies (Donoghue et al., 2006; Parkinson et al., 2008).

Representation of microstructural variation using quantities extracted from volumetric microCT datasets requires more work than the approaches described above. It can be as simple as plotting the mean linear attenuation coefficient per slice as a function of position along an important direction (assumed parallel to the tomography rotation axis). Figure 6.9 shows such a plot for a stack of SiC cloths, and the peaks and valleys clearly show the average centers of the cloths (Lee, 1993; Lee et al., 1998). Deformation of two fairly dense Al "foams" produced by powder metallurgy (46 percent and 37 percent

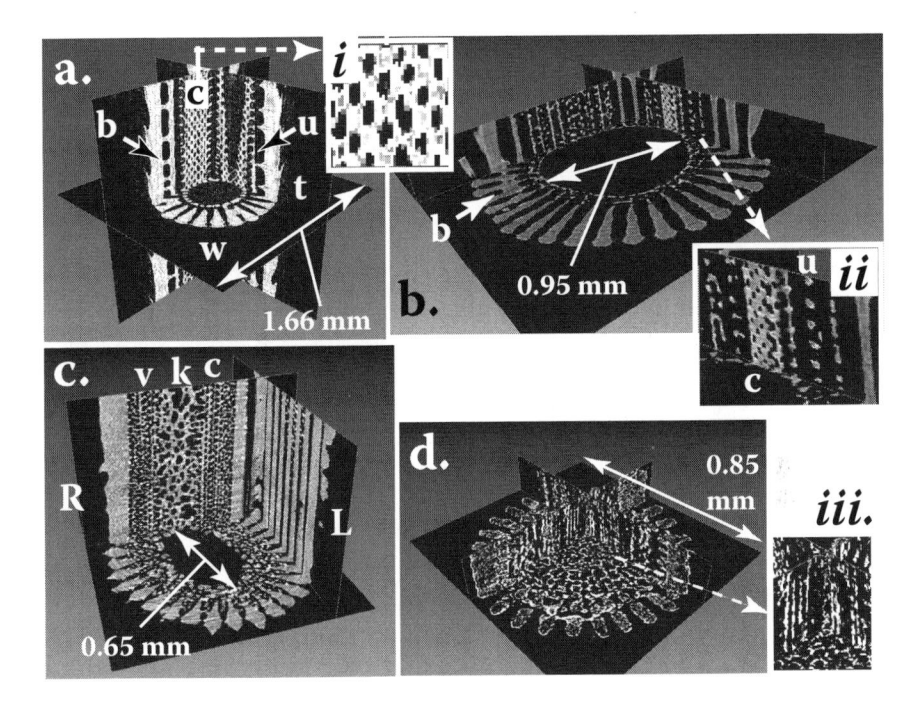

Experimental Design, Data Analysis, Visualization 131

FIGURE 6.5
Three orthogonal sections through reconstructed spines of (a) *Diadema setosum*, (b) *Centrostephanus rodgersii*, (c) *Echinothrix diadema*, and (d) *Lytechninus variegatus* showing three orthogonal slices through the spines. Insets *i*, *ii*, and *iii* are enlargements of the fine stereom indicated by the dashed lines. The wedges "w" of both *C. rodgersii* and *D. setosum* are linked by well-defined bridges "b" and capped by thorns "t" (upward oriented in (a), not visible in (b)). The axial distribution of bridges tends toward regular spacing in (a). In (b), *C. rodgersii* has at least one bridge every 250–300 μm, although the spacing is sometimes much less; bridges link five or more adjacent wedges in the same transverse plane and spiral around the spine axis. In *E. diadema*, however, the wedges are linked by many fine trabeculae "v" (c). In *C. rodgersii* and *D. setosum* the central cavities are bordered by a thin, nearly circular calcite cylinder that is pierced by a regular array of holes (insets *i* and *ii*). These perforated cylinders connect to the wedges through thin, regularly spaced radial bars "u." In contrast, *E. diadema*'s oval central cavity is bordered by irregular, coarse stereom "k" in which the perforated cylinder is buried (below "c" in (c)). In (d), the regular axial structure of the inner fine stereom in the *L. variegatus* spine can be appreciated only in 3D (inset *iii*). The data were collected at 2-BM, APS, with a 1K × 1K detector, 0.25° rotation steps, and: (a) 18 keV x-rays and reconstructed with 1.66-μm voxels, (b, c) 21 keV x-rays and 5-μm voxels, and (d) 15 keV x-rays and 1.3-μm voxels. (Reproduced from Stock et al. (2006).)

of full density) was quantified by measuring mean porosity as a function of slice number for compression along the tomographic rotation axis (Wang et al., 2006), much as what was done for the SiC cloths or for liquid foam (Lambert et al., 2005). Displacement of local maxima/minima in mean slice porosity could be followed through the different increments of deformation as the porosity diminished by up to a factor of two.

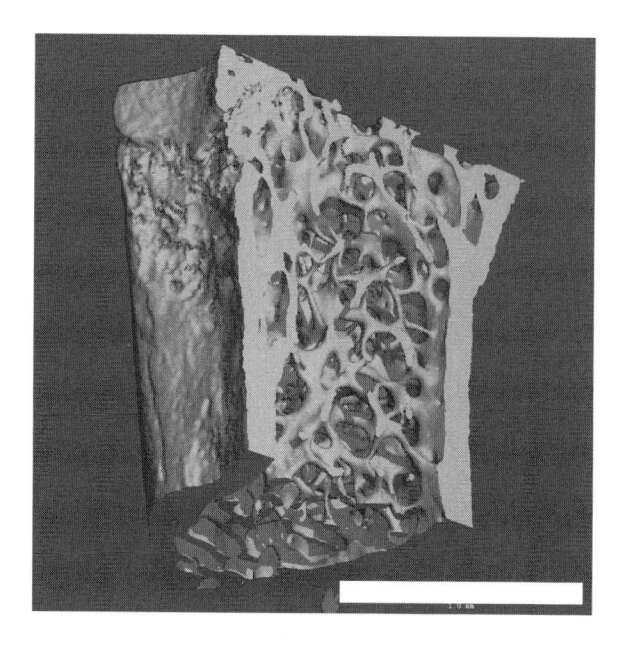

132   *MicroComputed Tomography: Methodology and Applications*

FIGURE 6.6
3D rendering of the trabecular bone in the mouse femur shown in Figures 6.2 and 6.4. The scale bar at the bottom right is 1 mm long.

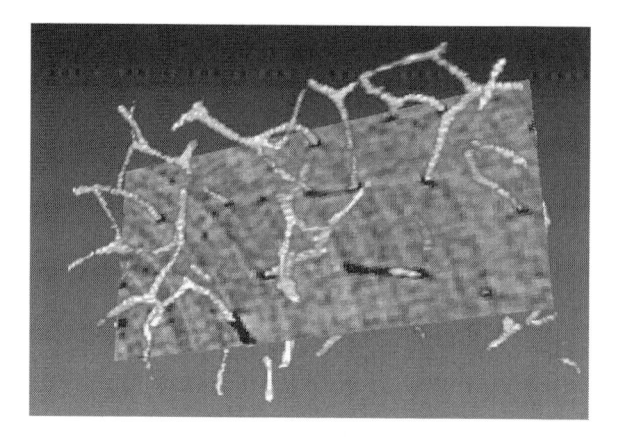

FIGURE 6.7 (SEE COLOR INSERT FOLLOWING PAGE 144.)
Small canals in rabbit femoral bone. Low-absorption voxels rendered as white, and higher-absorption voxels rendered transparent. A small section of one of the reconstructed slices within the volume is shown for background as a grayscale image. 21 keV, 1K × 1K reconstruction, 0.25° rotation increment, ~5-µm isotropic voxels. (S. R. Stock, K. Ignatiev, N. M. Rajamannan, F. De Carlo, 2-BM, APS, March 2003.)

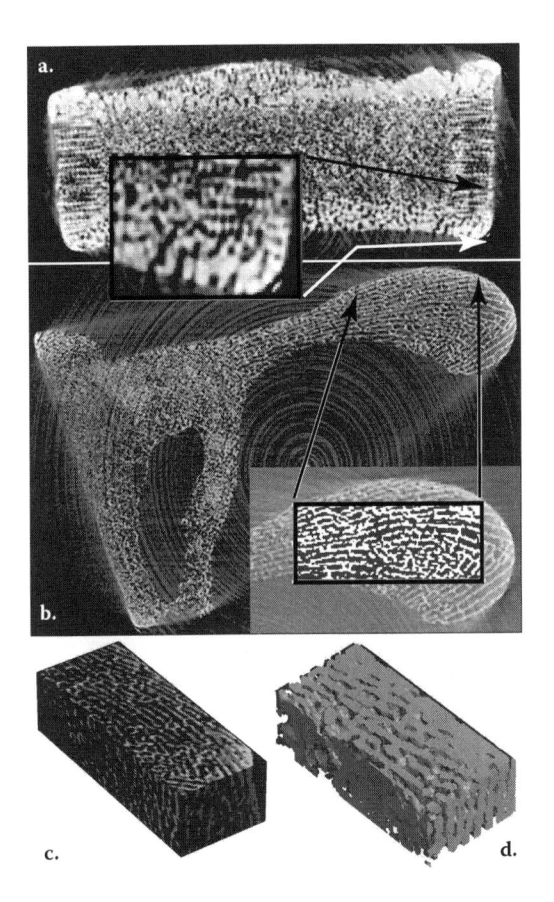

Experimental Design, Data Analysis, Visualization 133

FIGURE 6.8
Synchrotron microCT data for an interambulacral plate of *Lytechinus variegatus* (a) and a demipyramid of *Asthenosoma varium* (b–d). (a) Slice with the plate exterior at the top; the inset box shows the magnified section of the ossicle indicated by the arrows and the horizontal field of view is 550 voxels (2.75 mm). (b) Slice with an enlarged area inset. The black box defines the ROI for numerical microstructure evaluation; the pixels are binarized to either calcite (white) or void (black). The horizontal field of view is 687 voxels (3.44 mm). (c) Grayscale isometric views and (d) thresholded 3D rendering from within the ROI defined in (b); the volumes are both 37 voxels (0.185 mm) high. In (a–c), the lighter the pixel is, the higher the linear attenuation coefficient. In (d), the higher absorption voxels are shown solid and lower values are rendered transparent. 21 keV, 1K × 1K reconstruction, 0.25° rotation increment (Stock et al., 2004a).

Often, 3D surfaces are represented by the triangular elements of a mesh fitted to the surface. These surface elements can be used numerically for analyses other than visual representations: for each element, its vertices define its 3D position and the normal to the element. Silva and coworkers used this information and the normal directions of neighboring surface elements to quantify bone surface erosion in a rat model of rheumatoid arthritis (Silva et al., 2006). This approach should be very valuable in microCT studies in many disciplines.

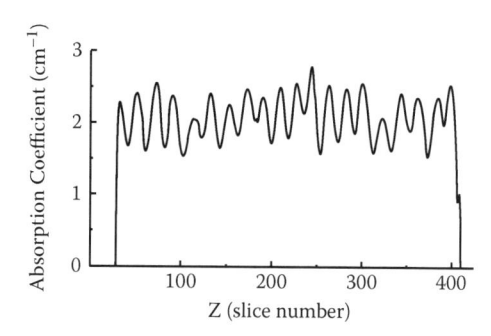

FIGURE 6.9
Mean linear attenuation coefficient per slice for a composite preform of SiC cloths. The 20 peaks correspond to the centers of the 20 cloths stacked perpendicular to the tomography rotation axis (Lee, 1993; Lee et al., 1998).

More complicated analysis is required to extract 3D variations of quantities such as porosity or crack opening, particularly when the quantity needs to be followed as a function of time (for crack opening, as a function of applied stress). It is best to illustrate this approach concretely, and the example used, crack opening as a function of position for different in situ loads (Guvenilir, 1995; Guvenilir et al., 1997), anticipates more detailed discussion in Chapter 11.

For rough cracks without significant tortuosity (i.e., the cracks do not have many parallel branches but do deviate considerably from a simple surface), crack opening versus position can be determined relatively simply as follows. Assume that the loading direction (z) is parallel to the tomography rotation axis and hence perpendicular to the reconstruction plane (containing coordinate axes **x** and **y**; Figure 6.10). The voxels along each column (x,y) have a range of values, and generally the crack will be at the position with the minimum voxel. In practice one also requires that the identified crack voxel be in the vicinity of the other crack voxels. The column consists of a variable number of partial voxels (Figure 6.10a), and the difference of the voxels' values from the mean value of the uncracked material represents partial voxels of opening (Figure 6.10b). These fractions can be added to give the total opening at that position (x,y) and can be projected along **z** onto the *x-y* plane to give a simple 2D representation of the variation of crack opening.

Figure. 6.11 shows the fatigue crack opening in a cylindrical Al specimen projected onto the *x-y* plane for four applied loads; the different colors show the local experimental opening (Guvenilir et al., 1997). This 2D simplification focuses attention on patterns of opening without the complication of the third spatial dimension. Similar 2D projected views were developed for intercloth channel widths within SiC/SiC cloth-layup composites and revealed patterns of opening that depended on the relative

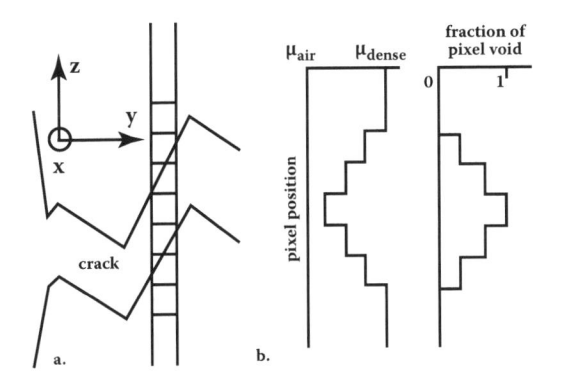

Experimental Design, Data Analysis, Visualization 135

FIGURE 6.10
Illustration of the analysis used to generate the distribution of crack openings shown in Figures 6.11 and 6.12. (a) A column of voxels (represented by squares) along the loading direction z (parallel to the tomography reconstruction axis). Crack openings occupying whole and partial voxels are shown for one (x,y) position. (b) Schematic of values of linear attenuation coefficient along a particular column (left) and corresponding partial voxels of open crack (right). The partial voxels are summed for each column and used in the 2D and 3D representations of the distribution of crack opening.

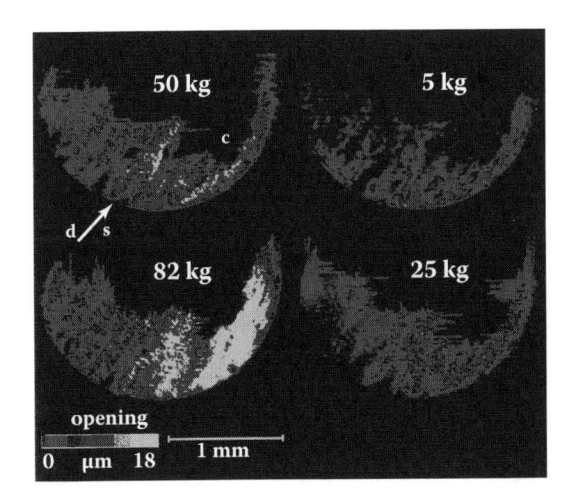

FIGURE 6.11 (SEE COLOR INSERT FOLLOWING PAGE 144.)
Crack opening in a sample of an Al alloy projected on the plane normal to the load axis. The color bar (lower left) indicates the total opening at each position, and the symbols are explained in the text in Chapter 11. Maps for four different applied forces (kgf) are shown. The corresponding stress intensities are given in parentheses. (Bottom left) 82 kgf (7.1 MPa√m). (Top left) 50 kgf (4.3 MPa√m). (Bottom right) 25 kgf (2.2 MPa√m). (Top right) 5 kgf (0.4 MPa√m) (Guvenilir et al., 1997).

displacements of the holes in the woven cloths on either side of the channel (Lee, 1993; Lee et al., 1998). These well-defined patterns (see Chapter 10) and their relationship to hole position were difficult to appreciate in numerical sections or in inverse 3D renderings of the specimen porosity.

The 2D projection of crack openings onto a plane (Figure 6.11) ignores the important third dimension characteristic of the very rough crack of this Al alloy. Figure 6.12, top left, shows a mesh map of one of the crack faces, and Figure 6.12, bottom left, top and bottom right, show fatigue crack opening (in color) at the correct positions on the 3D surface (Guvenilir et al., 1997). The contact or lack of contact between peaks and valleys of opposing crack faces is readily apparent and, as is discussed in Chapter 11, changes with applied stress.

Distance transform data of local structural thicknesses can be readily combined with 3D renderings using local coloring of the solid to emphasize local thickness. This sort of representation is used frequently in biomedical studies (e.g., trabecular bone or blood vessel networks). Figure 6.13 shows the voxels of cracks and voids rendered solid in a cement specimen that had been attacked by sulfate ions. The colors are the local thicknesses as determined by the distance transform method, and the results of the study of sulfate attack of cement, mentioned in the experiment design section above, are reviewed in the chapter on environmental interactions.

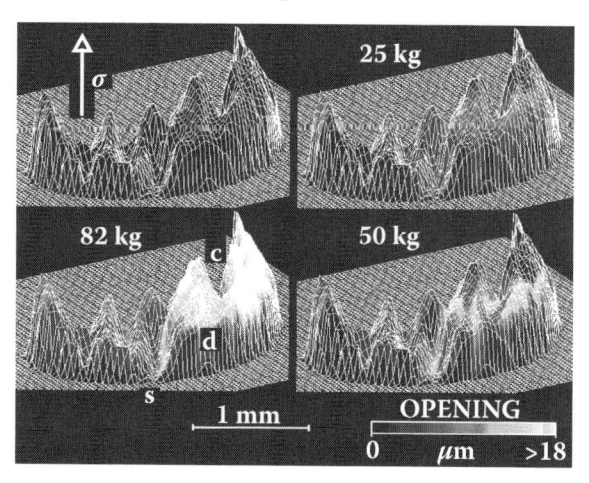

FIGURE 6.12 (SEE COLOR INSERT FOLLOWING PAGE 144.)
Crack opening in a sample of an Al alloy in 3D (see Figure 6.11 for the openings projected onto the plane normal to the load axis). The loading direction is indicated by "σ", the color bar (lower right) indicates the total opening at each position on the 3D mesh, and the other symbols are explained in the text in Chapter 11. Maps for three different applied forces (kgf) are shown. The corresponding stress intensities are given in parentheses. (Upper right) Mesh map showing the 3D crack position. (Lower left) Crack opening at 82 kgf (7.1 MPa√m). (Lower right) Opening at 50 kgf (4.3 MPa√m). (Upper right) Opening at 25 kgf (2.2 MPa√m) (Guvenilir et al., 1997).

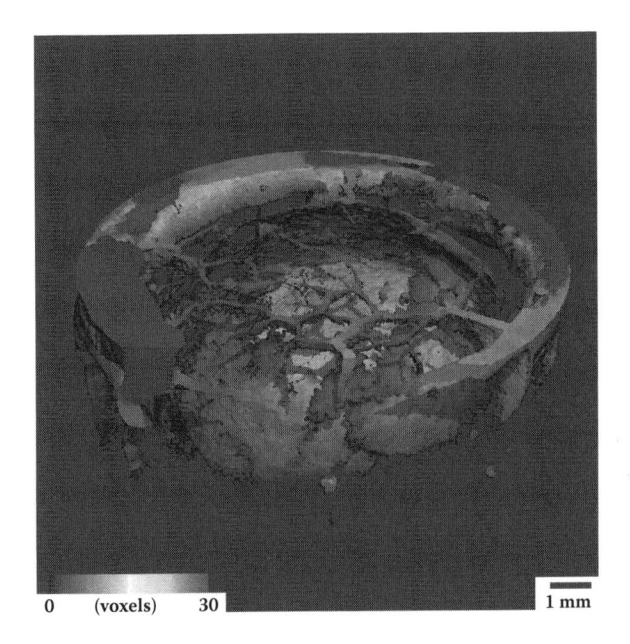

Experimental Design, Data Analysis, Visualization 137

FIGURE 6.13 (SEE COLOR INSERT FOLLOWING PAGE 144.)
Three-dimensional rendering of the open voxels within the interior of a Portland cement specimen that had suffered significant damage from sulfate ion attack. The different gray levels represent local thicknesses (in units of voxels, each of which was 37 μm on an edge) determined using the distance transform. (Unpublished image from the study described in Stock et al. (2002), Naik (2003), and Naik et al. (2006).)

A very effective use of 3D renderings is to use two different thresholds to represent different volumes of the specimen and to render the more extensive volume semi-transparent. Figure 6.14 shows a human heart valve in which significant calcification developed (Rajamannan et al., 2005). The 3D rendering shows the soft tissue as semi-transparent light blue and the more highly absorbing calcification as white.

Many recent microCT papers incorporate color images (if not in the hard copy, then in the online version of the journal). As the human eye distinguishes more levels of contrast in color images than in grayscale images, color is sometimes used to increase the dynamic range visible in slices, but this is used less frequently than one would expect. Color has been used primarily to present four- or more dimensional data or to label different discrete subvolumes within 3D renderings, and these types of color images will be increasingly important in descriptions of scientific and engineering studies. More than a few color figures have appeared, however, that could have been equally effective as grayscale images, but this, perhaps, is an overly pedantic observation.

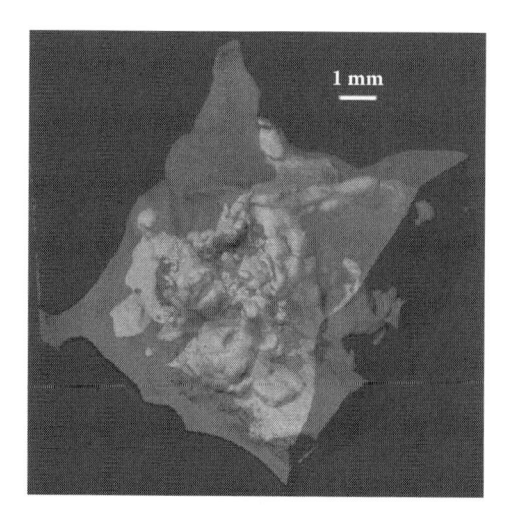

FIGURE 6.14 (SEE COLOR INSERT FOLLOWING PAGE 144.)
Rendering of calcified human heart valve employing two thresholds, a lower value for the soft tissue (semi-transparent gray) and a higher level for the calcification (white). The flat surface represents a numerical section through the volume. (Rajamannan, et al. (2005); © Lippincott, Williams & Wilkins, 2005, used with permission.)

Supplemental data — in particular, movies — posted on journal websites have become an increasingly important component of publications. Movies paging through a stack of slices, showing different perspectives on a rendered volume or removing outer layers of an object and exposing interior structure, are popular supplements. Earlier, microCT was used to create physical models with rapid prototyping manufacture (e.g., osteoporotic and normal trabecular bone (Borah et al., 2001)), but this is too expensive to do on a routine basis, and 3D renderings (spinning or viewing perspective under user control) can be produced without particular difficulty or expense.

Some groups maintain websites where extensive microCT or CT datasets are posted. For example, the digital morphology website (www.digimorph.org) includes anatomical data on a wide variety of animals, renderings and slices of which can be viewed in a variety of ways. The visible cement website maintained by NIST contains synchrotron microCT datasets and serves as a standard for benchmarking new analysis programs for this class of materials (Bentz et al., 2002).

References

Apostol, L. and F. Peyrin (2007). Connectivity analysis in very large 3D microtomographic images. *IEEE Trans Nucl Sci* **54**: 167–172.

Benaouli, A.H., L. Froyen, and M. Wevers (2000). Micro focus computed tomography of aluminium foams. *X-Ray Tomography in Materials Science*. J. Baruchel, J.Y. Buffière, E. Maire, P. Merle, and G. Peix (Eds.). Paris, Hermes Science: 139–154.

Benouali, A.H., L. Froyen, T. Dillard, S. Forest, and F. N'Guyen (2005). Investigation on the influence of cell shape anisotropy on the mechanical performance of closed cell aluminum foams using micro-computed tomography. *J Mater Sci* **40**: 5801–5811.

Bentz, D.P., S. Mizell, S. Satterfield, J. Devaney, W. George, P. Ketcham, J. Graham, J. Porterfield, D. Quenard, F. Vallee, H. Sallee, E. Boller, and J. Baruchel (2002). The visible cement data set. *J Res NIST* **107**: 137–148. See also visiblecement.nist.gov.

Borah, B., G.J. Gross, T.E. Dufresne, T.S. Smith, M.D. Cockman, P.A. Chmielewski, M.W. Lundy, J.R. Hartke, and E.W. Sod (2001). Three-dimensional micro-imaging (MRµI and µCT), finite element modeling, and rapid prototyping provide unique insights into bone architecture in osteoporosis. *Anat Rec* **265**: 101–110.

Borbély, A., P. Kenesei, and H. Biermann (2006). Estimation of the effective properties of particle-reinforced metal–matrix composites from microtomographic reconstructions. *Acta Mater* **54**: 2735–2744.

Boyd, S.K., S. Moser, M. Kuhn, R.J. Klinck, P.L. Krauze, R. Muller. and J.A. Gasser (2006). Evaluation of three-dimensional image registration methodologies for in vivo micro-computed tomography. *J Biomed Eng* **34**: 1587–1599.

Breunig, T.M., S.R. Stock, S.D. Antolovich, J.H. Kinney, W.N. Massey, and M.C. Nichols (1992). A framework relating macroscopic measures and physical processes of crack closure of Al–Li Alloy 2090. *Fracture Mechanics: Twenty-Second Symposium (Vol. 1)*. H.A. Ernst, A. Saxena, and D.L. McDowell (Eds.). Philadelphia, ASTM. *ASTM STP* **1131**: 749–761.

Breunig, T.M., S.R. Stock, A. Guvenilir, J.C. Elliott, P. Anderson, and G.R. Davis (1993). Damage in aligned fibre SiC/Al quantified using a laboratory x-ray tomographic microscope. *Composites* **24**: 209–213.

Brunke, O., S. Oldenbach, and F. Beckmann (2005). Quantitative methods for the analysis of synchrotron-µCT datasets of metallic foams, Eur. *Eur J Appl Phys* **29**: 73–81.

Burghardt, A.J., G.J. Kazakia, and S. Majumdar (2007). A local adaptive threshold strategy for high resolution peripheral quantitative computed tomography of trabecular bone. *J Biomed Eng* **35**: 1678–1686.

Caselles, V., R. Kimmel, and G. Sapiro (1997). Geodesic active contours. *Int J Comput Vision* **22**: 61–79.

Chow, P.L., D.B. Stout, E. Komisopoulou, and A.R. Chatziioannou (2006). A method of image registration for small animal, multi-modality imaging. *Phys Med Biol* **51**: 379–390.

Cullity, B.D. and S.R. Stock (2001). *Elements of X-ray Diffraction*. Upper Saddle River, NJ: Prentice-Hall.

Donoghue, P.C.J., S. Bengtson, X. Dong, N.J. Gostling, T. Huldtgren, J.A. Cunningham, C. Yin, F.P.Z. Yue, and M. Stampanoni (2006). Synchrotron X-ray tomographic microscopy of fossil embryos. *Nature* **442**: 680–683.

Elmoutaouakkil, A., G. Fuchs, P. Bergounhon, R. Péres, and F. Peyrin (2003). Three-dimensional quantitative analysis of polymer foams from synchrotron radiation x-ray microtomography. *J Phys D* **36**: A37–A43.

Fan, L. and C.W. Chen (1999). Integrated approach to 3D warping and registration from lung images. *Developments in X-ray Tomography II*. U. Bonse (Ed.). Bellingham, WA, SPIE. *SPIE Proc Vol* **3772**: 24–35.

Feeney, D.S., J.W. Crawford, T. Daniell, P.D. Hallett, N. Nunan, K. Ritz, M. Rivers, and I.M. Young (2006). Three-dimensional microorganization of the soil–root–microbe system. *Microb Ecol* **52**: 151–158.

Fu, X., M. Dutt, A.C. Bentham, B.C. Hancock, R.E. Cameron, and J.A. Elliott (2006). Investigation of particle packing in model pharmaceutical powders using x-ray microtomography and discrete element method. *Powder Technol* **167**: 134–140.

Gong, H., M. Zhang, L. Qin, and Y. Hou (2007). Regional variations in the apparent and tissue-level mechanical parameters of vertebral trabecular bone with aging using micro-finite element analysis. *Annal Biomed Eng* **35**: 1622–1631.

Gonzalez, R.C., R.E. Woods, and S.L. Eddins (2004). *Digital Image Processing Using MATLAB®*. Upper Saddle River, NJ: Pearson.

Guvenilir, A. (1995). Investigation into Asperity Induced Closure in an Al–Li Alloy Using X-ray Tomography, Ph.D. Thesis. Atlanta, Georgia Institute of Technology.

Guvenilir, A. and S.R. Stock (1998). High resolution computed tomography and implications for fatigue crack closure modeling. *Fatigue Fract Eng Mater Struct* **21**: 439–450.

Guvenilir, A., T.M. Breunig, J.H. Kinney, and S.R. Stock (1997). Direct observation of crack opening as a function of applied load in the interior of a notched tensile sample of Al–Li 2090. *Acta Mater* **45**: 1977–1987.

Guvenilir, A., T.M. Breunig, J.H. Kinney, and S.R. Stock (1999). New direct observations of crack closure processes in Al–Li 2090 T8E41. *Phil Trans Roy Soc (Lond)* **357**: 2755–2775.

Hangartner, T.N. (2007). Thresholding technique for accurate analysis of density and geometry in QCT, pQCT and microCT images. *J Musculoskelet Neuronal Interact* **7**: 9–16.

Hildebrand, T. and P. Rüegsegger (1997). A new method for the model independent assessment of thickness in three-dimensional images. *J Microsc* **185**: 67–75.

Hildebrand, T., A. Laib, R. Müller, J. Dequecker, and P. Rüegsegger (1999). Direct 3D morphometric analysis of human cancellous bone: Microstructural data from spine, femur, iliac crest can calcaneus. *J Bone Miner Res* **14**: 1167–1174.

Johnson, R.H., K.L. Karau, R.C. Molthen, and C.A. Dawson (1999). Quantification of pulmonary arterial wall distensibility using parameters extracted from volumetric microCT images. *Developments in X-ray Tomography II*. U. Bonse (Ed.). Bellingham, WA, SPIE. *SPIE Proc Vol* **3772**: 15–23.

Jones, A.S., A. Reztsov, and C.E. Loo (2007). Application of invariant grey scale features for analysis of porous minerals. *Micron* **38**: 40–48.

Jupe, A.C., S.R. Stock, P.L. Lee, N.N. Naik, K.E. Kurtis, and A.P. Wilkinson (2004). Phase composition depth profiles using spatially resolved energy dispersive x-ray diffraction. *J Appl Cryst* **37**: 967–976.

Kim, C.H., H. Zhang, G. Mikhail, D. von Stechow, R. Müller, H.S. Kim, and X.E. Guo (2007). Effects of thresholding techniques on µCT-based finite element models of trabecular bone. *J Biomed Eng* **129**: 481–486.

Lambert, J., I. Cantat, R. Delannay, A. Renault, F. Graner, J.A. Glazier, I. Veretennikov, and P. Cloetens (2005). Extraction of relevant physical parameters from 3D images of foams obtained by x-ray tomography. *Colloids Surf A* **263**: 295–302.

Lee, S.B. (1993). Nondestructive examination of chemical vapor infiltration of 0°/90° SiC/Nicalon composites, Ph.D. Thesis. Atlanta, Georgia Institute of Technology.

Lee, S.B., S.R. Stock, M.D. Butts, T.L. Starr, T.M. Breunig, and J.H. Kinney (1998). Pore geometry in woven fiber structures: 0°/90° plain-weave cloth lay-up preform. *J Mater Res* **13**: 1209–1217.

Lenoir, N., M. Bornert, J. Desrues, P. Bésuelle, and G. Viggiani (2007). Volumetric digital image correlation applied to X-ray microtomography images from triaxial compression tests on Argillaceous rock. *Strain* **43**: 193–205.

Levitz, P. (2007). Toolbox for 3D imaging and modeling of porous media: Relationship with transport properties. *Cement Concr Res* **37**: 351–359.

Lindquist, W. (2001). Quantitative analysis of three-dimensional x-ray tomographic images. *Developments in X-ray Tomography III*. U. Bonse (Ed.). Bellingham, WA, SPIE. *SPIE Proc Vol* **4503**: 103–115.

Liu, L. and E.F. Morgan (2007). Accuracy and precision of digital volume correlation in quantifying displacements and strains in trabecular bone. *J Biomech* **40**: 3516–3520.

McDonald, S.A., L.C.R. Schneider, A.C.F. Cocks, and P.J. Withers (2006). Particle movement during the deep penetration of a granular material studied by x-ray microtomography. *Scripta Mater* **54**: 191–196.

Mendoza, F., P. Verboven, H.K. Mebatsion, G. Kerckhofs, M. Wevers, and B. Nicolai (2007). Three-dimensional pore space quantification of apple tissue using x-ray computed tomography. *Planta* **226**: 559–570.

Montanini, R. (2005). Measurement of strain rate sensitivity of aluminium foams for energy dissipation. *Int J Mech Sci* **47**: 26–42.

Mulder, L., L.J. van Ruijven, J.H. Koolstra, and T.M.G.J. van Eijden (2007). The influence of mineralization on intrabecular stress and strain distribution in developing trabecular bone. *Annal Biomed Eng* **35**: 1668–1677.

Naik, N. (2003). Sulfate attack on Portland cement-based materials: Mechanisms of damage and long term performance, Ph.D. Thesis. Atlanta, Georgia Institute of Technology.

Naik, N.N., A.C. Jupe, S.R. Stock, A.P. Wilkinson, P.L. Lee, and K.E. Kurtis (2006). Sulfate attack monitored by microCT and EDXRD: Influence of cement type, water-to-cement ratio, and aggregate. *Cement Concr Res* **36**: 144–159.

Naik, N.N., K.E. Kurtis, A.P. Wilkinson, A.C. Jupe, and S.R. Stock (2004). Sulfate deterioration of cement-based materials examined by x-ray microtomography. *Developments in X-ray Tomography IV*. U. Bonse (Ed.). Bellingham, WA, SPIE. *SPIE Proc Vol* **5535**: 442–452.

Nakashima, Y., T. Nakano, K. Nakamura, K. Uesugi, A. Tsuchiyama, and S. Ikeda (2004). Three-dimensional diffusion of non-sorbing species in porous sandstone: Computer simulation based on x-ray microtomography using synchrotron radiation. *J Contam Hydrol* **74**: 253–264.

Nielsen, S.F., F. Beckmann, R.B. Godiksen, K. Haldrup, H.F. Poulsen, and J.A. Wert (2004). Measurement of the components of plastic displacement gradients in three dimensions. *Developments in X-Ray Tomography IV*. U. Bonse (Ed.). Bellingham, WA, SPIE. *SPIE Proc Vol* **5535**: 485–492.

Nielsen, S.F., H.F. Poulsen, F. Beckmann, C. Thorning, and J. Wert (2003). Measurements of plastic displacement gradient components in three dimensions using marker particles and synchrotron x-ray absorption microtomography. *Acta Mater* **51**: 2407–2415.

Oh, W. and W.B. Lindquist (1999). Image thresholding by indicator kriging. *IEEE Trans Patt Anal Mach Intell* **21**: 590–602.

Parkinson, D.Y., G. McDermott, L.D. Etkin, M.A. Le Gros, and C.A. Larabell (2008). Quantitative 3-D imaging of eukaryotic cells using soft X-ray tomography. *J Struct Biol*, **162**: 380–386.

Perilli, E., F. Baruffaldi, M. Visentin, B. Bordini, F. Traina, A. Cappello, and M. Viceconti (2007). MicroCT examination of human bone specimens: Effects of polymethylmethacrylate embedding on structural parameters. *J Microsc* **225**: 192–200.

Rajagopalan, S., L. Lu, M.J. Yaszemski, and R.A. Robb (2005). Optimal segmentation of microcomputed tomographic images of porous tissue-engineering scaffolds. *J Biomed Mater Res* **75A**: 877–887.

Rajamannan, N.M., T.B. Nealis, M. Subramaniam, S.R. Stock, K.I. Ignatiev, T.J. Sebo, J.W. Fredericksen, S.W. Carmichael, T.K. Rosengart, T.C. Orszulak, W.D. Edwards, R.O. Bonow, and T.C. Spelsberg (2005). Calcified rheumatic valve neoangiogenesis is associated with VEGF expression and osteoblast-like bone formation. *Circulation* **111**: 3296–3301.

Rajon, D.A., J.C. Pichardo, J.M. Brindle, K.N. Kielar, D.W. Jokisch, P.W. Patton, and W.E. Bolch (2006). Image segmentation of trabecular spongiosa by inspection of visual gradient magnitude. *Phys Med Biol* **51**: 4447–4467.

Ramaswamy, S., M. Gupta, A. Goel, U. Aaltosalmi, M. Kataja, A. Koponen, and B.V. Ramarao (2004). The 3D structure of fabric and its relationship to liquid and vapor transport. *Colloids Surf A* **241**: 323–333.

Rolland Du Roscoat, S., M. Decain, X. Thibault, C. Geindreau, and J.F. Bloch (2007). Estimation of microstructural properties from synchrotron x-ray microtomography and determination of the REV in paper materials. *Acta Mater* **55**: 2841–2850.

Rüegsegger, P., B. Koller, and R. Müller (1996). A microtomographic system for the nondestructive evaluation of bone architecture. *Calcif Tiss Int* **58**: 24–29.

Sheppard, A.P., C.H. Arns, A. Sakellariou, T.J. Senden, R.M. Sok, H. Averdunk, M. Saadatfar, A. Limaye, and M.A. Knackstedt (2006). Quantitative properties of complex porous materials calculated from x-ray µCT images. *Developments in X-ray Tomography V*. U. Bonse (Ed.). Bellingham, WA, SPIE. *SPIE Proc Vol* **6318**: 631811–1 to –15.

Showalter, C., B.D. Clymer, B. Richmond, and K. Powell (2006). Three-dimensional texture analysis of cancellous bone cores evaluated at clinical CT resolutions. *Osteopor Int* **17**: 259–266.

Silva, M.D., J. Ruan, E. Siebert, A. Savinainen, B. Jaffee, L. Schopf, and S. Chandra (2006). Application of surface roughness analysis on micro-computed tomographic images of bone erosion: Examples using a rodent model of rheumatoid arthritis. *Mole Imaging* **5**: 475–484.

Spowage, A.C., A.P. Shacklock, A.A. Malcolm, S.L. May, L. Tong, and A.R. Kennedy (2006). Development of characterization methodologies for macroporous materials. *J Porous Mater* **13**: 431–438.

Stock, S.R., T.A. Ebert, K. Ignatiev, and F.D. Carlo (2006). Structures, structural hierarchy and function in sea urchin spines. *Developments in X-ray Tomography V*. U. Bonse (Ed.). Bellingham, WA, SPIE. *SPIE Proc Vol* **6318**: 63180A–1 to –4.

Stock, S.R., K. Ignatiev, and F.D. Carlo (2004a). Very high resolution synchrotron microCT of sea urchin ossicle structure. *Echinoderms: München*. T. Heinzeller and J.H. Nebelsick (Eds.). London, Taylor and Francis: 353–358.

Stock, S.R., K.I. Ignatiev, S.A. Foster, L.A. Forman, and P.H. Stern (2004b). MicroCT quantification of in vitro bone resorption of neonatal murine calvaria exposed to IL–1 or PTH. *J Struct Biol* **147**: 185–199.

Stock, S.R., S. Nagaraja, J. Barss, T. Dahl, and A. Veis (2003). X-Ray microCT study of pyramids of the sea urchin *Lytechinus variegatus*. *J Struct Biol* **141**: 9–21.

Stock, S.R., N.N. Naik, A.P. Wilkinson, and K.E. Kurtis (2002). X-ray microtomography (microCT) of the progression of sulfate attack of cement paste. *Cement Concr Res* **32**: 1673–1675.

Su, R., G.M. Campbell, and S.K. Boyd (2007). Establishment of an architecture-specific experimental validation approach for finite element modeling of bone by rapid prototyping and high resolution computed tomography. *Med Eng Phys* **29**: 480–490.

Tan, J.C., J.A. Elliott, and T.W. Clyne (2006). Analysis of tomography images of bonded fibre networks to measure distributions of fiber segment length and fiber orientation. *Adv Eng Mater* **8**: 495–500.

Thompson, K.E., C.S. Willson, K.T.W Zhang, A.H. Reed, and L. Beenken (2006). Quantitative computer reconstruction of particulate materials from microtomography images. *Powder Technol* **163**: 169–182.

Toda, H., T. Ohgaki, K. Uesugi, M. Kobayashi, N. Kuroda, T. Kobayashi, M. Niinomi, T. Akahori, K. Makii, and Y. Aruga (2006a). Quantitative assessment of microstructure and its effects on compression behavior of aluminum foams via high-resolution synchrotron x-ray tomography. *Metall Mater Trans A* **37**: 1211–1219.

Toda, H., I. Sinclair, J.Y. Buffiere, E. Maire, T. Connolley, M. Joyce, K.H. Khor, and P. Gregson (2003). Assessment of the fatigue crack closure phenomenon in damage-tolerant aluminium alloy by in-situ high-resolution synchrotron x-ray microtomography. *Phil Mag* **83**: 2429–2448.

Toda, H., I. Sinclair, J.Y. Buffiere, E. Maire, K.H. Khor, P. Gregson, and T. Kobayashi (2004). A 3D measurement procedure for internal local crack driving forces via synchrotron x-ray microtomography. *Acta Mater* **52**: 1305–1317.

Toda, H., M. Takata, T. Ohgaki, M. Kobayashi, T. Kobayashi, K. Uesugi, K. Makii, and Y. Aruga (2006b). 3-D image-based mechanical simulation of aluminium foams: Effects of internal microstructure. *Adv Eng Mater* **8**: 459–467.

Verhulp, E., B. van Rietbergen, and R. Huiskes (2004). A three-dimensional digital image correlation technique for strain measurements in microstructures. *J Biomech* **37**: 1313–1320.

Waarsing, J.H., J.S. Day, J.C. v. d. Linden, A.G. Ederveen, C. Spanjers, N.D. Clerck, A. Sasov, J.A.N. Verhaar, and H. Weinans (2004). Detecting and tracking local changes in the tibiae of individual rats: A novel method to analyse longitudinal in vivo microCT data. *Bone* **34**: 163–169.

Wan, S.Y., E.L. Ritman, and W.E. Higgins (2002). Multi-generational analysis and visualization of the vascular tree in 3D microCT images. *Computers Biol Med* **32**: 55–71.

Wang, M., X.F. Hu, and X.P. Wu (2006). Internal microstructure evolution of aluminum foams under compression. *Mater Res Bull* **41**: 1949–1958.

Wilkinson, A.P., A.C. Jupe, K.E. Kurtis, N.N. Naik, S.R. Stock, and P.L. Lee (2004). Spatially resolved energy dispersive x-ray diffraction (EDXRD) as a tool for nondestructively providing phase composition depth profiles on cement and other materials. *Applications of X-Rays in Mechanical Engineering 2004*. New York, ASME: 49–52.

Zettler, R., T. Donath, J.F. d. Santos, F. Beckman, and D. Lohwasser (2006). Validation of marker material flow in 4mm thick friction stir welded Al 2024–T351 through computer microtomography and dedicated metallographic techniques. *Adv Eng Mater* **8**: 487–490.

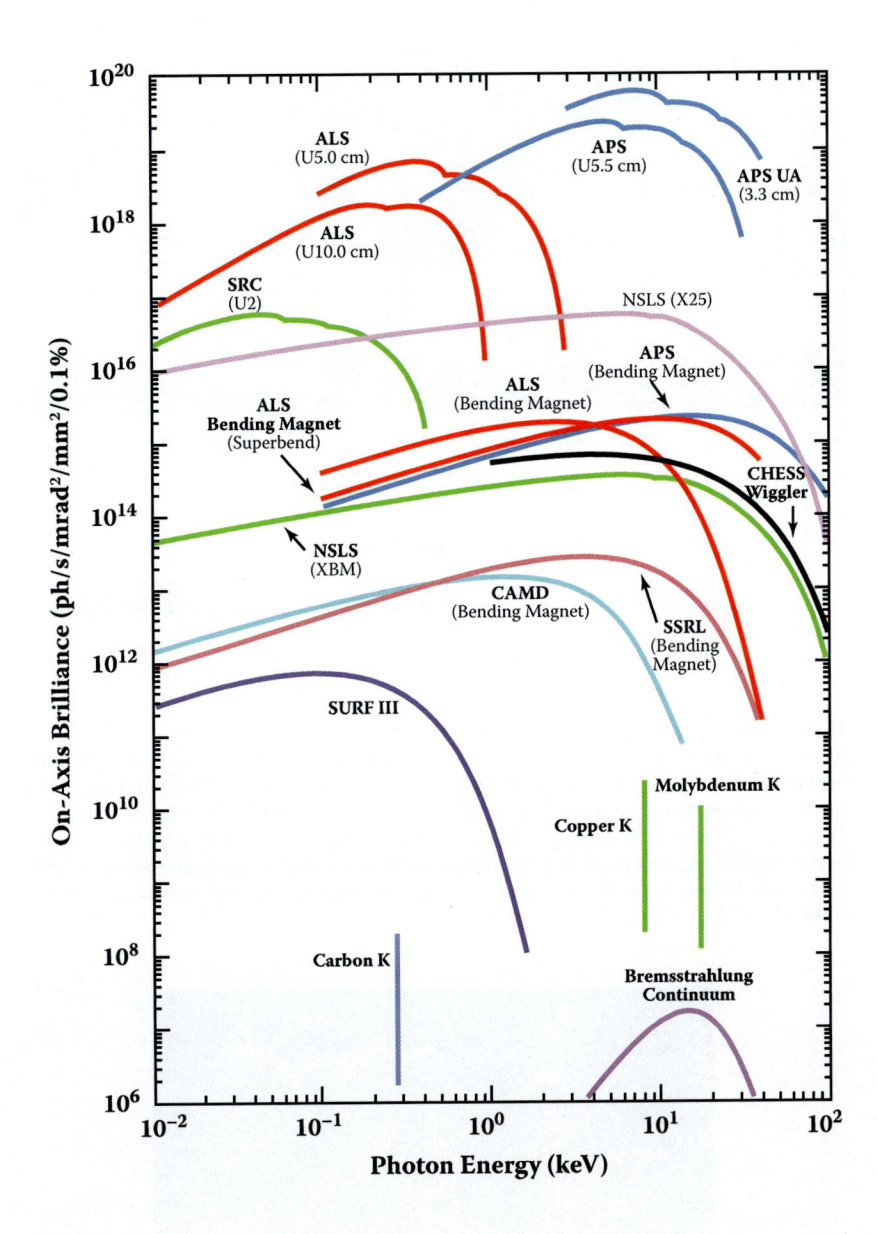

FIGURE 2.4
Relative intensities of x-ray tubes and of synchrotron radiation sources in the United States as a function of photon energy. The vertical axis plots brilliance, that is, intensity per unit time per unit area per unit solid angle per unit bandpass. (The figure was produced by Argonne National Laboratory, managed and operated by UChicago Argonne LLC for the U.S. Department of Energy under Contract No. DE-AC02-06CH11357, and is used with permission.)

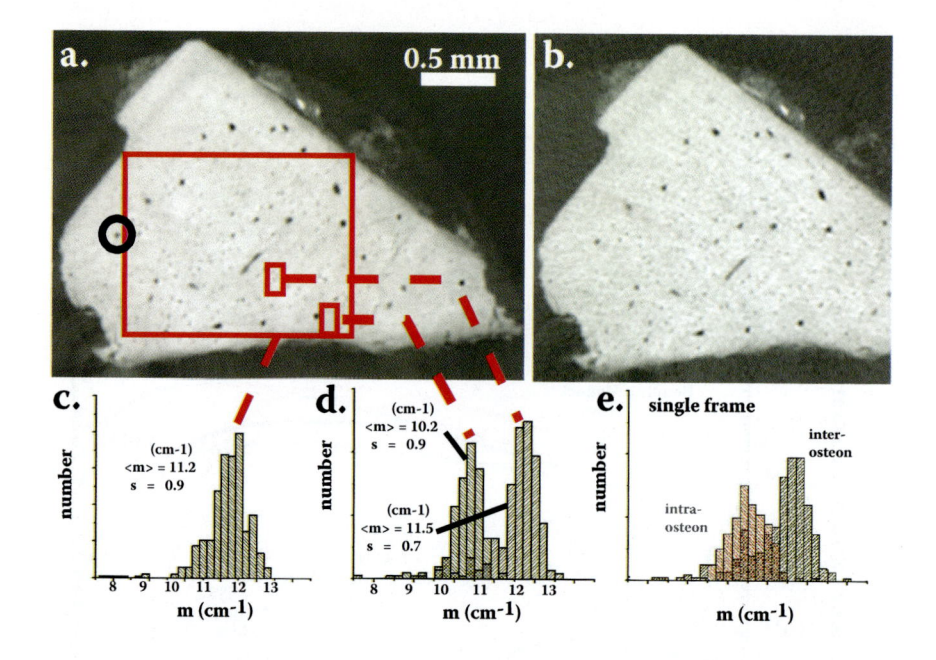

FIGURE 5.12
Counting time vs. contrast sensitivity in bone: (a) 32 frame average and (b) single frame. The mineral level within remodeled osteons (e.g., inside the black circle in (a)) is lower than that in the older bone between the osteons and is more clearly visible in (a) than in (b). (c) Histogram of linear attenuation coefficient values µ (mean <µ> and standard deviation σ) for the large boxed area in (a). (d) Histograms of the two small areas indicated in (a) (intra-osteonal and inter-osteonal bone) superimposed. (e) Histograms of the areas in (b) matching those in (d).

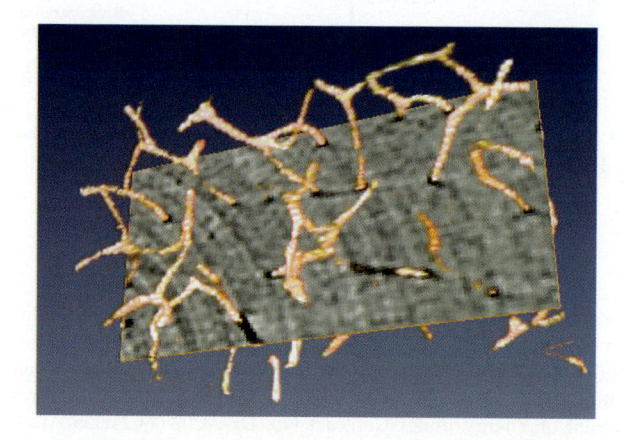

FIGURE 6.7
Small canals in rabbit bone. Low absorption voxels orange-white, higher absorption voxels transparent, slice section in gray.

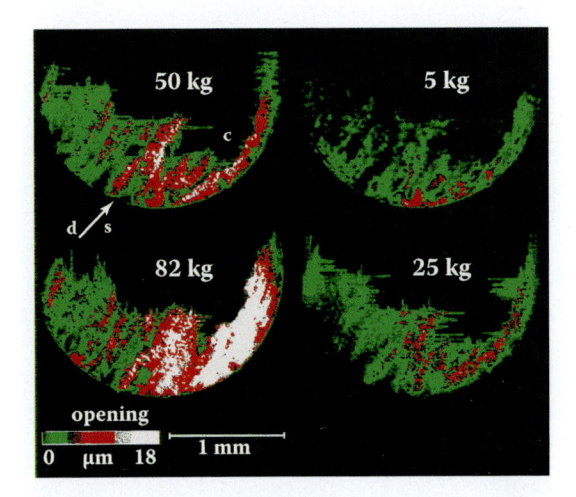

FIGURE 6.11
Crack opening in a sample of an Al alloy projected on the plane normal to the load axis. The color bar (lower left) indicates the total opening at each position, and the symbols are explained in the text in Chapter 11. Maps for four different applied forces (kgf) are shown. The corresponding stress intensities are given in parentheses. (Bottom left) 82 kgf (7.1 MPa√m). (Top left) 50 kgf (4.3 MPa√m). (Bottom right) 25 kgf (2.2 MPa√m). (Top right) 5 kgf (0.4 MPa√m) (Guvenilir et al., 1997).

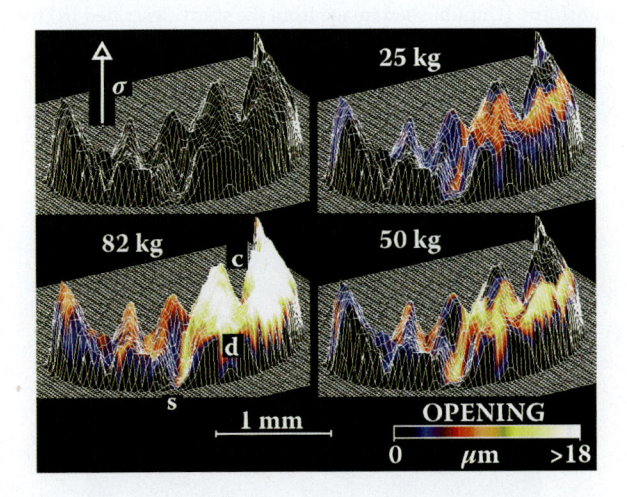

FIGURE 6.12
Crack opening in a sample of an Al alloy in 3D (see Figure 6.11 for the openings projected onto the plane normal to the load axis). The loading direction is indicated by "σ", the color bar (lower right) indicates the total opening at each position on the 3D mesh, and the other symbols are explained in the text in Chapter 11. Maps for three different applied forces (kgf) are shown. The corresponding stress intensities are given in parentheses. (Upper right) Mesh map showing the 3D crack position. (Lower left) Crack opening at 82 kgf (7.1 MPa√m). (Lower right) Opening at 50 kgf (4.3 MPa√m). (Upper right) Opening at 25 kgf (2.2 MPa√m) (Guvenilir et al., 1997).

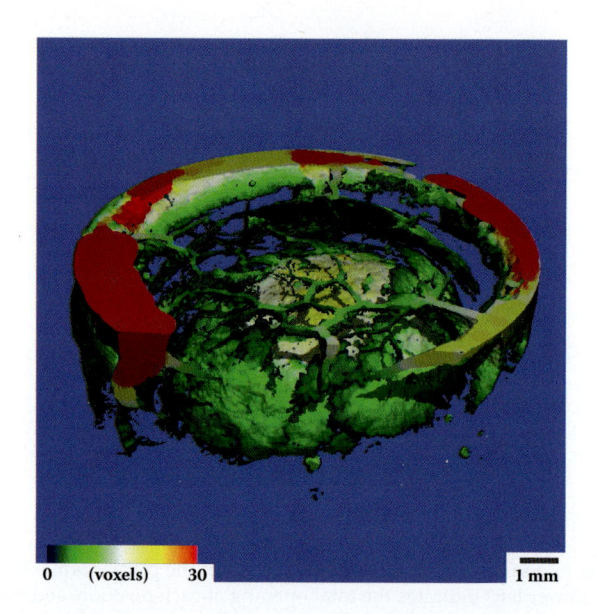

FIGURE 6.13
Three-dimensional rendering of the open voxels within the interior of a Portland cement specimen that had suffered significant damage from sulfate ion attack. The colors represent local thicknesses (in units of voxels, each of which was 37 µm on an edge) determined using the distance transform. (Unpublished image from the study described in Stock et al. (2002), Naik (2003), and Naik et al. (2006).)

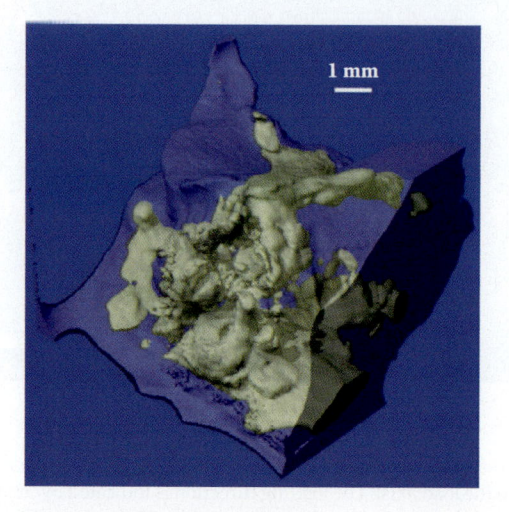

FIGURE 6.14
Rendering of calcified human heart valve employing two thresholds, a lower value for the soft tissue (semi-transparent light blue) and a higher level for the calcification (white). The flat surface represents a numerical section through the volume. (Rajamannan, et al. (2005); © Lippincott, Williams & Wilkins, 2005, used with permission.)

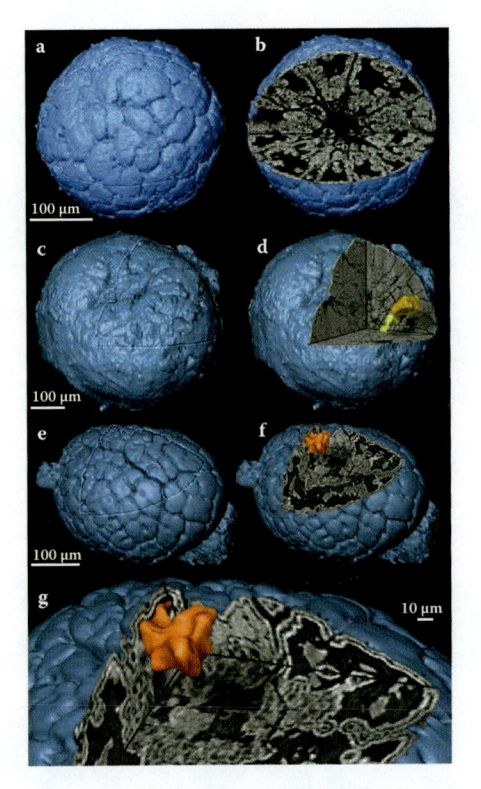

FIGURE 7.3
Fossil cleavage-stage embryos from the Lower Cambrian. Divisions between blastomeres (a,b) variably preserved, (c,d) well preserved (light orange/yellow shows one column), (e-g) not preserved (orange, cavity within diagenetic infilling) (Donoghue et al., 2006).

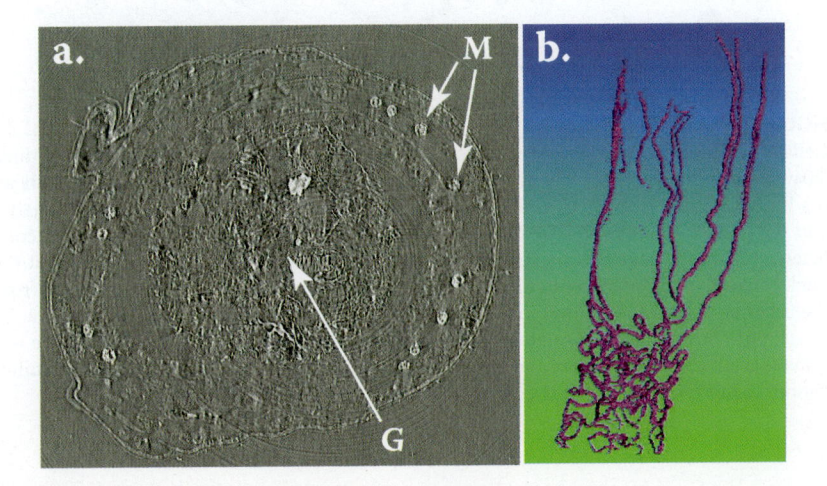

FIGURE 7.4
Silkworm (a) slice, and (b) 3D rendering of calcium oxalate-filled Malpighian tubes (pink).

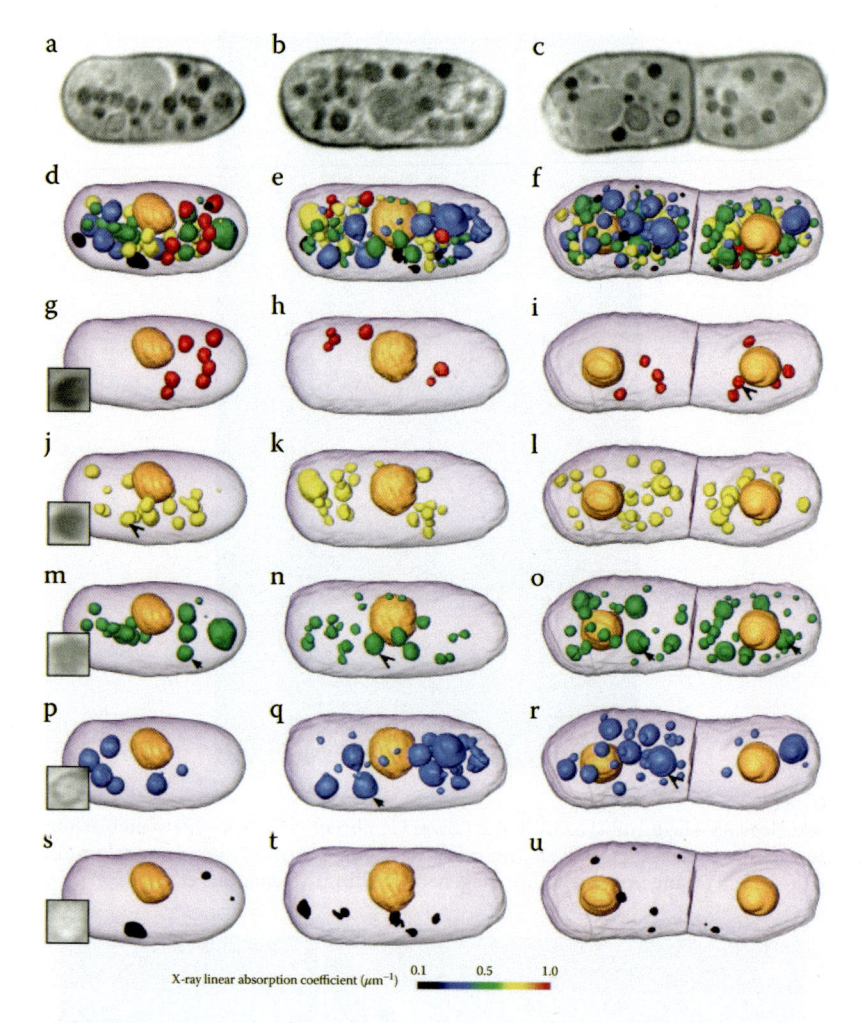

FIGURE 7.7
Cells of *Schizosaccharomyces pombe* imaged by soft x-ray tomography. (a)–(c) One-voxel-thick (20 nm) slices through the reconstructed volumes of four cells, with the darker gray indicating higher absorption. (d)–(f) Colored surfaces representing the boundaries of organelles and the plasma membrane. The colors correspond to the average linear attenuation coefficients inside the organelles (see the color bar at the bottom of the figure). Note that the nuclei should actually be colored blue but are shown as orange to differentiate them from the other structures. The length of the cell in (a) is 5 μm. (g)–(u) The same surfaces as in (d)–(f) but with each color isolated from the others. At the left side of each row there is a grayscale slice image of one of the organelles shown in that row; the arrows point to mitochondria and the wedges to other organelles (Parkinson et al., 2008).

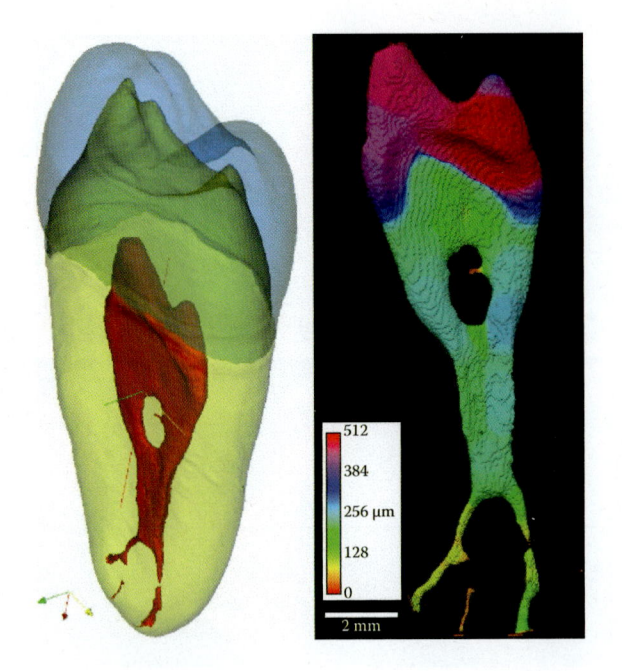

FIGURE 7.8
(Left) 3D tooth image: enamel (cyan), dentin (yellow), pulp (red). (Right) Local thicknesses map (Dougherty and Kunzelmann, 2007).

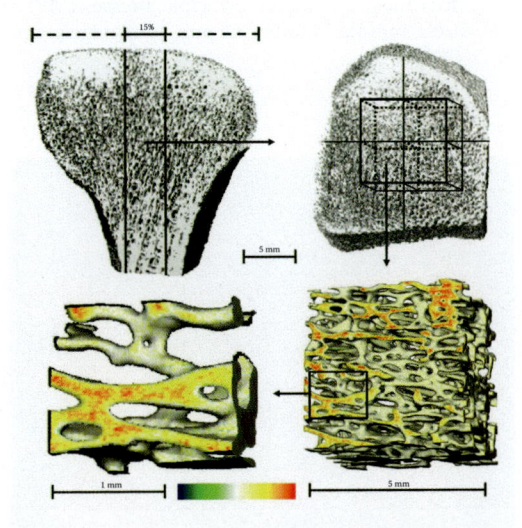

FIGURE 8.3
Left condyle of a 43 week old pig. (Upper left) Frontal cross-section. (Upper right) Medial view, sagittal slice (posterior on left) showing four subvolumes. (Lower right) Lower left subvolume (posteroinferior) magnified and (lower left) further magnification of trabeculae; color indicates increasing degree of mineralization from blue to red (Willems et al., 2007).

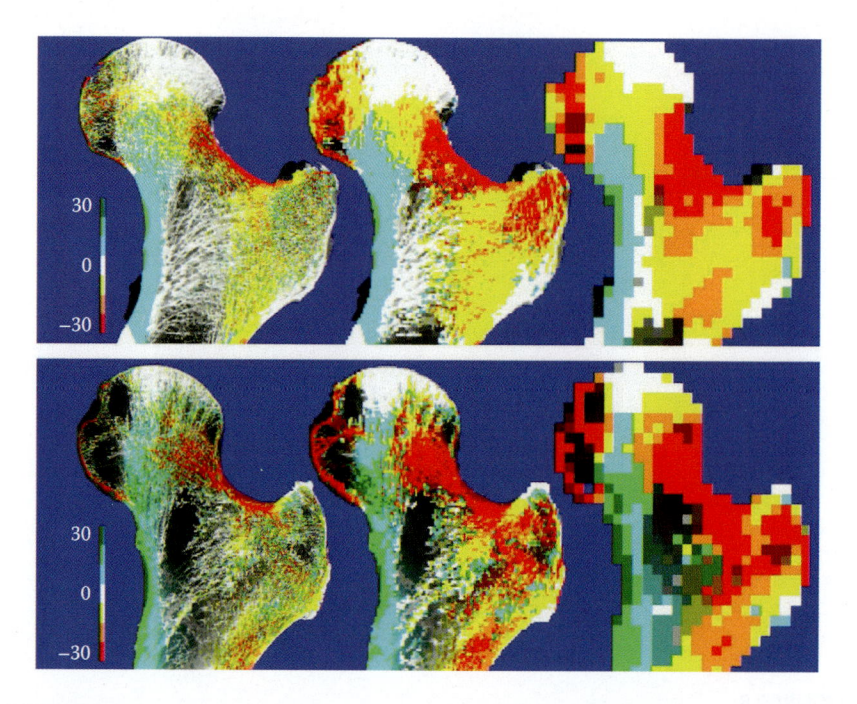

FIGURE 8.4
Finite element models at three resolutions for a healthy proximal femur (top row) and an osteoporotic femur (bottom row). Maximum principal stresses (MPa, in color). Left column 80 µm isotropic voxels, middle column 0.64 mm voxel edges, right column 3.04 mm edges (Verhulp et al., 2006).

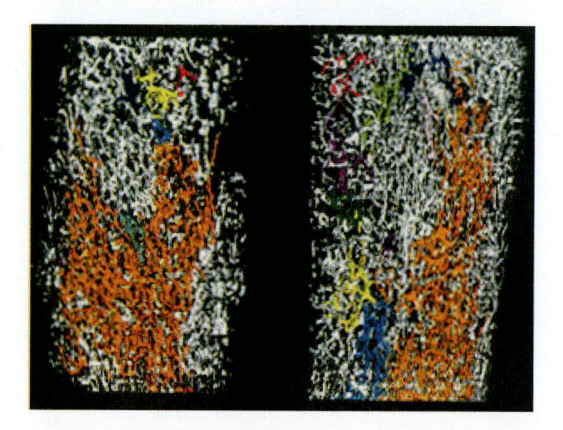

FIGURE 9.4
Haversian canals in a 3D rendering showing only the low-absorption voxels in 1 mm x 1 mm x 3 mm volume. The low-absorption voxels, which are interconnected, are shown in the same color (Ritman et al., 1997).

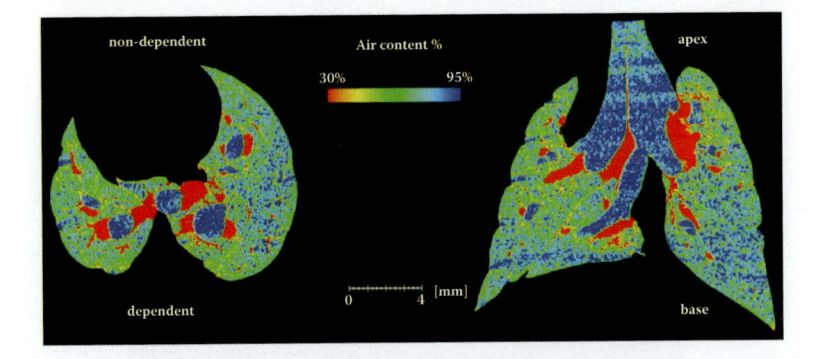

FIGURE 9.6
Axial (left) and coronal (right) 2D sections of a typical mouse lung showing regional air content differences represented by the colors shown in the color bar (Namati et al., 2006).

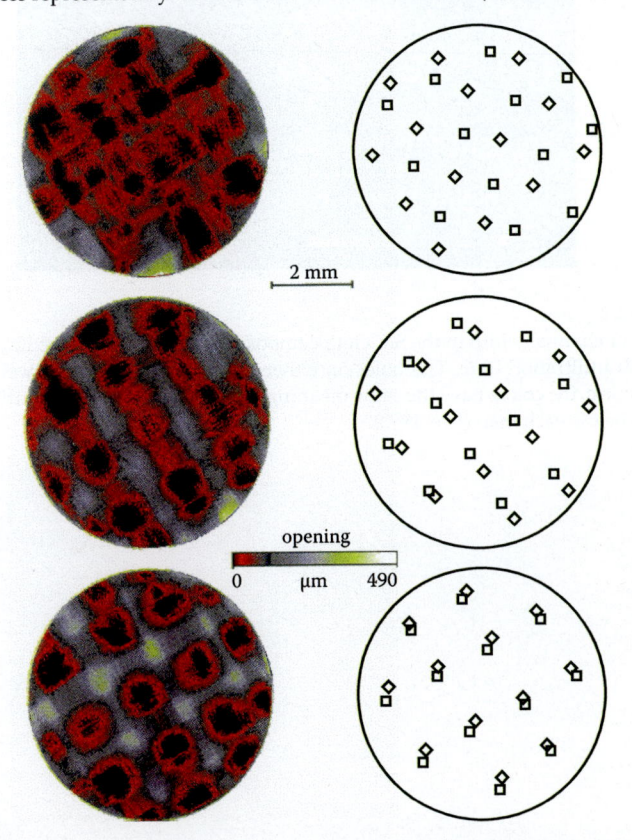

FIGURE 10.1
Projected channel width (left col.) vs. (right col.) positions of holes in SiC cloths bordering channel (squares, diamonds). Colors (bar) denote opening. (Top) Closed channel: holes displaced 45° to tows (fiber bundles). (Middle) Holes displaced along tow axes produce 1D pipes. (Bottom) Closely aligned holes and 2D pipes (Lee, 1993; Lee et al., 1998; © Mater. Res. Soc.).

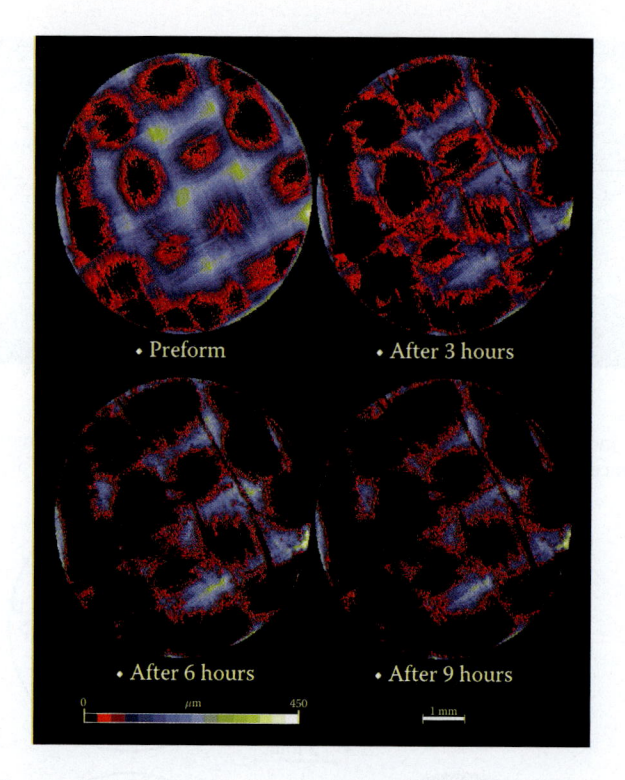

FIGURE 10.2
Variation of channel width in the SiC cloth composite described in Figure 10.1 as a function of SiC CVI infiltration time. The holes on either side of the channel are well aligned (2D array of pipes), the colors have the same meaning as in Figure 10.1, and the infiltration times are given below each map (Lee, 1993).

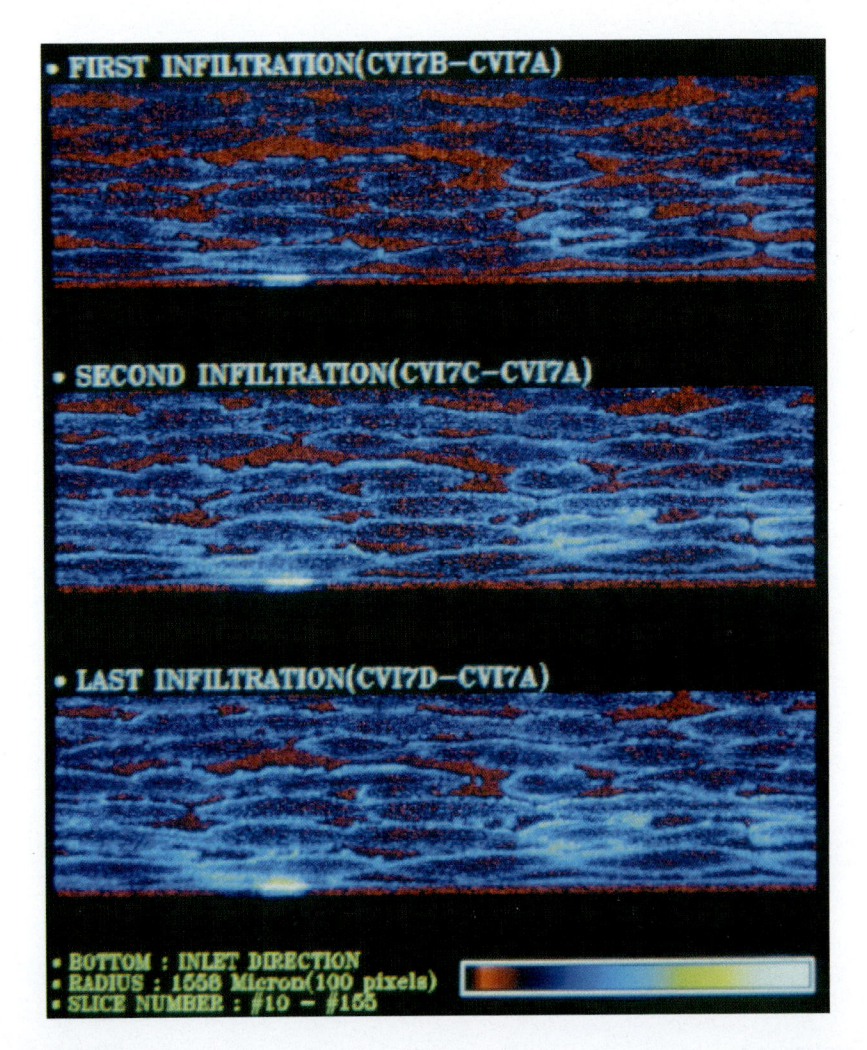

FIGURE 10.3
Circle cuts through a SiC/SiC cloth preform composite showing where SiC was deposited. (Top) Difference image showing SiC deposited in the first three hours of CVI. (Middle) Difference image showing the total SiC deposited after six hours. (Bottom) Difference image showing the total SiC deposited after nine hours. Red, blue, and white show increasing change from the preform. The open areas between tows, where there are no fibers onto which SiC can be deposited, remain red throughout. The gas enters the reaction chamber from the bottom, and the white shells seen on many tows show that the interior microporosity between individual ~15-μm-diameter fibers remains incompletely densified (Lee, 1993).

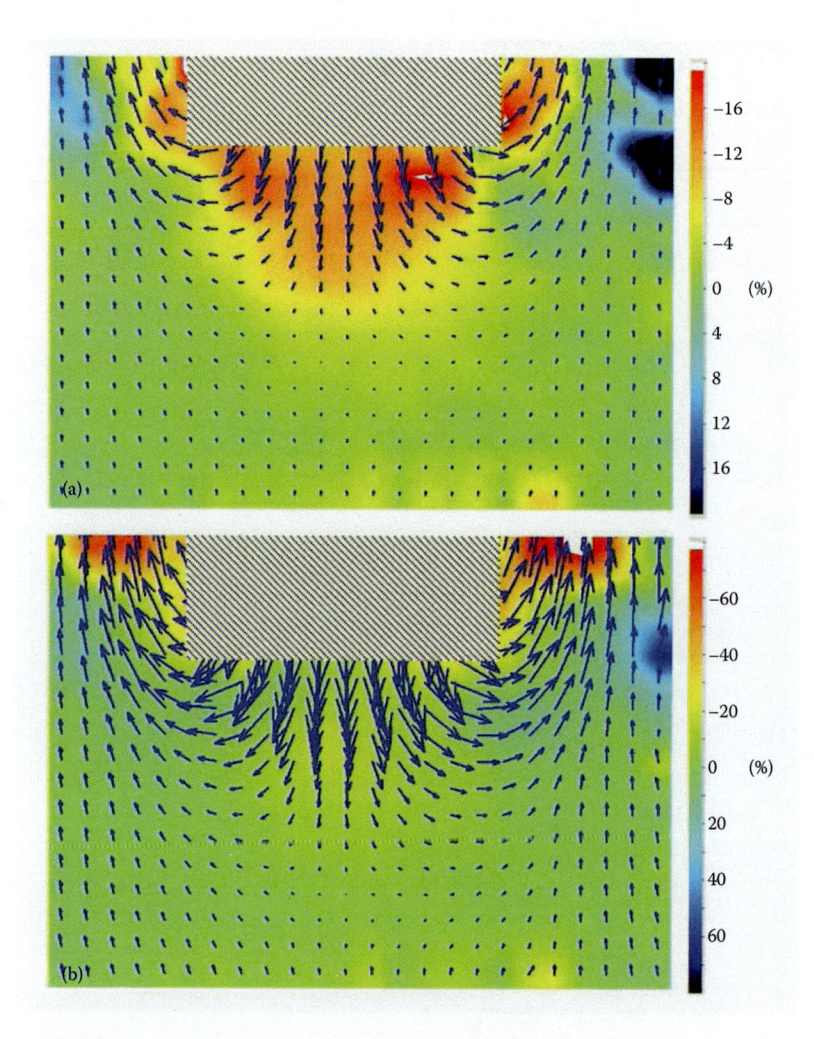

FIGURE 10.4
Powder displacement during compression. The arrows show particle displacement vectors around the downward moving punch calculated from image correlation over: (a) four 0.5 mm displacement steps (2.0 mm total) and (b) eight steps (4 mm total). The background color scale represents the dilational (volumetric) strain calculated from the particle displacements, effectively showing the change in density across the diametral section. The powder immediately under the punch is compacted (negative strain), but dilation (positive strain) is observed as the loosely packed powder is sheared, particularly as it flows around the corners of the punch and upward against the sides of the container. (Reprinted from McDonald et al. (2006); © 2006, with permission from Elsevier.)

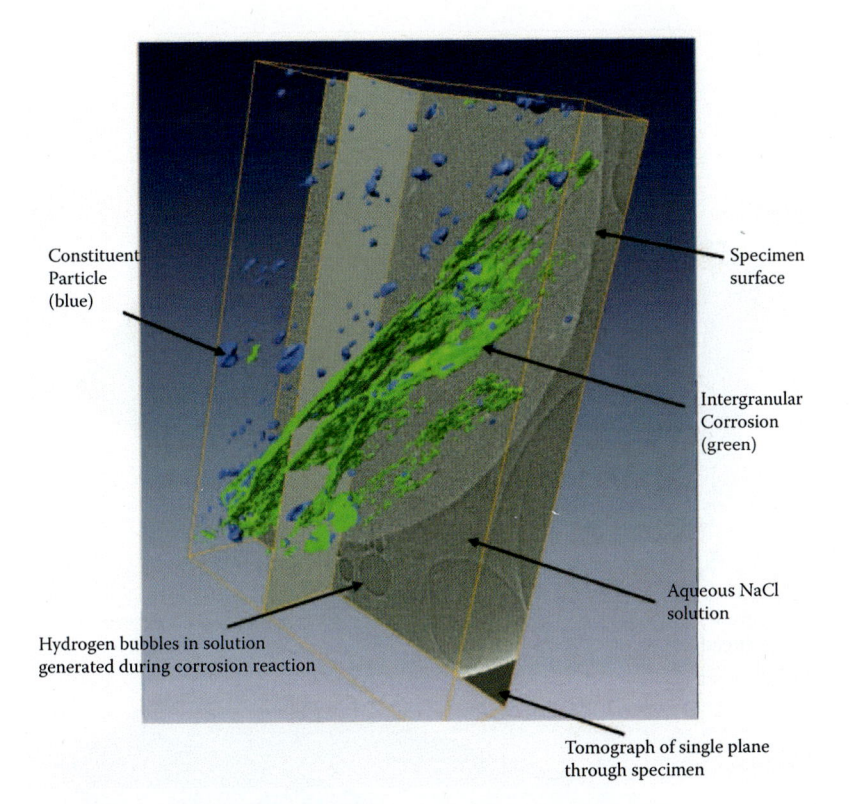

FIGURE 10.7
3D rendering of the heat-affected zone near a weld in AA2024 after 40 h exposure to 0.6 M NaCl. About 200 slices are used, showing the sample transparent except for intergranular corrosion (green) and intermetallic particles (blue). The 0.5-mm-diameter specimen was imaged in situ with 20.5 keV synchrotron x-radiation and reconstructed with 0.7-μm isotropic voxels. (Reprinted from Connolly et al. (2006); © 2006, with permission of Maney Publishing (http://www.ingentaconnect.com/content/maney/mst).)

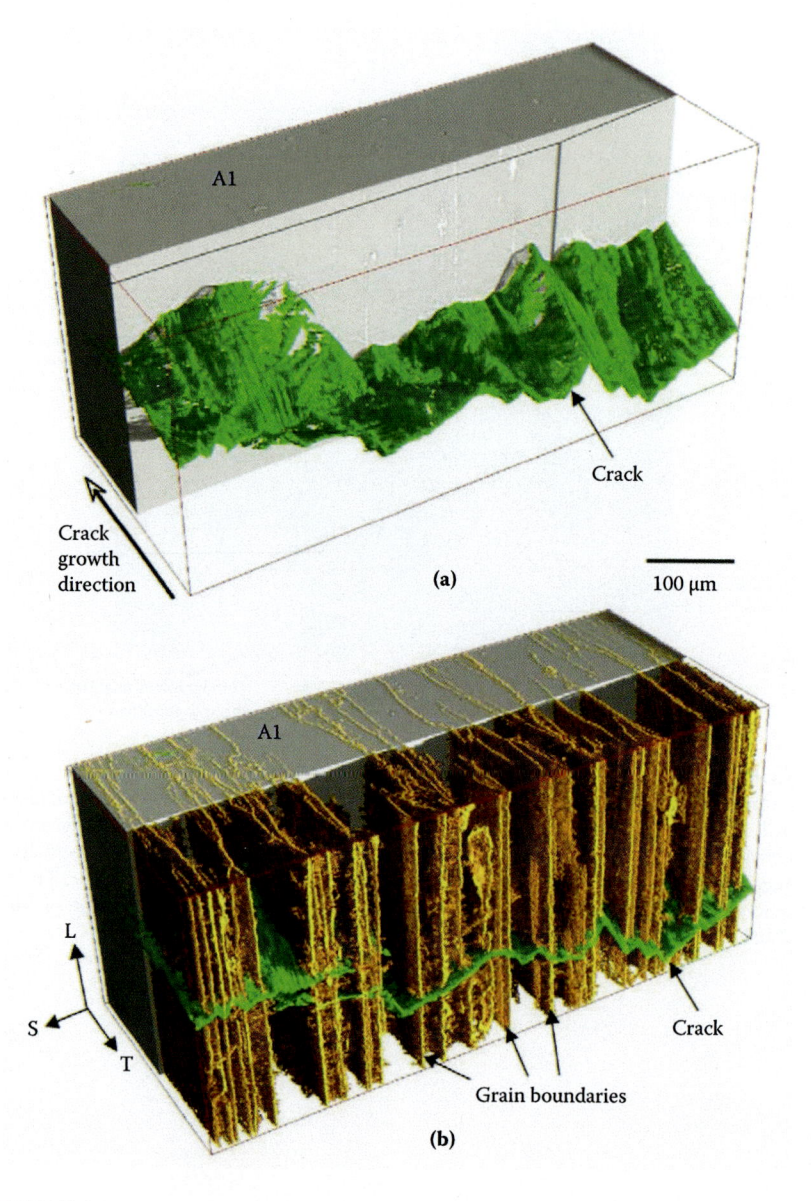

FIGURE 11.1
Three-dimensional representations of a fatigue crack in AA2024-T351. (a) Crack volume (green) and (b) grain boundaries decorated with Ga (gold). (Reprinted from Khor et al. (2006); © 2006, with permission from Elsevier.)

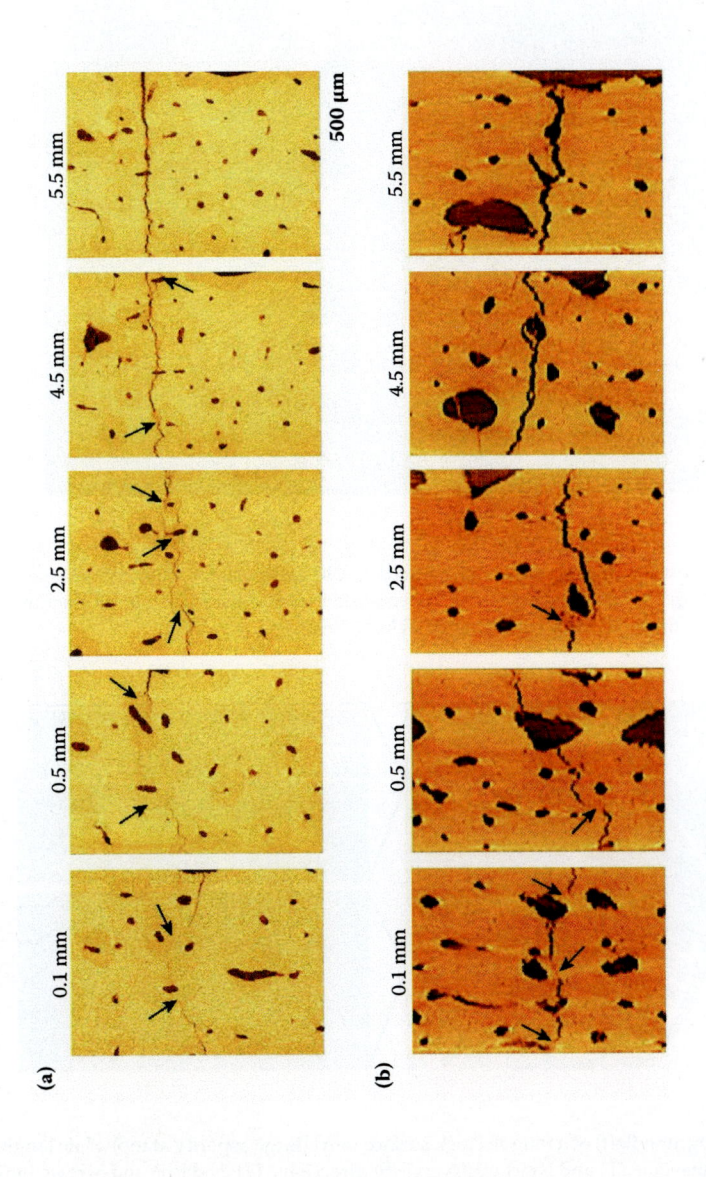

FIGURE 11.2
Synchrotron microCT sections through human cortical bone showing typical cracks in (a) young (34 years) and (b) aged (85 years) groups. The numbers above each column of images give the distance from the crack tip, and the black arrows indicate uncracked ligaments bridging the crack. The darker the shade of orange, the lower is the x-ray absorption and mineral content (Nalla et al., 2004).

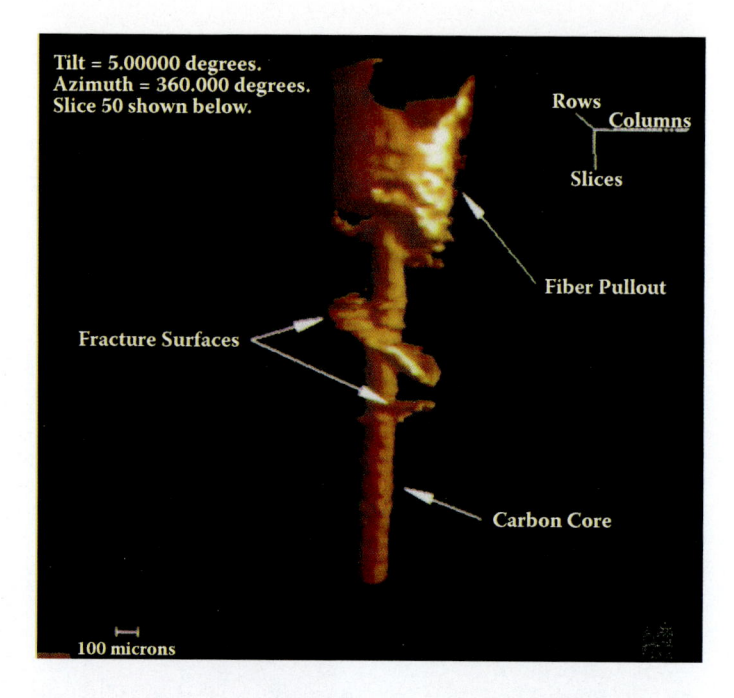

FIGURE 11.4
Surface of low absorption voxels in fractured Al-SiC composite. C fiber core extends down from fracture surface (top) where SiC fiber pulled from Al surface; spiral crack midway along fiber (Breunig, 1992; Kinney and Nichols, 1992).

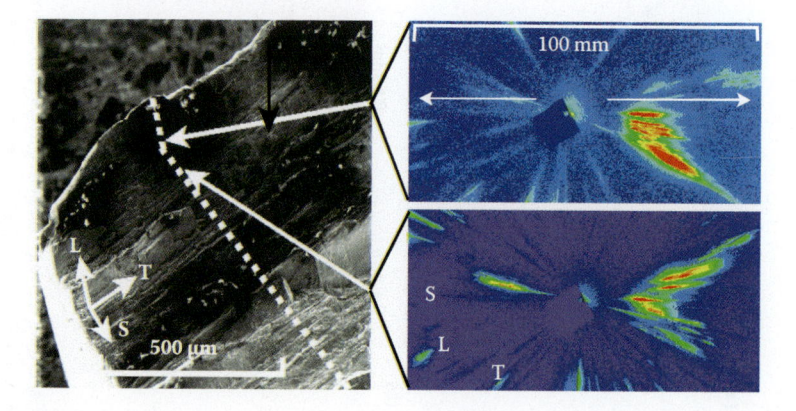

FIGURE 12.1
SEM fractograph (left) of a fatigue crack surface in Al (large asperity at top). Plate longitudinal (L), transverse (T) and short transverse (S) directions. Dashed line and arrows indicate positions of transmission microbeam Laue patterns (right). The streaks are 111 reflections (purple, dark blue, light blue, green, yellow, orange, red show increasing intensity) whose orientations change abruptly inside vs. outside of the asperity (Haase et al., 1998).

7

Simple Metrology and Microstructure Quantification

The subject of this chapter is metrology and microstructure characterization. Both subjects are covered elsewhere in Chapters 8 to 12, and the coverage here focuses on "simple" quantification, that is, microCT studies relying on elementary thresholding and on single (one time point) examination of each specimen. More complex interrogations (i.e., studies of cellular solids such as metallic, ceramic, polymeric, or liquid foams or trabecular bone; observations of cracking as a function of applied stress) are postponed until later chapters.

Distribution and morphology of phases is the subject of the first section of this chapter. Pharmaceutical materials are covered first, followed by geological materials. Examples of studies of monolithic engineered materials (metals, ceramics, and polymers) are the next subject. Manufactured composites are the fourth topic, and the fifth introduces biological tissues as phases.

Metrology refers to the techniques of measurement, and it is coupled with phylogeny, the study of the evolutionary relations between species, in the second half of this chapter. The reason for coupling these seemingly dissimilar topics is that quantification and comparison of dimensions or structural motifs underlie the experimental and analysis approaches. The order of the subsections is: industrial metrology, paleontology, invertebrates, and vertebrates.

This chapter may strike the reader as a bit of a potpourri better postponed until the closing pages of the book or as a miscellany whose topics are best folded among those of the other chapters. Instead, this chapter is collected together because the author sees a common theme of simple analysis approaches answering the questions at hand. It is important to never lose sight of the fact that the simplest approaches can yield important results and that better (more complex analyses) can be the worst enemy of good enough (i.e., a completed meaningful study). The commonality of analysis requirements and approaches is the organizing principle behind the subsequent chapters (at least as far as the author's perception extends), and there is no reason not to start thinking in these terms from the beginning of the coverage of microCT applications.

7.1 Distribution of Phases

Phases can be voids as well as different solid phases. The spatial distribution and relative amounts of different phases control macroscopic properties to a great extent. For many applications, microCT is the best method of obtaining 3D information or avoiding specimen preparation artifacts. There are many more studies that could have been included, but the reader should be able to gain an appreciation of this area of microCT from the examples below.

7.1.1 Pharmaceuticals

A relatively new application of lab microCT is evaluation of the spatial distribution of phases in pharmaceutical manufactured materials, that is, in solid dosage forms such as tablets and soft-gelatin capsules (Hancock and Mullarney, 2005). Thicknesses and interface character in multilayer tablets, microstructure of rapidly dissolving tablets produced by lyophilization (ice crystallization followed by drying), and particle sizes within controlled-release osmotic tablets are important characteristics directly measurable via microCT. Nondestructive microCT comparison of genuine and counterfeit tablets has the additional advantage of preserving the evidence in patent litigation or other legal proceedings.

Pores in pharmaceutical granules have also been studied with lab microCT (Farber et al., 2003), including quantification of localized density variations (Busignies et al., 2006), as has the spatial conformation of components in modified release tablets (Traini et al., 2008). Hydrophilic drugs dispersed in polymeric matrices are expected to have decreased release rates as diffusion path length increases, and microCT has been examined as a way of fine-tuning the pore structure to achieve more uniform drug release (Wang et al., 2007). The discrete element method (for simulating structural changes) and microCT have been compared to advance understanding of tablet formation by powder compaction (Fu et al., 2006). Magnetic particles are being investigated for controlled drug delivery, for example, to tumors, and microCT has been used to measure these particles' distribution in tissue of animal models (Brunke et al., 2005).

7.1.2 Geological Materials

In one synchrotron microCT study (Gualda and Rivers, 2006), quartz, magnetite, and sanidine size distributions were measured in pumice clasts (isolated crystals surrounded by a low-density matrix); this investigation was undertaken to address possible limitations of earlier work on the same material. Previous characterization used a crushing + sieving + winnowing procedure to quantify the size distributions (Gualda et al., 2004), avoiding stereology's well-known limitations in transforming 2D data into true measures of the 3D arrangements in the solid. Such processing, however, tends to cause significant loss of small crystals and frequently fragments larger crystals. This latter artifact is especially significant in that it obscures characterization of fragmentation generated in magmatic processes. As the steps involved in the analysis of the microCT data are characteristic of those often encountered in microCT studies, they are described here in some detail.

In phase quantification studies, the first image analysis step (after reconstruction) is image classification, that is, assigning each voxel in the 3D volume to a given phase. The second step is identifying individual grains, that is, clusters of voxels belonging to a single-phase particle. In the study cited in the previous paragraph (Gualda and Rivers, 2006), contrast sensitivity (256 gray levels) was adequate to quantify the volumes and size distributions for particles greater than five voxels diameter in the Bishop Tuff pumice clast; noise limited the investigators' ability to reliably identify smaller particles. Nonetheless, Gualda and Rivers (2006) found that the combination of contrast and spatial information allowed distinction between quartz and sanidine, despite the fact that the distribution of linear attenuation coefficients of the minority phase (sanidine) formed an indistinct shoulder of the quartz peak. The microCT results agreed with earlier results from destructive analysis (Gualda et al., 2004), namely, that the distribution of quartz particle sizes indicated action of magmatic fragmentation processes but that of magnetite was largely unaffected by the fragmentation process recorded by quartz.

The trade-off between spatial resolution and contrast sensitivity is clearly explained as it affects these results, but the authors do not provide details of the number of x-ray counts recorded in the 2 × 2 binned detector pixels (Gualda and Rivers, 2006), and one is unable to assess the numerical extent to which contrast for a given phase is the spread because of counting statistics. It would have been interesting if the investigators had investigated frame averaging for improving contrast sensitivity and increasing the small particle detection limit. Of further interest in this carefully done study is the documentation of exceptional volumes that diverge quite markedly from the rest of the sample: this simple result should serve as a caution to all investigators using microCT of small sections derived from larger objects (Gualda and Rivers, 2006).

Bubble (vesicle) characteristics in basalts provide insight into various processes in magma prior to, during, and after eruption; a distribution of bubble sizes, for example, can be due to multiple or continuous nucleation processes or differences in growth rates. Synchrotron microCT of five basalts from different locations showed bubbles were spheroidal and composed of 45 vol.% of lavas to 80 vol.% of scoria and were at least 90 percent interconnected (Song et al., 2001).

Anisotropy measurements of magnetic susceptibility and of elastic constants (from P-wave velocity) were combined with microCT in a study of Callovo–Oxfordian argillite (David et al., 2007). An unexpected finding was the presence of tubular structures of dense materials (pyrite).

Study of small-diameter inclusions and of materials of different voxel or subvoxel porosity content can be very important and challenging. Microdiamond content and size distribution in kimberlite are important indicators of the likelihood of finding coarser valuable diamonds. MicroCT has been applied to the problem of quantifying diamond content of drill-hole cores, and these authors report tomography research at De Beers (Schena et al., 2005). Zones of differing levels of porosity have been studied in mortar and are important in resistance of this construction material to environmental attack (Diamond and Landis, 2007).

7.1.3 Two or More Phase Metals, Ceramics, and Polymers

Pyun and coworkers used phase microCT to visualize the 3D morphology of polymer blends (Pyun et al., 2007). The material was polystyrene plus high-density polyethylene; the investigators visualized both phases by segmentation and, when one phase was dissolved, found that the interfacial S_V after various annealing times agreed with the results of mercury porosimetry.

Radio-opaque polymers are desirable for dental applications; Anderson and coworkers employed lab microCT to show barium methacrylate monomer did not blend well when diluted in methacrylate, whereas inhomogeneities could not be detected when tin methacrylate was used (Anderson et al., 2006). Void distribution determination in HY-100 steel is one of the few synchrotron microCT studies of very highly attenuating material (Everett et al., 2001). Catalytic conversion of natural gas to (clean) liquid fuels via Fe nanoparticles is of interest for lessening internal combustion-related pollution, and the spatial distribution of these nanoparticles has been studied with fluorescence microCT (Jones et al., 2005).

Toda and coworkers used nanoCT to image γ' Ag_2Al precipitates in an Al alloy (Toda et al., 2006). The $\{111\}_{Al}$ habit planes of the precipitates were visible in their 3D renderings (Figure 7.1). Solidification structures in an Al–Si alloy were seen via phase microCT, even though the difference in absorption contrast between Al and Si is very small (Figure 4.10).

7.1.4 Manufactured Composites

In early work on manufactured composites, the pore structure in degassed and nondegassed reaction-bonded silicon nitride/silicon carbide was studied with 10-µm voxels in the 1 mm × 1 mm cross-section (Stock et al., 1989), and, not surprisingly, the 15-µm diameter Nicalon fibers could not be resolved. Within a given cross-section of an Al powder processed composite, the content of 12-µm TiB_2 particles was found to deviate

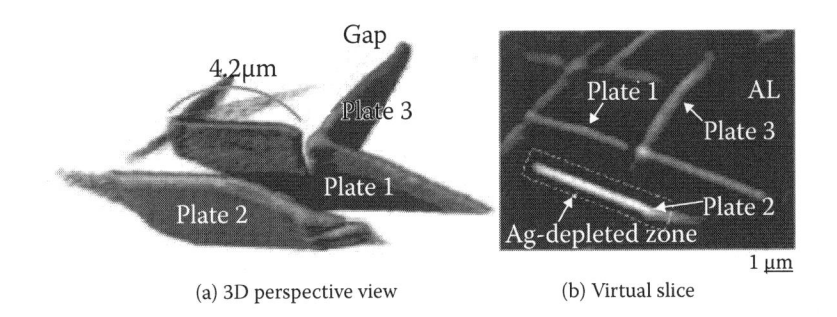

FIGURE 7.1
Platelike γ′ Ag$_2$Al precipitates imaged with a Fresnel zone plate nanoCT system. (a) 3D view of several precipitates segmented to only show the Ag-rich phase. (b) Slice through the plates shown in (a) and showing the Al matrix (gray) and Ag-depleted zones (black) around each plate (white). Imaging at 9.8 keV with ~90 nm isotropic voxels, with the precipitates being 2.5 to 7 voxels thickness (Toda et al. (2006); © 2006, American Institute of Physics).

substantially from the nominal 20 vol.% of reinforcement: contents as low as 10 vol.% were reported (Mummery et al., 1995). Images of a SiC monofilament Si$_3$N$_4$ (91 wt.% Si$_3$N$_4$, 6 wt.% Y$_2$O$_3$, 3 wt.% Al$_2$O$_3$) composite have been obtained with microCT (Hirano et al., 1989); in 111-µm-thick slices, the 140-µm-diameter fibers and their 30-µm-diameter cores are quite visible, and the radial variation of linear attenuation coefficient at 24.0 keV for the SiC fibers agrees with others' 21-keV measurements of similar fibers (Kinney et al., 1990).* Thermomechanical fatigue of ceramic as well as metal matrix composites has also been investigated (Baaklini et al., 1995).

Aluminum matrix composites reinforced by Al$_2$O$_3$ whiskers 2 to 4 µm in diameter and 50 to 80 µm in length or by Al$_2$O$_3$ fibers about 20 µm in diameter have been studied (Bonse et al., 1991). The sample diameter of the former composite was ~1 mm, whereas that of the latter, which was not reported, was apparently much less than 1 mm. The whisker-reinforced composite is a material normally used for diesel engine pistons, and its aluminum matrix contains significant levels of Si, Cu, Ni, and Fe. The resulting intermetallic phases were seen to form a three-dimensional network whose mesh size was on the order of 15 µm. These investigations also found that the individual Al$_2$O$_3$ fiber images and the distribution of fiber images for the second composite agreed with scanning electron micrographs and demonstrated spatial resolution in the reconstruction of 6 µm (MTF of 80 lines pair/mm at 20 percent contrast; Bonse et al. (1991)).

The distribution of Cr particles in alumina, determined via synchrotron microCT, was used as the input for a finite element calculation of residual stresses arising during cooling from the 1450°C processing temperature

---

* Figures 5.10 and 5.11 and associated text as well as the discussion in Chapter 11 cover these other results on SiC fibers in Al matrices.

(Geandier et al., 2003). Hydrostatic stresses were found in the alumina near the Cr particles. In a separate paper, Geandier and coworkers compared calculations with actual residual stress measurements (Geandier et al., 2002). Such residual stress measurements are readily performed, nondestructively, using synchrotron high-energy x-ray diffraction (Haeffner et al., 2005). Similar input was used for FEM of dynamic response of porous (and epoxy-infiltrated) shape memory alloy specimens (Qidwai and DeGiorgi, 2004).

Calculations of macroscopic properties have been performed using actual particle spatial and orientation distributions measured in an Al–20 vol.% Al_2O_3 composite. Finite element modeling (FEM) and mean field and multiscale modeling were used to compute various elastic properties; good agreement with experimental moduli measurements was obtained (Borbely et al. 2003, 2006; Kenesei et al. 2006a,b), and a somewhat larger fraction of particles fractured in the interior of the specimen compared to a zone 2–3 particle diameters from the surface. Sanchez et al. (2006) quantified graphite volume fraction in Al matrix composites and used the microCT-determined spatial distribution of graphite to calculate flow of Al into the graphite preform and to simulate the spatial distribution of strains from deformation of the solid composite.

As Heggli et al. (2005) discuss in conjunction with their microCT data on a graphite/Al composite, accurate models depend on employing representative volume elements that are sufficiently large to be representative of the material on a macroscopic scale (but are small enough to be tractable numerically); these authors concluded that reasonably accurate predictions of resistivity could be obtained using ensemble averaging over a sufficient number of small models. Titanium dioxide–polymer composites (for bioimplants) were compacted with radial density gradients observed (based on linear attenuation coefficients from averaged voxel values), but it is unclear whether the reported variation was not, in fact, an artifact of beam hardening, Ti being quite absorbing (De Santis et al., 2007).

The distribution of different carbon-based phases in a composite was studied with phase microCT (Coindreau et al., 2003). The refractive index decrements of resin, carbon fibers, and deposited carbon were clearly different and allowed clear segmentation of the different phases. Absorption contrast would have revealed only porosity. Carbon fibers in a carbon matrix composite have been the subject of another phase microCT study (Martín-Herrero and Germain, 2007). Phase contrast provides enough contrast to differentiate the fibers from the matrix (Figure 7.2), something that would never be seen with absorption contrast.

Quite a few other microCT composite studies have appeared and are covered elsewhere in this volume.

FIGURE 7.2
Three orthogonal sections through a C/C composite from phase microCT. The volume enclosed by the planes is 0.0033 mm^3, and the voxel size equals 0.745 µm. Phase contrast shows the C fibers (dark gray in the image) and the carbon matrix deposited on the fibers (light gray) (Martín-Herrero and Germain, 2007).

7.1.5 Biological Tissues as Phases

Different biological tissue types can be regarded as phases in the materials sense, and the number of studies employing microCT of tissue specimens, scaffolds for implants, and the like dwarfs those of engineering materials. Mineralized tissues such as bone (apatite, calcium phosphate, plus collagen) or echinoderm ossicles (calcite, calcium carbonate) can take the form of a cellular solid (i.e., plates and struts surrounded by soft tissue), and it is sensible to review these studies in the chapter on cellular solids. Likewise, studies of blood vessel and airway networks are covered in the porous solids chapter. MicroCT studies of dense mineralized tissues (cortical bone, sea urchin teeth) are distributed within other subsections, but pathological calcifications have also been studied.

Several examples of pathological calcification have already been presented. Figure 4.6 shows a synchrotron and matched lab microCT slice of a mouse model of osteoarthritis; the highly porous bone on the right side resulted from the treatment. Calcifications as part of juvenile dermatomyositis have been studied (Stock et al., 2004b, 2004c). Kidney stones have also been the subject of studies, both their microstructure and their location within the kidney (Stock et al., 2004c). Williams and coworkers studied calcium oxalate/apatite calculi attached to renal papilla (Williams et al., 2006). Figure 6.14 shows a semitransparent 3D rendering of a human heart valve that became heavily calcified (Rajamannan et al., 2005a). Quantitative comparisons based on lab microCT data have shown statistically significant differences in the volume of calcification between control and disease-affected heart valves (Rajamannan et al., 2005b). Other microCT studies of calcification gone wrong abound.

MicroCT of soft tissue is mostly done with phase contrast in order to differentiate between different soft tissue types, although, as recent data on brain tissue (Müller et al., 2006a) and on fixed lung sections (Shimizu et al., 2000) showed, this is not always necessary for certain tissue types. Tumor tissue has been well-differentiated from healthy tissue (Momose et al., 1998a,b; Momose and Hirano, 1999; Momose et al., 1999), and different normal tissue structures (e.g., mammary ducts) can also be made out (Takeda et al., 1998). An example where in vivo absorption microCT was used to study soft tissue features (lung tumors in a mouse model) is typical of such studies: a priori information was necessary to differentiate tumors from other features (large blood vessels; Weber et al., 2004). Respiratory gating was used in this study of the accuracy of microCT characterization, and further details are covered in the section on accuracy. Injection of contrast agents that concentrate in the soft tissue type of interest has been used to good effect to image murine liver tumors, but success depends on adequate contrast enhancement (Weber et al., 2004), a process that may be difficult to control.

Not all medical x-ray applications require full 3D information, and considerations of dose limitation or specimen geometry often dictate that simple radiography be employed. It is worth a brief mention of the few phase-radiographic imaging studies, then, because this modality may become more important clinically in the future. Digital-phase mammography has received attention because sensitivity per unit dose to tumor cells or tumor precursors such as microcalcifications is much greater than with conventional mammography (Arfelli et al., 1998; Kotre and Birch, 1999; Yu et al., 1999). Diffraction-enhanced radiography of cartilage in disarticulated as well as in intact joints is quite promising (Mollenhauer et al., 2002; Li et al., 2003; Muehleman et al., 2003, 2004), although the technical challenges of covering the FOV for human joints such as the ankle are considerable.

7.2 Metrology and Phylogeny

MicroCT is frequently used to measure internal or external dimensions and shapes both in manufactured and biological objects. The abilities to view structures in 3D and to take internal measurements along the proper 3D directions are very important.

7.2.1 Industrial Metrology

High-definition inspection of fuel injector components is being investigated in the automobile industry (Bauer et al., 2004). Stop action, high-temporal resolution microCT of evolution of fuel spray (5.1-μs temporal and 150-μm spatial resolution) has been achieved using the pulsed nature of the APS storage ring (Liu et al., 2004; Im et al., 2007). These experiments relied on the reproducibility of the spray from injection to injection.

7.2.2 Paleontology and Archeology

Tafforeau and coworkers reviewed application of synchrotron microCT for nondestructive 3D studies of paleontological specimens (Tafforeau et al., 2006). MicroCT of structures in fossil fish (Dominguez et al., 2002) and in ossicles of fossil sea urchins (Stock and Veis, 2003) has been reported. The internal structures of fossil embryos from the lower Cambrian were preserved by early stages of diagenesis, and the 3D structures of blastomeres (cells formed by the early division of the fertilized egg) revealed important information about the initial diversification of metazoan animals (Donoghue et al., 2006). Figure 7.3 shows the preserved blastomere structure in cleavage stage embryos using a combination of 3D renderings and sectioning planes. Mazurier et al. (2006) focused on fossilized trabecular bone.

High-resolution CT has found considerable application in anthropology and primatology. In primatology, a new species of great ape from ~10 Ma ago was established and, through analysis of its dentition and dentinoenamel junction (DEJ), related to the gorilla clade (Suwa et al., 2007). Enamel thickness is an important diagnostic characteristic for hominoids, and, given that small sample sizes (fragments, small number of specimens) are the rule rather than the exception, enamel and dentin volume quantification with microCT (i.e., establishing these phylogenic characters numerically) is particularly important (Gantt et al., 2006). McErlain and coworkers found that microCT could reveal dental anatomy (attrition of enamel, internal 3D caries structure) in a 500-year-old tooth (McErlain et al., 2004).

MicroCT of the cranial vault of an Oligocene anthropoid (~29–30 Ma) revealed previous estimates of the endocranial volume were too large, and this study suggested greater caution is needed in phylogenic analyses of these and related animals (Simons et al., 2007). MicroCT was to be a very useful diagnostic tool for studying a wide range of pathologies in a collection of historical skulls (Rühli et al., 2007). Lab microCT was used to study the symphyseal ontogeny (mandible) of a subfossil lemur (Ravosa

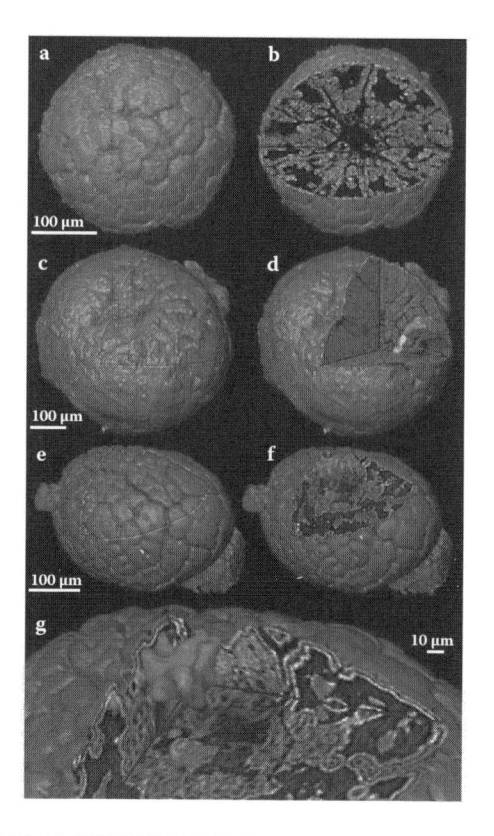

FIGURE 7.3 (SEE COLOR INSERT FOLLOWING PAGE 144.)
Fossil cleavage-stage embryos from the Lower Cambrian. (a),(b) Divisions between adjacent blastomeres are variably preserved on this embryo's surface and within. (c),(d) Divisions between nearly all blastomeres are preserved to their full extent (the light features show a column of blastomeres). (e)–(g) Divisions between blastomeres are not preserved and the light structure is a rendering of a cavity within the diagenetic infilling (Donoghue et al., 2006). (Note: The structures are much more visible in the color version.)

et al., 2007). Synchrotron microCT has been used to study, with very high spatial resolution, the dentin, enamel, and DEJ of *Homo neanderthalensis* (Macchiarelli et al., 2006) and early *Homo sapiens* (Smith et al., 2007).

7.2.3 Invertebrates and Micro-Organisms

Insects are ideally suited for imaging with x-ray phase contrast, and phase contrast microCT for this application has been recently reviewed (Betz et al., 2007). Although considerable understanding has been gained on phenomena such as insect respiration using real-time phase radiography to quantify tracheal volume changes (Westneat et al., 2003; Socha et al., 2007), microCT is needed to gain a complete understanding of this 3D process.

Recording a number of different radiographs from each projection (sampling the different respiratory stages) and using postexperiment processing to combine views at the equivalent respiration state is an approach used with animals, and one supposes this should work with insects. Structures in small insects were imaged in an SEM-based microCT system (Tanisako et al., 2006). Neurons were stained in a study of the *Drosophila* central nervous system using microCT (Mizutani et al., 2007).

Silkworms eat mulberry leaves that contain crystals of calcium oxalate and have evolved specialized organs for processing this unpalatable material. These Malpighian tubules lie outside the worm's gut, extend about one-half the length of the worm, and, at certain stages of life, are filled with calcium oxalate crystals. Figure 7.4a shows a slice of the gut "G" and a number of Malpighian tubules "M" surrounding it from a 13-day-old larva preserved in concentrated ethanol. Phase-enhanced edge contrast allows other soft tissue structure to be observed. Figure 7.4b shows a 3D rendering of high-absorption voxels (calcium oxalate crystals) within a stack of ~2K slices.

Changes in snail shells during growth were studied through in vivo microCT (Postnov et al., 2006). Synchrotron microCT and SEM were compared in another study of the formation of embryonic and juvenile snail shells (Marxen et al., 2008).

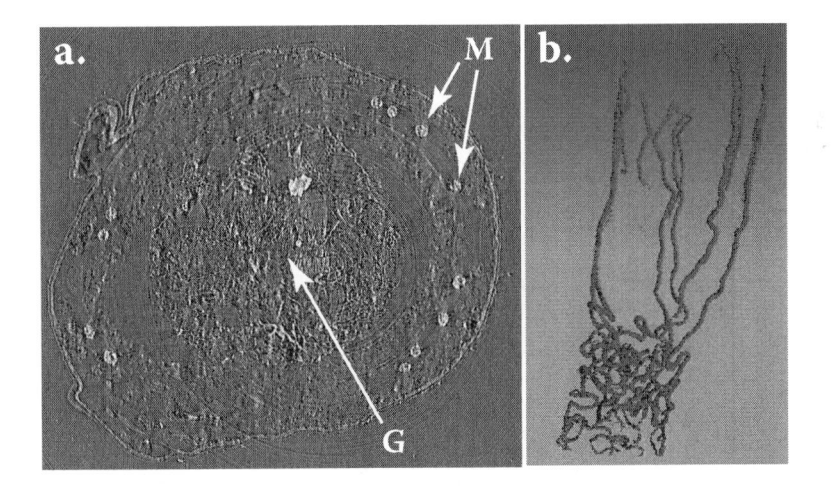

FIGURE 7.4 (SEE COLOR INSERT FOLLOWING PAGE 144.)
Silkworm slice and 3D rendering of calcium oxalate–filled Malpighian tubes. (a) Slice, 480 voxels horizontal field of view. Gut "G" and Malpighian tubules "M" filled with high-absorption calcium oxalate crystals (compared to the rest of the larva and surrounding fluid) are labeled. (b) 3D rendering showing the high-absorption voxels, that is, the Malpighian tubes. The larva's rectum is at the bottom of the ~2.5-mm-long rendering, and the wavy structure of the tubules is natural. 14.6 keV, 2K × 2K reconstruction, 1.4-µm voxels, 0.125° rotation increment (S. R. Stock, M. A. Webb, A. J. Wyman, K. I. Ignatiev, and F. De Carlo, 2-BM, APS, December 2007).

Structures of terrestrial plants and of marine plants and animals have also been measured with microCT. In grain kernels, for example, infestations and fissures could be readily detected (Dogan, 2007). Localization of iron in seeds of *Arabidopsis* was studied with fluorescence microCT and related to the presence of a specific transporter VIT-1 (Kim et al., 2006). Fluorescence microCT was also used to study metal storage mechanisms in the metal hyperaccumulator *Alyssum murale*, a plant developed as a commercial crop for phytoremediation or for phytomining Ni from metal-enriched soils (Tappero et al., 2007). In the stalk of the (marine) horsetail, microCT showed silica occurs in a thin continuous layer of the entire outer epidermis and is highly concentrated in the knob regions of the long epidermis cells (Sapei, et al., 2007).

Sea urchins form complex geometric structures, generally of single crystal calcite, and, in their hierarchy of structures, appear to employ as wide a range of strengthening strategies as has been employed in human-engineered composites. Three main types of calcite structures are produced by sea urchins: porous stereom, a structure whose topology resembles that of trabecular bone and which is discussed in Chapter 8 and also shown in Figure 5.5; a dense structure of plates and prisms constituting the teeth; and open, highly regular microarchitecture of certain spines (Figure 6.5 and the following paragraph). The literature on the microarchitecture of sea urchin mineralized tissue is quite extensive and scattered, but, aside from very brief comments relating to microCT, this interesting subject is left unexamined.

Sea urchin spines protect the animal's body from predators and high-energy, rough surf environments, and the spines of different phylogenic families have different characteristic architectures as well as a wide range of lengths and diameters. In addition to stereom, spines can contain radial wedges or a dense cortical shell. A wide range of sea urchins has adopted the strategy of employing long, hollow spines capable of rapid reorientation, and the regularity of some of these structures is remarkable. Reports of structural analyses have appeared recently (Stock et al., 2006, 2007), and numerical analysis of the structures revealed by microCT may shed light onto why certain characteristic structures have persisted over millions of years, providing insight into functional advantages conferred by certain structures.

Figure 6.5 shows the complex 3D spine structure of one type of sea urchin: the radial wedges and hollow center concentrate mass where it will be most effective in resisting bending, and the bridges between wedges and the perforated central cylinder link the wedges. This structure is a single crystal, and the exquisite biological control of mineralization geometry and crystallography at ocean temperatures is even more remarkable because the calcite's high Mg content grows at equilibrium only about several hundred °C. Several studies suggest that spatial constraint of the mineralization space controls single-crystal calcite growth in sea urchin ossicles (Park and

Meldrum, 2002; Aizenburg et al., 2003). Materials processing employing this strategy and suitable macromolecular crystallization adjuncts appears to be a very attractive route to improved cellular solids.

Sea urchin teeth differ from the urchin's other skeletal elements in that the tooth does not consist of stereom (see the section on cellular materials) or of wedges and bridges but, rather, of a dense, single crystal calcite structure suited to rasping food from the surfaces of rocks. As sea urchin teeth contain their entire developmental history, they have received considerable attention as a biomineralization model. At the tooth's aboral end, deep within the urchin, little mineral is present, but, by the adoral end (i.e., the cutting edge), mineral has filled virtually all of the tooth's volume. Within the tooth, myriad reinforcement strategies are employed to make a functional structure from calcite, an otherwise wretched structural material (Stock et al., 2004a).

Sea urchin teeth have cross-sections shaped like "U" or "T" (e.g., the slice shown in Figure 5.13 shows the leg of a "T"-shaped tooth; note that sand dollars, closely related to the regular sea urchins discussed here, have teeth with diamond cross-sections), but brevity requires the discussion be limited to the "T"-shaped or camarodont teeth. The leg of the "T" is called the keel, and the bar across the top is termed the flange. The five teeth mounted in the five pyramids (each consisting of two demipyramids; see Chapter 8) of the oral apparatus have their flanges tangent to the oral cavity and their keels running radially from the flange to the axis of the jaw structure. Each tooth functions as a T-girder, a structure providing considerable resistance to bending per unit mass, with the flange loaded in compression and the keel loaded in tension.

The microstructure of the mature flange seems well adapted to compression loading and the keel to tensile loading. The flange consists of a stack of plates paralleling the upper and lower surfaces of the bar of the "T" and close to parallel to the tooth's axis, and the interior of the flange contains thin needles running along the tooth's axis and extending into the keel, where they widen into much larger-diameter prisms. The flange plates and needles are compressed end-on during eating, a geometry allowing high resistance to wear and to catastrophic failure. The keel, loaded in tension when the flange is compressed on hard surfaces, is essentially a composite structure reinforced by fibers (prisms) aligned along the tensile loading axis.

In addition to the prisms within the center of the keel, teeth of sea urchins such as *Lytechinus variegatus* also have carinar process plates running along the flanks of the keel. This is different from the stirodont tooth structure pictured in Figure 5.13, where the carinar process plates are absent. Synchrotron microCT observations of carinar process plate orientation in *L. variegatus* suggested that these plates (on the sides of the keel) serve to prevent deflection (and fracture) of the keel along a secondary bending axis (Figure 7.5), that is, a situation that might be encountered

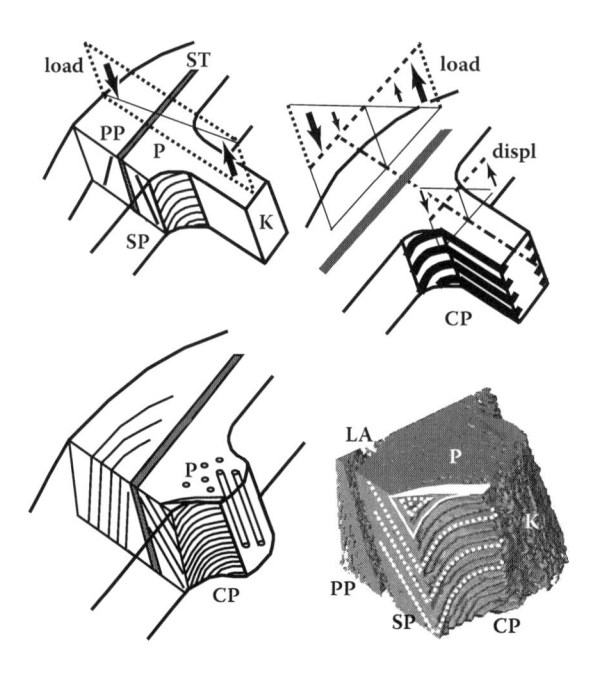

FIGURE 7.5
Function of *Lytechinus variegatus* (Echinoida, Toxopneustida) carinar process plates in resisting transverse tooth bending. In all drawings and the rendering, the cutting edge of the tooth is up, and CP denotes the carinar process plates, K the keel, P the prisms, and LA the low absorption portion of the tooth seen in slices. (Upper left) Primary bending to which the tooth is exposed (compression in the flange and tension in the keel). The stone part ST (center of the flange), the primary plates PP, and secondary plates SP are shown schematically. (Upper right) Possible displacements in the keel in response to uneven flange loading during scraping on an uneven surface, that is, transverse bending about a second axis. (Lower left) Geometry of the reinforcing elements of the structure. (Lower right) MicroCT rendering of a portion of the tooth in the same orientation as the schematics and illustrating the relative orientations and locations of plates and prisms. (Reprinted from Stock et al. (2003); © 2004, with permission from Elsevier.)

when only one side of the flange is in contact with a hard substrate (Stock et al., 2003). Different camarodonts have different arrangements of carinar process plates (Figure 7.6), although all of those studied to date have plates arranged to resist the bending hypothesized for *L. variegatus* (Stock et al., 2004a). Appreciation of the possible function of the carinar process plates occurred only after (microCT-derived) 3D datasets were available.

Soft x-ray nanoCT has been used to study the internal structure of cells (Larabell and Le Gros, 2004; Parkinson et al., 2008). The contrast between different organelles was great enough to differentiate structures. The 3D rendering reproduced in Figure 7.7 shows the nuclei in orange, for example; dimensions, volumes, and numbers of organelles were tabulated from the volumetric data (Parkinson et al., 2008).

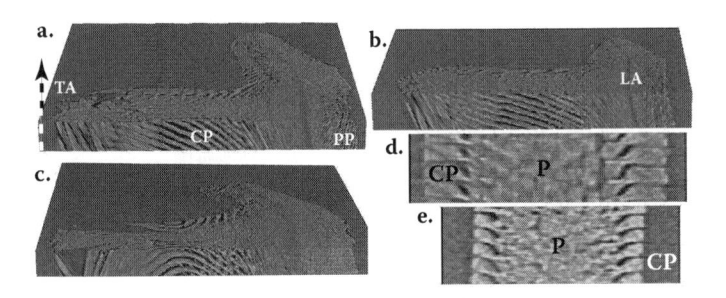

Simple Metrology and Microstructure Quantification 159

FIGURE 7.6
Carinar process plates in different camarodont teeth shown in 3D orthogonal sections and planes transverse to two keels. Each image extends 50 voxels (~85 μm) vertically (i.e., along tooth axis TA). LA, PP, CP, P are the low absorption area of the keel, primary plates, carinar process plates, and prisms, respectively. (a), (d) *Strongylocentrotus franciscanus* (Echinoida, Strongylocentrotus). (b), (e) *Hetereocentrotus trigonarius* (Echinoida, Echinometrida). (c) *Paracentrotus lividus* (Echinoida, Echinida) (Stock et al. 2004a).

7.2.4 Vertebrates

MicroCT of trabecular structures has been used as the basis for improved dosimetry estimates for different bone sites (Kramer et al., 2006, 2007). MicroCT was used as the input for quantifying skeletal chord lengths for estimating doses received by the cells in the bone marrow during nuclear medicine therapy (Shah et al., 2005). Those interested in this subject are directed to the references of these papers for others' application of microCT to dosimetry.

MicroCT of the auditory apparatus and associated blood vessels has received attention (Shibata et al., 1999; Vogel et al., 2001; Muller, 2004; Gea et al., 2005; Müller et al., 2006b). Yin and coworkers have examined guinea pig cochleae with both diffraction-enhanced microCT and histology (Yin et al., 2007). Scala vestibuli pressure and 3D stapes velocities were analyzed for the gerbil using microCT-generated 3D solid structures (Deacraemer et al., 2007). MicroCT scanning is also being investigated as a tool for evaluating the surgical positioning of cochlear implant electrodes and the damage produced by the electrode placement (Postnov et al., 2006); note that samples were cut from the surrounding temporal bone and were not imaged in human patients in vivo.

Comparison of trabecular structure across different mouse genetic strains (Martín-Badosa et al., 2003) and of femoral heads of two different primates with very different sizes and loading environments (MacLatchy and Müller, 2002) were other studies with a metrology component. Skeletal microCT atlases of different mouse strains also have appeared (Jacob and Chole, 2006; Perlyn et al., 2006; Chan et al., 2007; Olafsdottir et al., 2007). Comparison of genetic strain-average structure with aligned microCT

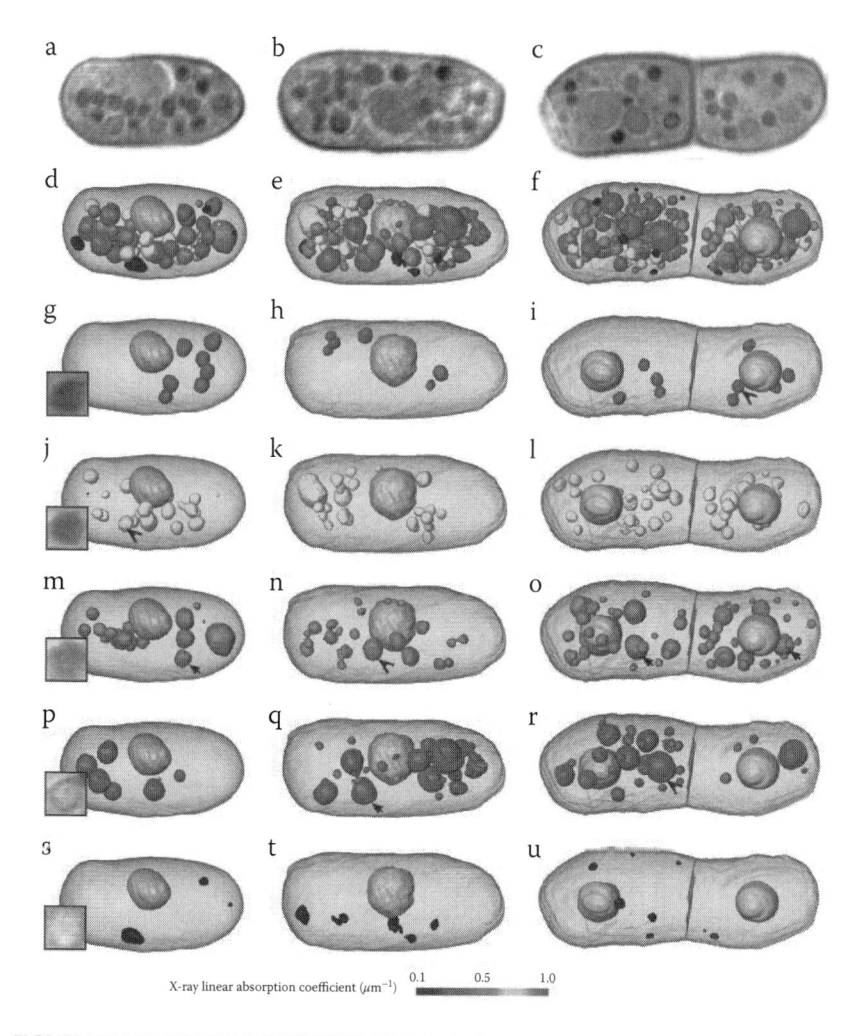

160   *MicroComputed Tomography: Methodology and Applications*

FIGURE 7.7 (SEE COLOR INSERT FOLLOWING PAGE 144.)
Cells of *Schizosaccharomyces pombe* imaged by soft x-ray tomography. (a)–(c) One-voxel-thick (20 nm) slices through the reconstructed volumes of four cells, with the darker gray indicating higher absorption. (d)–(f) Shaded surfaces representing the boundaries of organelles and the plasma membrane. The shades correspond to the average linear attenuation coefficients inside the organelles (see the gray scale bar at the bottom of the figure). Note that the nuclei should actually be colored blue but are shown as orange to differentiate them from the other structures. The length of the cell in (a) is 5 µm. (g)–(u) The same surfaces as in (d)–(f) but with each shade isolated from the others. At the left side of each row there is a grayscale slice image of one of the organelles shown in that row; the arrows point to mitochondria and the wedges to other organelles (Parkinson et al., 2008).

and MRI 3D datasets illustrate how the atlases can be used to correlate features of the central nervous system with those of the surrounding bones (Chan et al., 2007).

Studies of fish skeletons as biomineralization models have appeared (Neues et al., 2006, 2007), the latter study being of particular interest because a normal wild type and mutant are compared. MicroCT has been used in a study of the sonic organ of the singing midshipman fish, although MRI was the main imaging modality (Forbes et al., 2006); it would be interesting to study this organ in vivo using x-ray phase imaging.

Pneumatization (i.e., extension of pulmonary air sacs into marrow cavities) is an important adaptation of birds. MicroCT was used to study cortical and trabecular bone in pneumatic and apneumatic thoracic vertebrae (two species of ducks) and found significant differences in microarchitecture (Fajardo et al., 2007).

Nielsen et al. (1997) provided a three-dimensional rendering of the root canal in an endodontically prepared human maxillary molar, and the tissue loss (i.e., increase in volume of the root canal space) due to the endodontic procedure was clear even in the 127 µm × 127 µm voxels of the reconstructions. Dowker et al. (1997) obtained similar images of the human upper third molar, upper central incisor, and upper lateral incisor with and without files inserted in the root canal; these data consisted of isotropic ~39-µm voxels and showed considerable detail such as unfilled space in the canal sealant and the presence of dentin debris. Dowker et al. also indicated that the three-dimensional datasets will be used to illustrate the variety and complexity of root canals during computer-assisted dental training. Another library of microCT images for teaching 3D dental structures also has been reported (Seifert et al., 2004). It is interesting to contemplate whether x-ray microCT of engineering materials can provide analogous computer-assisted learning opportunities for undergraduate and graduate engineering or science students.

Future microCT applications are expected to include imaging accessory canals and investigating fluid transport into root-filled teeth; one anticipates improvements in clinical procedures will soon be a direct consequence. Peters et al. (2000) used lab microCT to analyze root canal geometry by adapting tools developed for histomorphometry of trabecular bone. Macromorphology of human tooth roots (Plotino et al., 2006) and distance transform illustration of local diameter of the tooth pulp (Dougherty and Kunzelmann, 2007) are also of interest. Figure 7.8 shows images from the latter study: 3D tri-level segmentation of enamel, dentin, and pulp (left) and 3D local pulp widths determined by the distance transform method.

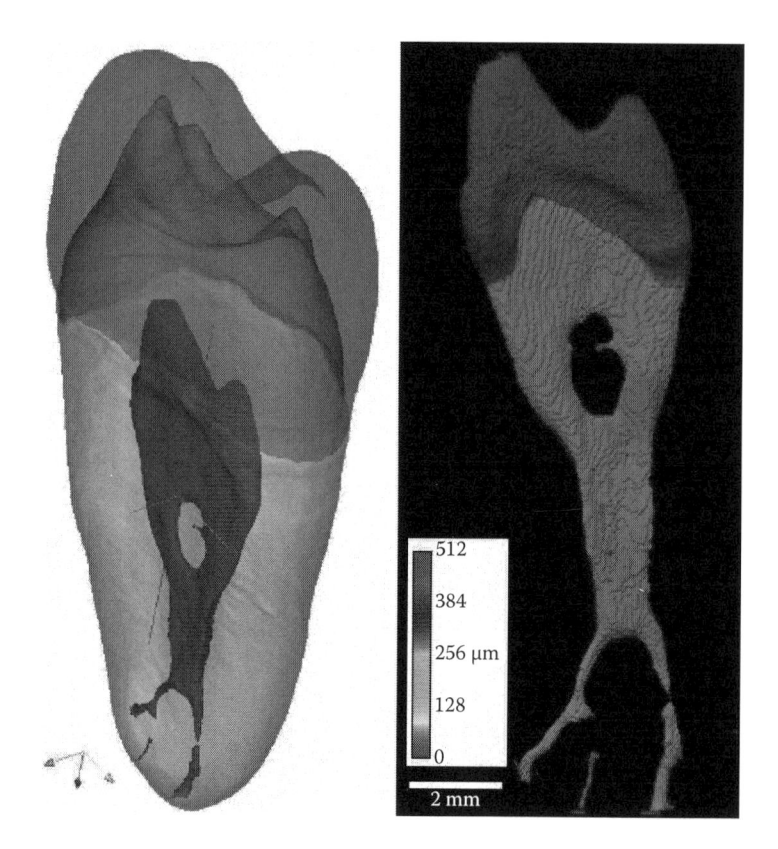

FIGURE 7.8 (SEE COLOR INSERT FOLLOWING PAGE 144.)
(Left) 3D rendering of a tooth segmented into enamel (medium gray), dentin (light gray), and pulp (dark gray). (Right) Local thicknesses (gray scale) of the pulp of the tooth in (a) determined with the distance transform method. 324 × 254 × 669 voxels. (Dougherty and Kunzelmann (2007); © 2007 Microscopy Society of America. Reprinted courtesy of the Microscopy Society and Cambridge University Press.)

References

Aizenburg, J., D.A. Muller, J.L. Grazul, and D.R. Hartmann (2003). Direct fabrication of large micropatterned single crystals. *Science* **299**: 1205–1208.

Anderson, P., Y. Ahmed, M.P. Patel, G.R. Davis, and M. Braden (2006). X-ray microtomographic studies of novel radio-opaque polymeric materials for dental applications. *Mater Sci Technol* **22**: 1094–1097.

Arfelli, F., M. Assante, V. Bonvicini, A. Bravin, G. Canatore, E. Castelli, L.D. Palma, M.D. Michiel, R. Longo, A. Olivo, S. Pani, D. Pontoni, P. Poropat, M. Prest, A. Rashevsky, G. Tromba, A. Vacchi, E. Vallazza, and F. Zanconati (1998). Low dose phase contrast x-ray medical imaging. *Phys Med Biol* **43**: 2845–2852.

Baaklini, G.Y., R.T. Bhatt, A.J. Eckel, P. Engler, M.G. Castelli, and R.W. Rauser (1995). X-ray microtomography of ceramic and metal matrix composites. *Mater Eval* **53**: 1040–1044.

Bauer, W., F.T. Bessler, E. Zabler, and R.B. Bergmann (2004). Computer tomography for nondestructive testing in the automotive industry. *Developments in X-ray Tomography IV*. U. Bonse (Ed.). Bellingham, WA, SPIE. *SPIE Proc Vol* **5535**: 464–472.

Betz, O., U. Wegst, D. Weide, M. Heethoff, L. Helfen, W.K. Lee, and P. Cloetens (2007). Imaging applications of synchrotron phase-contrast microtomography in biological morphology and biomaterials science. I. General aspects of the technique and its advantages in the analysis of millimetre-sized arthropod structure. *J Microsc* **227**: 51–71.

Bonse, U., R. Nusshardt, F. Busch, R. Pahl, J.H. Kinney, Q.C. Johnson, R.A. Saroyan, and M.C. Nichols (1991). X-ray tomographic microscopy of fibre-reinforced materials *J Mater Sci* **26**: 4076–4085.

Borbely, A., H. Biermann, O. Hartmann, and J. Buffiere (2003). The influence of the free surface on the fracture of alumina particles in an Al-Al2O3 metal-matrix composite. *Compu Mater Sci* **26**: 183–188.

Borbély, A., P. Kenesei, and H. Biermann (2006). Estimation of the effective properties of particle-reinforced metal–matrix composites from microtomographic reconstructions. *Acta Mater* **54**: 2735–2744.

Brunke, O., S. Odenbach, C. Fritsche, I. Hilger, and W.A. Kaiser (2005). Determination of magnetic particle distribution in biomedical applications by x-ray microtomography. *J Mag Mag Mater* **289**: 428–430.

Busignies, V., B. Leclerc, P. Porion, P. Evesque, G. Couarraze, and P. Tchoreloff (2006). Quantitative measurements of localized density variations in cylindrical tablets using x-ray microtomography. *Euro J Pharm Biopharm* **64**: 38–50.

Chan, E., N. Kovacevic, S.K.Y. Ho, R.M. Henkelman, and J.T. Henderson (2007). Development of a high resolution three-dimensional surgical atlas of the murine head for strains 129S1/SvImJ and C57Bl/6J using magnetic resonance imaging and micro-computed tomography. *Neurosci* **144**: 604–615.

Coindreau, O., G. Vignoles, and P. Cloetens (2003). Direct 3D microscale imaging of carbon-carbon composites with computed holotomography. *Nucl Instrum Meth B* **200**: 308–314.

David, C., P. Robion, and B. Menéndez (2007). Anisotropy of elastic, magnetic and microstructural properties of the Callovo-Oxfordian argillite. *Phys Chem Earth* **32**: 145–153.

De Santis, R., M. Catauro, L. Di Silvio, L. Manto, M.G. Raucci, L. Ambrosio, and L. Nicolais (2007). Effects of polymer amount and processing conditions on the in vitro behaviour of hybrid titanium dioxide/polycaprolactone composites. *Biomater* **28**: 2801–2809.

Deacraemer, W.F., O. de La Rochefoucauldi, W. Dong, S.M. Khanna, J.J.J. Dirckx, and E.S. Olson (2007). Scala vestibuli pressure and three-dimensional stapes velocity measured in direct succession in gerbil. *J Acoust Soc Am* **121**: 2774–2791.

Diamond, S. and E. Landis (2007). Microstructural features of a mortar as seen by computed microtomography. *Mater Struct* **40**: 989–993.

Dogan, H. (2007). Nondestructive imaging of agricultural products using x-ray microtomography. *Microsc Microanal* **13**(Suppl 2): 1316CD.

Dominguez, P., A.G. Jacobson, and R.P.S. Jefferies (2002). Paired gill slits in a fossil with a calcite skeleton. *Nature* **417**: 841–844.

Donoghue, P.C.J., S. Bengtson, X. Dong, N.J. Gostling, T. Huldtgren, J.A. Cunningham, C. Yin, F.P.Z. Yue, and M. Stampanoni (2006). Synchrotron X-ray tomographic microscopy of fossil embryos. *Nature* **442**: 680–683.

Dougherty, R.P. and K.H. Kunzelmann (2007). Computing local thickness of 3D structures with ImageJ. *Microsc Microanal* **13**(Suppl 2): 1678CD.

Dowker, S.E.P., G.R. Davis, and J.C. Elliott (1997). X-ray microtomography: non-destructive three-dimensional imaging for in vitro endodontic studies. *Oral Surg Oral Med Oral Pathol Oral Radiol Endod* **83**: 510–516.

Everett, R.K., K.E. Simmonds, and A.B. Geltmacher (2001). Spatial distribution of voids in HY-100 steel by x-ray tomography. *Scripta Mater* **44**: 165–169.

Fajardo, R.J., E. Hernandez, and P.M. O'Connor (2007). Post cranial skeletal pneumaticity: A case study in the use of quantitative microCT to assess vertebral structure in birds. *J Anat* **211**: 138–147.

Farber, L., G. Tardos, and J.N. Michaels (2003). Use of x-ray tomography to study the porosity and morphology of granules. *Powder Technol* **132**: 57–63.

Forbes, J.G., H.D. Morris, and K. Wang (2006). Multimodal imaging of the sonic organ of *Porichthys notatus*, the singing midshipman fish. *Mag Res Imaging* **24**: 321–331.

Fu, X., M. Dutt, A.C. Bentham, B.C. Hancock, R.E. Cameron, and J.A. Elliott (2006). Investigation of particle packing in model pharmaceutical powders using x-ray microtomography and discrete element method. *Powder Technol* **167**: 134–140.

Gantt, D.G., J. Kappleman, R.A. Ketcham, M.E. Alder, and T.H. Deahl (2006). Three-dimensional reconstruction of enamel thickness and volume in humans and hominoids. *Eur J Oral Sci* **114**(suppl 1): 360–364.

Gea, S.L.R., W.F. Decraemer, and J.J.J. Dirckx (2005). Region of interest micro-CT of the middle ear: A practical approach. *J X-Ray Sci Technol* **13**: 137–147.

Geandier, G., A. Hazotte, S. Denis, A. Mocellin, and E. Maire (2003). Microstructural analysis of alumina chromium composites by x-ray tomography and 3-D finite element simulation of thermal stresses. *Scripta Mater* **48**: 1219–1224.

Geandier, G., P. Weisbecker, S. Denis, A. Hazotte, A. Mocellin, J.L. Lebrun, and E. Elkaim (2002). X-ray diffraction analysis of residual stresses in alumina-chromium composites and comparison with numerical simulations. *Mater Sci Forum* **404–407**: 547–552.

Gualda, G.A.R. and M. Rivers (2006). Quantitative 3D petrography using x-ray tomography: Application to Bishop Tuff pumice clasts. *J Volcanol Geothermal Res* **154**: 46–62.

Gualda, G.A.R., D.L. Cook, R. Chopra, L. Qin, A.T.A. Jr., and M. Rivers (2004). Fragmentation, nucleation and migration of crystals and bubbles in the bishop tuff rhyolitic magma. *Trans Roy Soc Edin Earth Sci* **95**: 375–390.

Haeffner, D.R., J.D. Almer, and U. Lienert (2005). The use of high energy x-rays from the Advanced Photon Source to study stresses in materials. *Mater Sci Eng A* **399**: 120–127.

Hancock, B.C. and M.P. Mullarney (2005). X-ray microtomography of solid dosage forms. *Pharm Technol* **29**: 92–100.

Heggli, M., T. Etter, P. Wyss, P.J. Uggowitzer, and A.A. Gusev (2005). Approaching representative volume element size in interpenetrating phase composites. *Adv Eng Mater* **7**: 225–229.

Hirano, T., K. Usami, and K. Sakamoto (1989). High resolution monochromatic tomography with x-ray sensing pickup tube. *Rev Sci Instrum* **60**: 2482–2485.

Im, K.S., K. Fezzaa, Y.J. Wang, X. Liu, J. Wang, and M.C. Lai (2007). Particle tracking velocimetry using fast x-ray phase-contrast imaging. *Appl Phys Lett* **90**: 091919.

Jacob, A. and R.A. Chole (2006). Survey anatomy of the paranasal sinuses in the normal mouse. *Laryngoscope* **116**: 558–563.

Jones, K.W., H. Feng, A. Lanzirotti, and D. Mahajan (2005). Synchrotron x-ray microprobe and computed microtomography for characterization of nanocatalysts. *Nucl Instrum Meth B* **241**: 331–334.

Kenesei, P., H. Biermann, and A. Borbely (2006a). Estimation of elastic properties of particle reinforced metal-matrix composites based on tomographic images. *Adv Eng Mater* **8**: 500–506.

Kenesei, P., A. Klohn, H. Biermann, and A. Borbely (2006b). Mean field and multiscale modeling of a particle reinforced metal-matrix composite based on microtomographic investigations. *Adv Eng Mater* **8**: 506–510.

Kim, S.A., T. Punshon, A. Lanzirotti, L. Li, J.M. Alonso, J.R. Ecker, J. Kaplan, and M.L. Guerinot (2006). Localization of iron in Arabidopsis seed requires the vacuolar membrane transporter VIT1. *Science* **314**: 1295–1298.

Kinney, J.H., S.R. Stock, M.C. Nichols, U. Bonse, T.M. Breunig, R.A. Saroyan, R. Nusshardt, Q.C. Johnson, F. Busch, and S.D. Antolovich (1990). Nondestructive investigation of damage in composites using x-ray tomographic microscopy. *J Mater Res* **5**: 1123–1129.

Kotre, C.J. and I.P. Birch (1999). Phase contrast enhancement of x-ray mammography: A design study. *Phys Med Biol* **44**: 2853–2866.

Kramer, R., H.J. Khoury, J.W. Vieira, and I. Kawrakow (2006). Skeletal dosimetry in the MAX06 and the FAX06 phantoms for external exposure to photons based on vertebral 3D-microCT images. *Phys Med Biol* **51**: 6265-6289.

Kramer, R., H.J. Khoury, J.W. Vieira, and I. Kawrakow (2007). Skeletal dosimetry for external exposure to photons based on μCT images of spongiosa from different bone sites. *Phys Med Biol* **52**: 6697–6716.

Larabell, C.A. and M.A. Le Gros (2004). X-ray tomography generates 3-D reconstructions of the yeast, *Saccharomyces cerevisiae*, at 60-nm resolution. *Mol Biol Cell* **15**: 957–962.

Li, J., Z. Zhong, R. Lidtke, K.E. Kuettner, C. Peterfy, E. Aliyeva, and C. Muehleman (2003). Radiography of soft tissue of the foot and ankle with diffraction enhanced imaging. *J Anat* **202**: 463–470.

Liu, X., J. Liu, X. Li, S.K. Cheong, D. Shu, J. Wang, M.W. Tate, A. Ercan, D.R. Schuette, M.J. Renzi, A. Woll, and S.M. Gruner (2004). Development of ultrafast computed tomography of highly transient fuel sprays. *Developments in X-Ray Tomography IV*. U. Bonse (Ed.). Bellingham, WA, SPIE. *SPIE Proc Vol* **5535**: 21–28.

Macchiarelli, R., L. Bondioli, A. Debenath, A. Mazurier, J.F. Tournepiche, W. Birch, and C. Dean (2006). How Neaderthal molar teeth grew. *Nature* **444**: 748–751.

MacLatchy, L. and R. Müller (2002). A comparison of the femoral head and neck trabecular architecture of Galago and Perodicticus using microcomputed tomography. *J Human Evol* **43**: 89–105.

Martín-Badosa, E., D. Amblard, S. Nuzzo, A. Elmoutaouakkilo, L. Vico, and F. Peyrin (2003). Excised bone structures in mice: Imaging at three-dimensional synchrotron microCT. *Radiol* **229**: 921–928.

Martín-Herrero, J. and C. Germain (2007). Microstructure reconstruction of fibrous C/C composites from X-ray microtomography. *Carbon* **45**: 1242–1253.

Marxen, J.C., O. Prymak, F. Beckmann, F. Neues, and M. Epple (2008). Embryonic shell formation in the snail *Biomphalaria glabrata*: A comparison between scanning electron microscopy (SEM) and synchrotron radaition micro computer tomography. *J Mollus Stud* **74**: 19–25.

Mazurier, A., V. Volpato, and R. Macchiarelli (2006). Improved noninvasive microstructural analysis of fossil tissues by means of SR-microtomography. *Appl Phys A* **83**: 229–233.

McErlain, D.D., R.K. Chhem, R.N. Bohay, and D.W. Holdsworth (2004). Microcomputed tomography of a 500-year-old tooth: Technical note. *Can Assoc Radiol J* **55**: 242–245.

Mizutani, R., A. Takeuchi, T. Hara, K. Uesugi, and Y. Suzuki (2007). Computed tomography imaging of the neuronal structure of *Drosophila* brain. *J Synchrotron Rad* **14**: 282–287.

Mollenhauer, J., M. Aurich, Z. Zhong, C. Muehleman, A.A. Cole, M. Hasnah, O. Oltulu, K.E. Kuettner, A. Margulis, and L.D. Chapman (2002). Diffraction enhanced x-ray imaging of articular cartilage. *Osteoarthritis Cartilage* **10**: 163–171.

Momose, A. and K. Hirano (1999). The possibility of phase contrast X-ray microtomography. *Jpn J Appl Phys* **38** Suppl 1: 625–629.

Momose, A., T. Takeda, Y. Itai, J. Tu, and K. Hirano (1999). Recent observations with phase-contrast computed tomography. *Developments in X-ray Tomography II*. U. Bonse (Ed.). Bellingham, WA, SPIE. *SPIE Proc Vol* **3772**: 188–195.

Momose, A., T. Takeda, Y. Itai, A. Yoneyama, and K. Hirano (1998a). Perspective for medical applications of phase contrast x-ray imaging. *Medical Applications of Synchrotron Radiation*. C.U.M. Ando (Ed.). Tokyo, Springer: 54–61.

Momose, A., T. Takeda, Y. Itai, A. Yoneyama, and K. Hirano (1998b). Phase contrast tomographic imaging using an x-ray interferometer. *J Synchrotron Rad* **5**: 309–314.

Muehleman, C., L.D. Chapman, K.E. Kuettner, J. Rieff, J.A. Mollenhauer, K. Massuda, and Z. Zhong (2003). Radiography of rabbit articular cartilage with diffraction enhanced imaging. *Anat Rec* **272A**: 392–397.

Muehleman, C., S. Majumdar, A.S. Issever, F. Arfelli, R.H. Menk, L. Rigon, G. Heitner, B. Reime, J. Metge, A. Wagner, K.E. Kuettner, and J. Mollenhauer (2004). X-ray detection of structural orientation in human articular cartilage. *Osteoarthritis Cartilage* **12**: 97–105.

Müller, B., M. Germann, D. Jeanmonod, and A. Morel (2006a). Three-dimensional assessment of brain tissue morphology. *Developments in X-Ray Tomography V*. U. Bonse (Ed.). Bellingham, WA, SPIE. *SPIE Proc Vol* **6318**: 631803–1 to –8.

Müller, B., A. Lareida, F. Beckmann, G.M. Diakov, F. Kral, F. Schwarm, R. Stoffner, A.R. Gunkel, R. Glueckert, A. Schrott-Fischer, J. Fischer, A. Andronache, and W. Freysinger (2006b). Anatomy of the murine and human cochlea visualized at the cellular level by synchrotron radiation based microcomputed tomography. *Developments in X-Ray Tomography V*. U. Bonse (Ed.). Bellingham, WA, SPIE. *SPIE Proc Vol* **6318**: 631805–1 to –9.

Muller, R. (2004). A numerical study of the role of the tragus in the big brown bat. *J Acous Soc Am* **116**: 3701–3712.
Mummery, P.M., B. Derby, P. Anderson, G.R. Davis, and J.C. Elliott (1995). X-ray microtomographic studies of metal matrix composites using laboratory x-ray sources. *J Microsc* **177**: 399–406.
Neues, F., W.H. Arnold, J. Fischer, F. Beckmann, P. Gaengler, and M. Epple (2006). The skeleton and pharyngeal teeth of zebrafish (*Danio rerio*) as a model of biomineralization in vertebrates. *Mat-wiss Werkstofftech* **37**: 426–431.
Neues, F., R. Goerlich, J. Renn, F. Beckmann, and M. Epple (2007). Skeletal deformations in medaka (*Oryzias latipes*) visualized by synchrotron radiation micro-computer tomography (SRµCT). *J Struct Biol* **160**: 236–240.
Nielsen, R. B., A.M. Alyassih, D.D. Peters, D.L. Carnes, and J. Lancester (1997). Microcomputed tomography: An advanced system for detailed endodontic research. Oral *Surg Oral Med Oral Pathol Oral Radiol Endod* **83**: 510–516.
Olafsdottir, H., T.A. Darvann, N.V. Hermann, E. Oubel, B.K. Ersboll, A.F. Frangi, P. Larsen, C.A. Perlyn, G.M. Morriss-Kay, and S. Kreiborg (2007). Computational mouse atlases and their application to automatic assessment of craniofacial dysmorphology caused by the Crouzon mutation $Fgfr2^{C342Y}$. *J Anat* **211**: 37–52.
Park, R.J. and F.C. Meldrum (2002). Synthesis of single crystals of calcite with complex morphologies. *Adv Mater* **14**: 1167–1169.
Parkinson, D.Y., G. McDermott, L.D. Etkin, M.A. Le Gros, and C.A. Larabell (2008). Quantitative 3-D imaging of eukaryotic cells using soft X-ray tomography. *J Struct Biol*: **162**: 380–386.
Perlyn, C.A., V.B. DeLeon, C. Babbs, D. Glover, L. Burell, T. Darvann, S. Kreiborg, and G. Morriss-Kay (2006). The cranialfacial phenotype of the Crouzon mouse: Analysis of a model for syndromic craniosynostosis using three-dimensional microCT. *Cleft Palate–Craniofac J* **43**: 740–748.
Peters, O.A., A. Laib, P. Rüegsegger, and F. Barbakow (2000). Three-dimensional analysis of root canal geometry by high resolution computed tomography. *J Dent Res* **79**: 1405–1409.
Plotino, G., N.M. Grande, R. Pecci, R. Bedini, C.H. Pameijer, and F. Somma (2006). Three-dimensional imaging using microcomputed tomography for studying tooth macromorphology. *J Am Dent Assoc* **137**: 1555–1561.
Postnov, A., A. Zarowski, N. de Clerck, F. Vanpoucke, F.E. Offeciers, D. Van Dyck, and S. Peeters (2006). High resolution microCT scanning as an innovatory tool for evaluation of the surgical positioning of cochlear implant electrodes. *Acta Oto-Laryngologica* **126**: 467–474.
Pyun, A., J.R. Bell, K.H. Won, B.M. Weon, S.K. Seol, J.H. Je, and C.W. Macosko (2007). Synchrotron X-ray microtomography for 3D imaging of polymer blends. *Macromol* **40**: 2029–2035.
Qidwai, M.A. and V.G. DeGiorgi (2004). Numerical assessment of the dynamic behavior of hybrid shape memory alloy composite. *Smart Mater Struct* **13**: 134–145.
Rajamannan, N.M., T.B. Nealis, M. Subramaniam, S.R. Stock, K.I. Ignatiev, T.J. Sebo, J.W. Fredericksen, S.W. Carmichael, T.K. Rosengart, T.C. Orszulak, W.D. Edwards, R.O. Bonow, and T.C. Spelsberg (2005a). Calcified rheumatic valve neoangiogenesis is associated with VEGF expression and osteoblast-like bone formation. *Circulation* **111**: 3296–3301.

Rajamannan, N.M., M. Subramanium, S. Stock, F. Caira, and T.C. Spelsberg (2005b). Atorvastatin inhibits hypercholesterolemia-induced calcification in the aortic valves via the Lrp5 receptor pathway. *Circulation* **112** (suppl I): I–229 to I–234.

Ravosa, M.J., S.R. Stock, E.L. Simons, and R. Kunwar (2007). MicroCT analysis of symphyseal ontogeny in a subfossil lemur (Archaeolemur). *Int J Primatol* **28**: 1385–1396.

Rühli, F.J., G. Kuhn, R. Evison, R. Müller, and M. Schultz (2007). Diagnostic value of micro-CT in comparison with histology in the qualitative assessment of historical human skull bone pathologies. *Am J Phys Anthropol* **133**: 1099–1111.

Sanchez, S.A., J. Narciso, F. Rodriguez-Reinoso, D. Bernard, I.G. Watson, P.D. Lee, and R.J. Dashwood (2006). Characterization of lightweight graphite based composites using x-ray microtomography. *Adv Eng Mater* **8**: 491–495.

Sapei, L., N. Gierlinger, J. Hartmann, R. Nöske, P. Strauch, and O. Paris (2007). Structural and analytical studies of silica accumulations in *Equisetum hyemale*. *Anal Bioanal Chem* **389**: 1249–1257.

Schena, G., S. Favretto, L. Santoro, A. Pasini, M. Bettuzzi, F. Casali, and L. Mancini (2005). Detecting microdiamonds in kimberlite drill-hole cores by computed tomography. *Int J Miner Process* **75**: 173–188.

Seifert, A., M.J. Flynn, K. Montgomery, and P. Brown (2004). Visualization of x-ray microtomography data for a human tooth atlas. *Medical Imaging 2004: Visualization, Image-Guided Procedures, and Display, 2004.* R. Galloway, Jr. (Ed.). Bellingham, WA, SPIE. *SPIE Proc Vol* **5367**: 747–757.

Shah, A.P., D.A. Rajon, D.W. Jokisch, P.W. Patton, and W.E. Bolch (2005). A comparison of skeletal chord length distributions in the adult male. *Health Phys* **89**: 199–215.

Shibata, T., S. Matsumoto, and T. Nagano (1999). Tomograms of the arterial system of the human fetal auditory apparatus obtained by very high resolution microfocus x-ray CT and 3D reconstruction. *Acta Anat Nippon* **74**: 545–553.

Shimizu, K., J. Ikezoe, H. Ikura, H. Ebara, T. Nagareda, N. Yagi, K. Umetani, K. Uesugi, K. Okada, A. Sugita, and M. Tanaka (2000). Synchrotron radiation microtomography of the lung specimens. *Medical Imaging 2000: Physics of Medical Imaging.* J. T. Dobbins III and J.M. Boone (Eds.). Bellingham, WA, SPIE. *SPIE Proc Vol* **3977**: 196–204.

Simons, E.L., E.R. Seiffert, T.M. Ryan, and Y. Attia (2007). A remarkable female cranium of the early Oligocene anthropoid *Aegyptopithecus zeuxis* (Catarrhini, Propliopithecidae). *PNAS* **104**: 8731–8736.

Smith, T.M., P. Tafforeau, D.J. Reid, R. Grun, S. Eggins, M. Boutakiout, and J.J. Hublin (2007). Earliest evidence of modern human life history in North African early *Homo sapiens*. *PNAS* **104**: 6128–6133.

Socha, J.J., M.W. Westneat, J.F. Harrison, J.S. Waters, and W.K. Lee (2007). Real-time phase-contrast x-ray imaging: a new technique for the study of animal form and function. *BMC Biol* **5**: 6.

Song, S.R., K.W. Jones, W.B. Lindquist, B.A. Dowd, and D.L. Sahagian (2001). Synchrotron x-ray computed tomography: Studies on vesiculated rocks. *Bull Volcanol* **63**: 252–263.

Stock, S.R. and A. Veis (2003). Preliminary microfocus x-ray computed tomography survey of echinoid fossil microstructure. *Applications of X-ray Computed Tomography in the Goesciences*. F. Mees, R. Swennen, M. van Geet, and P. Jacobs (Eds.). London, Geological Society of London. Special Publication **215**: 225–235.

Stock, S.R., F. De Carlo, X. Xiao, and T.A. Ebert (2007). Bridges between radial wedges (septs) in two diadematid spine types. *Echinoderms: Durham*: in press.

Stock, S.R., T.A. Ebert, K. Ignatiev, and F.D. Carlo (2006). Structures, structural hierarchy and function in sea urchin spines. *Developments in X-ray Tomography V*. U. Bonse (Ed.). Bellingham, WA, SPIE. *SPIE Proc Vol* **6318**: 63180A–1 to –4.

Stock, S.R., A. Guvenilir, T.L. Starr, J.C. Elliott, P. Anderson, S.D. Dover, and D.K. Bowen (1989). Microtomography of silicon nitride / silicon carbide composites. *Ceram Trans* **5**: 161–170.

Stock, S.R., K.I. Ignatiev, T. Dahl, A. Veis, and F.D. Carlo (2003). Three-dimensional microarchitecture of the plates (primary, secondary and carinar process) in the developing tooth of *Lytechinus variegatus* revealed by synchrotron x-ray absorption microtomography (microCT). *J Struct Biol* **144**: 282–300.

Stock, S.R., K. Ignatiev, F. De Carlo, J.D. Almer, and A. Veis (2004a). Multiple mode x-ray study of 3-D tooth microstructure across different sea urchin families. *Proceedings 8th International Conference on Chemistry and Biology of Mineralized Tissue*. W.J. Landis and J. Sodek (Eds.). Toronto, University of Toronto Press: 107–110.

Stock, S.R., K. Ignatiev, P.L. Lee, K. Abbott, and L.M. Pachman (2004b). Pathological calcification in Juvenile Dermatomyositis (JDM): MicroCT and synchrotron x-ray diffraction reveal hydroxyapatite with varied microstructures. *Conn Tiss Res* **45**: 248–256.

Stock, S. R., N.M. Rajamannan, E.R. Brooks, C.B. Langman, and L.M. Pachman (2004c). Pathological calcifications studies with microCT. *Developments in X-ray Tomography IV*. U. Bonse (Ed.). Bellingham, WA, SPIE. *SPIE Proc Vol* **5535**: 424–431.

Suwa, G., R.T. Kono, S. Katoh, B. Asfaw, and Y. Beyene (2007). A new species of great ape from the late Miocene epoch in Ethiopia. *Nature* **448**: 921–924.

Tafforeau, P., R. Boistel, E. Boller, A. Bravin, M. Brunet, Y. Chaimanee, P. Cloetens, M. Feist, J. Hoszowska, J.J. Jaeger, R.F. Kay, V. Lazzari, L. Marivaux, A. Nel, C. Nemoz, X. Thibault, P. Vignaud, S. Zabler, H. Riesemeier, P. Fratzl, and P. Zaslansky (2006). Applications of X-ray synchrotron microtomography for non-destructive 3D studies of paleontological specimens. *Appl Phys A* **83**: 195–202.

Takeda, T., A. Momose, E. Ueno, and Y. Itai (1998). Phase contrast x-ray CT image of breast tumor. *J Synchrotron Rad* **5**: 1133–1135.

Tanisako, A., S. Tsuruta, A. Hori, A. Okumura, C. Miyata, C. Kuzuryu, T. Obi, and H. Yoshimura (2006). MicroCT of small insects by projection x-ray microscopy. *Developments in X-Ray Tomography V*. U. Bonse (Ed.). Bellingham, WA, SPIE. *SPIE Proc Vol* **6318**: 63180B–1 to –8.

Tappero, R., E. Peltier, M. Gräfe, K. Heidel, M. Ginder-Vogel, K.J.T. Livi, M.L. Rivers, M.A. Marcus, R.L. Chaney, and D.L. Sparks (2007). Hyperaccumulator *Alyssum murale* relies on a different metal storage mechanism for cobalt than for nickel. *New Phyt* **175**: 641–654.

Toda, H., K. Uesugi, A. Takeuchi, K. Minami, M. Kobayashi, and T. Kobayashi (2006). Three-dimensional observation of nanoscopic precipitates in an aluminum alloy by microtomography with Fresnel zone plate optics. *Appl Phys Lett* **89**: 143112.

Traini, D., G. Loreti, A.S. Jones, and P.M. Young (2008). X-ray computed microtomography for the study of modified release systems. *Microsc Anal* **22**: 13–15.

Vogel, U., F. Beckmann, T. Zahnert, and U. Bonse (2001). Microtomography of the human middle and inner ear. *Developments in X-ray Tomography III*. U. Bonse (Ed.). Bellingham, WA, SPIE. *SPIE Proc Vol* **4503**: 146–155.

Wang, Y., H.I. Chang, D.F. Wertheim, A.S. Jones, C. Jackson, and A.G.A. Coombes (2007). Characterisation of the macroporosity of polycaprolactone-based biocomposites and release kinetics for drug delivery. *Biomater* **28**: 4619–4627.

Weber, S.M., K.A. Peterson, B. Durkee, C. Qi, M. Longino, T. Warner, J.F.T Lee, and J.P. Weichert (2004). Imaging of murine liver tumor using microCT with a hepatocyte-selective contrast agent: Accuracy is dependent on adequate contrast enhancement. *J Surg Res* **119**: 41–45.

Westneat, M.W., O. Betz, R.W. Blob, K. Fezzaa, W.J. Cooper, and W.K. Lee (2003). Tracheal respiration in insects visualized with synchrotron x-ray imaging. *Science* **299**: 558–560.

Williams Jr., J.C., B.R. Matlaga, S.C. Kim, M.E. Jackson, A.J. Sommer, J.A. McAteer, J.E. Lingeman, and A.P. Evan (2006). Calcium oxalate calculi found attached to the renal papilla: Preliminary evidence for early mechanisms in stone formation. *J Endourol* **20**: 885–890.

Yin, H.X., T. Zhao, B. Lo, H. Shu, Z.F. Huang, X.L. Gao, P.P. Zhu, Z.Y. Wu, and S.Q. Luo (2007). Visualization of guinea pig cochleae with computed tomography of diffraction enhanced imaging and comparison with histology. *J X-ray Sci Technol* **15**: 73–84.

Yu, Q., T. Takeda, K. Umetani, E. Ueno, Y. Itai, Y. Hiranaka, and T. Akatsuka (1999). First experiment by two-dimensional digital mammography with synchrotron radiation. *J Synchrotron Rad* **6**: 1148–1152.

8
Cellular or Trabecular Solids

The subject in this chapter is cellular solids, that is, materials with significant volume fractions of empty, fluid-filled, or soft-tissue-filled space. This usage of *cellular* is, of course, quite different from what a microbiologist might intuitively envision. The common features between engineered materials such as foams and natural cellular materials such as wood or cancellous bone require similar analyses approaches. MicroCT studies of these materials are, therefore, collected in this chapter.

The following section gives background on cellular materials. Section 8.2 describes microCT studies of static cellular structures, and Section 8.3 expands coverage to studies of temporally evolving, nonbiological cellular materials. Section 8.4 describes some of the many microCT results on mineralized tissues such as cancellous (trabecular) bone, and Section 8.5 highlights a few studies on scaffolds being developed for cell in-growth, that is, as biomedical implants.

8.1 Cellular Solids

Following the definition of Gibson and Ashby (1999), a cellular solid consists of an interconnected network of solid struts (rods) or plates that form the edges and faces of cells, respectively. In other words, faces separate two cells, and edges are common to three or more subvolumes (cells) of the larger structure. In 3D, the cells are polyhedra filling space, and such materials are often termed foams. If the cells connect through open faces (the only material being struts at the cell edges), the material is termed an open cell foam; if the faces are solid, sealing off adjacent cells, it is a closed cell foam. Many cellular materials are produced by plants and animals, including wood, cork, and bone. Engineered (by humans) cellular solids employ all classes of materials including composites and are used for thermal insulation, packaging (energy absorption), structures (for high specific strengths), buoyancy (marine), and scaffolding for cell growth. The microstructural characteristics of cellular solids are difficult to quantify except with noninvasive, 3D methods such as microCT. In what follows, for purposes of simplicity, the solid phase is discussed as if it occupied a relatively small fraction of the total volume.

As noted by Maire et al. (2003a,b), cellular solids are often very challenging to analyze with tomographic techniques, particularly with respect to the different levels within the hierarchy of structural scales influencing the material's performance, specifically the scale of the constitutive material and the scale of cellular microstructure. The interplay among voxel size, contrast sensitivity, and field of view is particularly prominent in materials such as foams. If the distribution of cell sizes is important, then large voxel sizes may be required for the field of view to span the half-dozen or more cells required to represent the structure adequately, and an instrument optimized for these parameters (perhaps an industrial or medical peripheral CT system) might prove more efficient than a microCT system. If, on the other hand, features within the cell walls are of central importance, then microCT or even nanoCT on small sections of the material is required. Both scales sometimes can be studied productively on the same instrument, and several investigations have employed either two extreme resolution/field of view combinations on a single instrument (Salvo et al., 2004), two or more systems (Peyrin et al., 1998; Brunke et al., 2004), or local tomography techniques.

A structure such as a cellular solid (i.e., a complex mixture of empty space and solid) requires several parameters in order to describe its microstructure, specifically, how much material is present and how the material is distributed spatially. These microstructural characteristics have largely been defined in studies of cancellous bone, and unbiased methods of measuring these quantities are products of these studies, for example, Odgard (2001). The amount of material is given by the volume fraction of solid V_V. In certain applications, the mean cell size and the distribution of cell sizes of a foam must be specified in order for macroscopic properties to be predicted. Also important is the 3D distribution of solid material, the components of which can be in the form of plates and struts (rods); these are characterized by quantities such as the surface area per unit volume S_V and mean thickness <Th> of the structural elements. Accurate values of S_V and <Th> cannot be derived from isolated slices unless one can assume the individual structural elements are all plates or all rods: this is one circumstance where a dataset of contiguous slices is essential. As discussed in detail elsewhere (Odgard, 1997), one cannot simply make measurements of apparent thickness in the individual slices of a volume without risk of introducing significant, unpredictable bias into the data.

Additional quantities of importance include the connectivity density, Conn.D, the structural anisotropy, and the structure model index SMI. Connectivity reports the number of redundant trabeculae in the structure, that is, trabeculae that can be cut without increasing the number of separate parts of the structure, and Conn.D is calculated by dividing the connectivity by the examination volume. A more complete discussion of Euler numbers, connectivity, and edge effects is beyond the scope of this review (see Odgard (2001)), as these topics are not emphasized in the examples below. Anisotropy can be defined by main directions (perpendiculars to symmetry planes in the structure) and by numbers quantifying the concentration of directions around the main direction (Odgard, 2001). The fabric tensor compactly describes orthotropic architectural anisotropy via a 3 × 3 matrix of eigenvectors giving main directions and eigenvalues the degree of concentration around main directions (Cowin, 1985). Alternatively, the degree of anisotropy DA can be computed using the mean intercept length (MIL) method (Odgard, 1997, 2001). The structure model index relates the convexity of the structure to a model type and allows one to determine, for example, whether a given structure is more rodlike or more platelike (Hildebrand and Rüegsegger, 1997). An array of ideal (flat) plates has SMI = 0, a set of ideal cylindrical rods has SMI = 3, and a set of spheres has SMI = 4. If sufficient "air bubbles" are present within the structure, SMI < 0.

8.2 Static Cellular Structures

Before covering evolving cellular structures and biological/biomedical structures in detail, it is useful to consider some of the wide variety of cellular materials to which microCT has been applied. Although there are a few reports of plant-related cellular structures, some of which are mentioned in the following paragraph, the bulk of this section is composed of inorganic and engineered polymeric materials.

Seeds often have a cellular structure, and, based on analysis of the cellular structure, phase-contrast microCT has been used to study the phylogeny of fossil seeds from the Cretaceous (Friis et al., 2007). The cortical flesh of the apple fruit can be regarded as a cellular material, and microCT quantified differences in pore structure between two cultivars, despite considerable variability within the pore spaces of each cultivar (Mendoza et al., 2007). MicroCT has shown differences in porosity in banana slices resulting from different drying techniques (Léonard et al., 2008).

Porosity and grain structure in wood have been studied (Illman and Dowd, 1999; Steppe et al., 2004; Vetter et al., 2006). More recent phase-contrast microCT of spruce wood quantified total porosity, tracheid diameter, cell wall thickness, and pit diameter and suggested that subvoxel-sized features could be determined indirectly using watershed segmentation (Trtik et al., 2007). Figure 8.1 shows three orthogonal planes through a reconstructed volume of spruce wood; a wide range of pore dimensions is present and the axes of these cylindrical structures are primarily along the plant's axis (vertical in Figure 8.1) but also in other directions. In other words, many different levels of structural hierarchy contribute to the

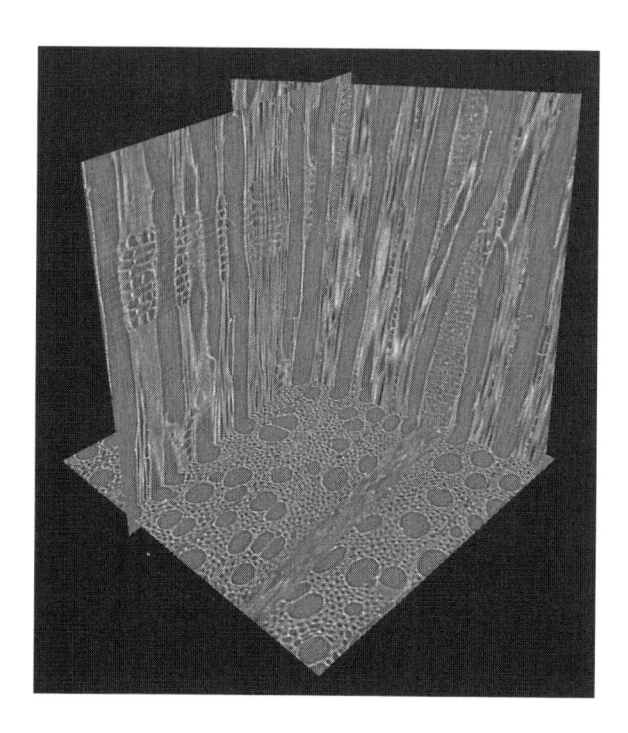

FIGURE 8.1
Synchrotron microCT of wood. Three orthogonal sections are shown with the plant's axis (i.e., the tree trunk axis) vertical. The lighter the pixel is, the more absorbing the corresponding voxel. (Unpublished synchrotron microCT data, T. E. Wilkes, K. T. Faber, F. De Carlo, S. R. Stock, 2-BM, APS, June 2007; see also Wilkes (2008).)

twin tasks of biological viability and of weight-bearing plus resistance to bending deformation. More details are found elsewhere (Wilkes, 2008).

Figure 8.2 shows a 3D rendering of an open cell Al foam, and microCT is very valuable for studying processes such as chemical milling of metal foams (Matsumoto et al., 2007). The elements in Figure 8.2 are mostly struts and not plates. In comparison with trabecular bone (e.g., Figure 6.6), the structure appears much more angular and transitions seem to be more abrupt.

Open-cell aluminum foams have been used as substrates/materials sources for zeolite catalyst growth, and lab microCT was used to characterize the starting cell diameters and strut thicknesses as well as to show that the zeolite film had homogeneous thickness (Scheffler et al., 2004). Pd–Ag/SiO$_2$ xerogel catalysts supported on Al$_2$O$_3$ foams were also characterized by microCT (Blacher et al., 2004). Synchrotron microCT was used to characterize the internal cellular structure within the walls of hollow fibers used for filtration (Remigy and Meireles, 2006) and the complex 3D porosity in a polymeric microfiltration membrane (Remigy et al., 2007), a study that would have been difficult to perform with other methods. Synchrotron microCT of replication processed open-celled aluminum

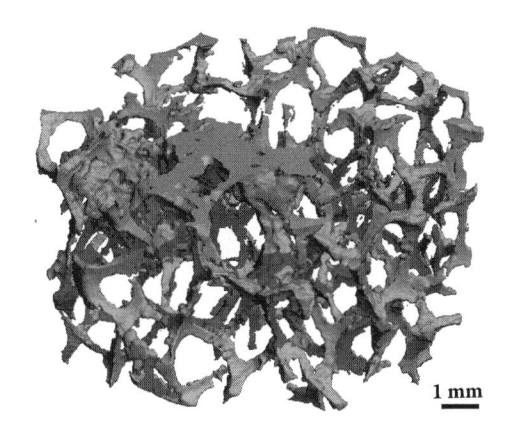

FIGURE 8.2
3D rendering of an open cell Al foam showing only the solid. (Unpublished lab microCT data, Y. Matsumoto, A. H. Brothers, D. C. Dunand, and S. R. Stock, October 2005; see also Matsumoto et al. (2007).)

foams was used to measure the internal architecture of these microcellular materials (Goodall et al., 2007). Cell shape did not affect uniaxial mechanical properties as long as the cells were roughly equiaxed; spherical and angular particle-derived foam properties differed, otherwise. Permeability in the Darcy regime of fluid flow, however, was relatively insensitive to details of pore shape.

Bubble-size distributions in simulated or actual volcanic silicate foams have been quantified with synchrotron microCT, and 2D and 3D measures of the distributions were compared and related to models of bubble growth (Robert et al., 2004). Pore initiation by blowing agents in metal foams has been studied by synchrotron microCT, and pores in prealloyed Al powders nucleated around the blowing agent particles, whereas they tended to nucleate around Si particles in Al–Si powder blends (Helfen et al., 2005). In a better developed, closed-cell foam produced by the blowing agent route, a large number of 3D parameters were quantified for pores (volume fraction, equivalent radius, surface area, sphericity), for the distribution of cell wall material (via granulometry) between cell faces and edges, and for Ti particle geometric parameters and size distribution (Brunke et al., 2005). The 3D spatial distribution of Ti particles was quantified and related to foam morphology in a simple representation, projection of the mass within each slice along the length of the cylindrical sample, that sufficed to show the principal foaming direction (Brunke et al., 2005). Different foaming times also altered the distribution of cell wall material (Brunke et al., 2004).

X-ray tomography (using a fairly large voxel size, 90 μm, dictated by the required field of view for a material with cells up to 5-mm diameter) of different structural metal foam types revealed differences in cell anisotropy (distribution of cell volumes and of aspect ratios quantified as equivalent ellipsoids (Benaouli et al., 2000, 2005); stereographic projections showing distribution of cell axes vs. orientation) that correlated with altered mechanical properties (Benouali et al., 2005). Defects in the cell walls (corrugations, holes, and cracks), however, substantially reduced Young's modulus and strength from theoretical values (Benouali et al., 2005). Another lab microCT report of a metal foam (Maire and Buffière, 2000) was followed by detailed FEM of foam deformation (Maire et al., 2003b); see below.

Synchrotron microCTs of polyvinylchloride foam (an example where cell diameters were smaller than 0.2 mm) showed that a higher processing temperature produced a higher volume fraction of porosity, larger equivalent cell diameters, decreased wall thicknesses, and decreased degree of anisotropy (Elmoutaouakkil et al., 2003). Interestingly, selection of small subvolumes produced the same results as for larger volumes of interest. Polymeric foams for hydrogen storage applications have also been examined with microCT.

Glass foams can be produced from silicate wastes (power generation waste plus bottle fragments plus a SiC foaming agent) and used for thermal or acoustic insulation and the like, a process of considerable interest given the amount of this waste that has been accumulating for years. Lab microCT has been used to relate foam structure and properties with processing temperature (Wu et al., 2006). Wu and coworkers studied lightweight glass–ceramic foams derived from silicate wastes by varying processing temperature and amount of foaming agent (Wu et al., 2007). Good agreement was obtained between experimental conductivity data and simulations based on microCT-derived finite element analysis, and the foams behaved mechanically in a manner typical of brittle foams.

8.3 Temporally Evolving, Nonbiological Cellular Structures

The 3D evolution of liquid foams over tens of hours has been studied by synchrotron microCT (Lambert et al., 2005), and the progress of drainage and bubble coarsening posed particular problems in this study. Although the minimum time between each set of projections for a single reconstruction was 8 min (10-mm field of view, 10-μm voxels, 50-ms exposure per projection, and 900 radiographs per dataset), nearly instantaneous changes such as bubbles bursting or moving produced occasional reconstruction artifacts. Despite an inability to image a significant fraction of thin film area, segmentation of the image into a set of bubbles could be done automatically, using a distance transformation plus watershed algorithm.

Even with boundary bubbles removed, the subvolume contained ~2 × 10³ bubbles at the start of the experiment, an order of magnitude increase over the number of bubbles that could be analyzed by other techniques such as MRI or optical tomography. A large portion of the distribution of bubble volumes could be characterized as exponential at a given time. The authors of this study also examined the number of faces per bubble, a quantity important in coarsening kinetics (Lambert et al., 2005).

Rapid synchrotron microCT (<30 s per reconstructed volume) has also been applied to bubble formation in bread, that is, during the process of the rising of the dough where the volume fraction of porosity increases from ~0.1 to ~0.7 (Babin et al., 2006). As dough rises over a period of 2–3 hr with relatively stable cell walls, unlike soap bubbles, the changes during each data collection increment were negligible. The first stage consisted of nucleation and growth of many similarly sized spherical bubbles. Bubble interaction and growth by coalescence characterized the second stage, and the third stage was dominated by coalescence-driven formation of large, irregularly shaped voids. Tracking individual bubbles during the first stage of rising allowed the kinetics of bubble growth to be compared to models. Baking of dough was also examined (Babin et al., 2006).

Mechanical deformation of cellular materials is generally performed incrementally either in an external testing apparatus or in an in situ load frame specially constructed for microCT. Although numerous apparatus (for applying loads in situ during microCT) have been built, their designs differ little from those developed in the first half of the 1990s and described in the chapter on microCT of deformed or cracked specimens, for example, Breunig et al. (1992, 1993, 1994) and Hirano et al. (1995).

With microCT characterization of foams, there is concern that the results of interrupted testing differ from those obtained during continuous loading. The lab microCT studies of Nazarian et al., however, found the results of stepwise compression of several types of cellular materials agreed well with data from conventional continuous testing (Nazarian and Muller, 2004; Nazarian et al., 2005).

MicroCT observation of deformation of cellular solids is frequently compared to FEM of the deformation based on the initial (microCT-determined) structure. In principle this allows the step-by-step FEM predictions of the deformation process to be compared with experiment and assumptions in the FEM to be refined. The steps involved in meshing microCT for FEM are covered elsewhere (Keaveny, 2001; van Rietbergen, 2001; van Rietbergen and Huiskes, 2001; Maire et al., 2003a,b; Youssef et al., 2005; Plougonven et al., 2006), and the reader is advised to search the bone literature for further discussion of approaches for cellular material.

Several microCT studies of elastomeric foams have appeared and include both open- and closed-cell materials (Kinney et al., 2001; Elliott et al., 2002; Youssef et al., 2005; Plougonven et al., 2006). These studies were aimed at improving understanding of deformation processes underlying the three stages of the typical elastomeric foam stress–strain curve (linear elastic stage I at lowest stresses transitioning into a plateau region; stage II, where little increase in stress produces large increases in strain and finishing in a densification regime; and stage III, of rapidly rising stress). Kinney and coworkers studied 1-mm-thick, 15-mm-diameter silica-reinforced (~25 wt.%) polysiloxane foam pads (Kinney et al., 2001). Synchrotron microCT of this closed-cell material was performed at four deformation levels from 0 to 35 percent strain, and analysis concentrated on the fraction of surface connected pores and finite element modeling of the structure. Because the pore diameters and mean cell wall thicknesses were ~1/3 and ~1/10 the specimen thickness, it is difficult to know whether the results can be applied to other geometries.

Plougonven et al. studied a polypropylene foam subjected to alternating cycles of interrupted shocks followed by synchrotron microCT imaging (Plougonven et al., 2006; Viot et al., 2007); this simulates energy absorption behavior during crashes. The investigators concentrated on behavior at the mesoscopic scale of grains (1–3-mm-diameter within the 10-mm-diameter specimen), and the hierarchy of structural scales of pores in the foam complicated the analysis. Phase contrast effects were suppressed before filtering suppressed noise, and a distance transform operation defined the cell walls. The relative densities of the different grains were then computed as a function of deformation. A similar approach combined with numerical modeling was used to study shock loading of an Al foam (Bourne et al., 2008).

Shape memory polymer foams can be compacted as much as 80 percent and still experience full strain recovery over multiple cycles. MicroCT of one such foam revealed bending, buckling, and cell collapse with increasing compression, consistent with results of numerical simulations (Di Prima et al., 2007).

In an open-celled polyurethane foam, Elliott and coworkers imaged the structure at 14 strains with the observations clustered at the critical regions of the stress–strain curve (i.e., the transition between linear elastic and plateau regimes, the transition between plateau and densification regions, and in the final stages of densification; Elliott et al. (2002)). Their focus was on strut deformation processes, and they employed local tomography to observe the central ~7 mm of a 25 mm diameter, 25 mm high specimen. The initial stages of compression (below 4 percent strain, in the linear elastic regime) were taken up by small amounts of bending in struts that were both longer than average and inclined to the compression axis. More severe bending and reorientation of struts within a localized deformation band accommodated strains up to ~23 percent. The densification regime began after multiple collapse bands formed and impinged on one another. Node and strut models of the deformation were applied, but further improvements were found to be necessary (Elliott et al., 2002). Brydon et al. (2005) considered, in much greater detail, numerical modeling of densification of open-celled carbon foams, whose structure had been measured by synchrotron microCT.

The microCT study of a closed-cell polyurethane foam stands in contrast to that described in the previous paragraph (Youssef et al., 2005). Synchrotron microCT (local reconstruction) was also performed on the central portion of a larger specimen (central ~2 mm of a 6-mm-diameter specimen) at four strains (maximum of 20 percent), and the investigators performed a thorough sensitivity analysis of FEM parameters and used FEM to fill in the gaps between microCT measurements. Direct comparison of FEM-derived strain distributions in the actual foam structure showed cell wall rupture occurred at local strain concentrators.

Metal foams often fail at significantly lower stresses than expected from theory, and this has motivated microCT imaging of foams after discrete increments of compression (Bart-Smith et al., 1998; Kádár et al., 2004; Montanini, 2005; Toda et al., 2006a,b; Wang et al., 2006). Even simple analysis methods and modest spatial resolution can provide informative results. Deformation of two fairly dense Al "foams" produced by powder metallurgy (46 percent and 37 percent of full density), for example, was quantified by measuring mean porosity as a function of slice number (compression along the tomographic rotation axis), much as what was done earlier for density gradients in chemical vapor infiltration (Lee, 1993) or more recently for liquid foam (Lambert et al., 2005). Displacement of local maxima/minima in mean slice porosity could be followed through the different increments of deformation as the porosity diminished by up to a factor of two. Correlation coefficients for cross-sections between two deformation states were never lower than 0.55, and the data confirm the phenomenon that the higher the cross-section porosity is, the greater the deformation of that cross-section. Similar approaches were used by others for an Al foam (Desischer et al., 2000) and for compression of a glass wool (Badel et al., 2003).

Medical CT (in plane voxel sizes a significant fraction of a millimeter and in out-of-plane voxels even larger) can reveal some details of wall deformation of closed-cell Al foams and can identify sites of weakness (Bart-Smith et al., 1998), but higher spatial resolution is required to produce data that can be incorporated usefully into numerical or analytical models. MicroCT of indentation-tested, low-density Al foams (between 0.06 and 0.17 percent of full density) revealed the deformation zone was confined to a spherical cap immediately under the indenter, with only occasional buckling of cell walls farther than one cell diameter from the zone directly under the indenter. Finite element modeling based on the reconstruction agreed with analytical expressions for the size of the deformation zone (Kádár et al., 2004).

Significant deformation occurs during the manufacture of open-cell Ni foams used as battery electrodes, and synchrotron microCT of these foams has proven very useful in characterizing tension and compression damage (Dillard et al., 2005). In tension, bending, stretching, and alignment of struts were observed, and the particulars depended on the initial anisotropy of the foam. Compression led to strain localization due to buckling of struts, and this differed from the large rotations observed under compression in polyurethane (Elliott et al., 2002).

Repeated observations (after different compression increments) with synchrotron microCT revealed damage accumulation in the cell walls and plateau borders of an Al foam (Ohgaki et al., 2006; Toda et al., 2006a,b); local tomography was required to provide adequate spatial resolution within the volume of interest and allow the specimen tested to have a large-enough cross-section to be representative of the foam. Displacement of thousands of micropores within the solid was tracked automatically for each increment of deformation, and a microstructure gauge formalism was used to calculate maps of the 3D strain tensor components. Characteristics of pores and the microcracks that nucleated from them were quantified, and the distribution of strain was discussed in terms of these features. The microCT-derived 3D structure was imported into FEM (Toda et al., 2006a,b). Similar data is required in applications in medicine, for example, displacements in CT images of lungs through image warping (Fan and Chen, 1999).

Compression of a low-porosity, closed-cell amorphous metal foam was observed with synchrotron microCT at 14 strains during loading and 3 strains during unloading (Demetriou et al., 2007). Yielding proceeded by percolation of an elastic buckling instability. Plots of individual cell displacements versus cell coordinate along the loading direction showed the width of the collapse zone.

Cellular structures can also be embedded in another solid phase. Madi and coworkers used synchrotron microCT to study a refractory consisting of zirconia grains coated by a softer amorphous (glassy) phase that formed the walls of the cellular system (Madi et al., 2007). Finite element modeling of the resulting structure was used to investigate the influence of the amount of glassy phase on stiffness.

Lab microCT of two syntactic metal foam types (preforms of hollow ceramic spheres infiltrated by liquid metal of two different Al alloys) was used to characterize deformation modes and to interpret features in stress–strain curves (Balch et al., 2005). The foam with commercial purity Al showed barreling over a large part of its area with uniform, rather modest plastic deformation of the matrix coupled with sphere fracture and a thin crush band. The foam with 7075–T6 matrix had damage concentrated in two intersecting and much thicker crush bands oriented ~45° to the compression axis; sharp stress drops after the peak stress resulted from the formation of the crush bands.

Adrien et al. (2007) compared syntactic foams containing hollow glass microspheres and one of three polymeric matrices (relatively stiff epoxy, relatively compliant polypropylene, and polyurethane) using synchrotron microCT with 0.7-μm voxels. The thickness of the spheres was rather constant, but their diameters were strongly distributed. During stepwise confined compression in the dry state, strong adherence was observed between the glass spheres and polyurethane, but it was weaker between the other two materials and the spheres. In the foams with more compliant matrices, damage was more homogeneous and mainly affected the larger spheres, whereas in the stiffer material (epoxy) damage localized in bands.

Examining the behavior of the foam and of the foam–sheet interface in Al-sandwich material required microCT at two very different scales (Salvo et al., 2004). Nonetheless, microCT allowed strains to be calculated based on measurements at four deformation states. The interface appeared to be sound, and the foam in the sandwich behaved as did similar free-standing foams examined by this group (Maire et al., 2003a,b).

8.4 Mineralized Tissue

Mineralized tissue is an important naturally occurring biomaterial that has been studied extensively with microCT. Indeed, a number of commercial lab microCT systems were optimized for studying the mineralized tissue of widest clinical interest, bone. Several books provide useful background on bone and its microarchitecture (Martin et al., 1998; Cowin, 2001; Majumdar and Bay, 2001; Currey, 2002).

This section begins with the microCT studies of porous mineralized tissue structures other than bone. Vertebrate bone and tooth are based on nanocrystallites of carbonated apatite (cAp), and sea urchins and other echinoderms use calcite. Biopsies of human tissue or whole limbs of small animal models continue to be a major fraction of the microCT literature; it is no longer practical to discuss all of these studies or, for that matter, to even list them, and more than a few examples follow in the second through fourth subsections. Section 8.4.2 describes the motivations for studying cancellous bone and covers the older studies first in order to give the reader a historical perspective, admittedly, at the risk of fragmenting certain topics. Section 8.4.3 provides a sampling of more recent studies on bone growth and aging, specifically on structural changes in cancellous bone produced by aging or by treatment of conditions such as osteoporosis. Section 8.4.4 considers mechanical testing and numerical modeling of mechanical properties. The reader should note that discussion of "dense" cAp-based mineralized tissue (cortical bone and tooth) is postponed until Chapters 10 and 11 in order to retain focus on cellular materials.

8.4.1 Echinoderm Stereom

The calcite stereom of sea urchins (mentioned briefly in Section 7.2.3) has many different fabrics (Smith, 1980), that is, dimensions and arrangements of struts and plates, and synchrotron microCT has high enough resolving power to study many types of stereom. Figure 6.8 shows a slice of one of the many plates constituting the test (protective globe enclosing the sea urchin's internal organs) and of a demipyramid (part of the jaw structure; Stock et al., 2004a). Within the region indicated in Figure 6.8b, BV/TV = 0.487, <Th> = 19.5 µm, <Sp> = 20.7 µm, and SMI = 0.22; that is, the structure consisted primarily of plates. Determination of specific stereom fabric type (e.g., galleried, labyrinthic, perforate) from segmented volumes will require further development of analysis tools.

MicroCT imaging of cellular mineralized tissues is useful even when the microstructure cannot be resolved. The finest stereom found in echinoderm ossicles can have submicrometer dimensions, for example, and nanoCT would be required. Interpretation of linear attenuation coefficients as partial voxels of calcite in a demipyramid of a sea urchin was discussed in Section 6.2.6.

8.4.2 Cancellous Bone: Motivations for Study and the Older Literature

Aging populations are characteristic of countries in the developed world, and improving treatment of osteoporosis in the aged has motivated, at least in part, microCT studies of bone. Osteoporosis is a disease of increased bone fragility seen in both elderly females and males, and the classical view is that decreased bone mineral density (BMD) explains fracture susceptibility. This paradigm fails in a significant fraction of patients, and the first refinement to this model is the proviso that decreased bone competence may also be the product of inferior bone microarchitecture, that is, impaired networks of struts and plates of the trabecular bone, a major constituent of vertebrae and the femoral head, both major risk sites for osteoporosis-related fractures.

Quite some time ago, it was realized that microCT was an ideal method for characterizing bone microarchitecture. Early microCT studies of trabecular bone, that is, those in the years before commercial microCT systems became widespread, had to establish the reliability of the technique. These studies focused on imaging trabeculae (Layton et al., 1988; Feldkamp et al., 1989; Dedrick et al., 1991, 1993; Goldstein et al., 1993; Bonse et al., 1994; Elliott et al., 1994; Bonse and Busch, 1996; Hildebrand et al., 1997; Kinney et al., 1997; Kirby et al., 1997; Ladd and Kinney, 1997; Müller and Hayes, 1997; Peyrin et al., 1997) and on establishing robust methodologies for threshold definition and for topological analysis (Engelke et al., 1997) — loss of trabecular connectivity was hypothesized as a major portion of the microarchitectural portion of bone quality deterioration.

Earlier serial sectioning of cancellous bone and unbiased estimates of connectivity did not reveal a simple relationship between connectivity and bone volume fraction (Odgaard and Gunderson, 1993). Using microCT, Feldkamp and coworkers investigated anisotropy and connectivity within 8-mm cubes of human cancellous bones (Layton et al., 1988). Goulet et al. (1994) suggest that mechanical influences seem to alter the amount of bone by changing trabecular thickness, whereas hormonal or chemical influences act to affect the number of trabeculae. It is also almost certain that the correlations between experimentally measured macroscopic properties and modeling based on the actual microstructure would not have been demonstrated without x-ray microCT. In theory, one could measure elastic response and then use exhaustive serial sectioning to define the trabecular network for use in three-dimensional modeling; in practice, the labor required prohibits this approach.

Quantitative (conventional) x-ray CT and dual energy absorptivity (i.e., quantitative radiography) have demonstrated that density measurements of human femora explained no more than 30 to 40 percent of observed variance in modulus and 50 to 60 percent of the variance in ultimate stress and that the orientation of cancellous cubes in the principal compressive trabeculae region was a significant contributor to mechanical properties independent of the bone density (Cody et al., 1996), a direct indication that trabecular microstructure is important in bone strength. Changes in trabecular morphology (thickness, relative density, bone fraction, and separation) were studied in a guinea pig model (Layton et al., 1988; Dedrick et al., 1991) and in a canine model of osteoarthritis (Kuhn et al., 1990); of particular interest in the canine study was that osteoarthritis was induced in one knee of the dog and the other knee was left unaffected and served as a control; earlier microCT work showed less than 5 percent difference in mean bone morphology variables between legs. Kinney and coworkers imaged trabecular bone architecture of rat tibia in vivo before and five weeks after ovariectomy-induced osteoporosis and found a major loss in trabecular bone following ovariectomy (OVX; Kinney et al. (1992, 1995)). More important was the change from an interconnected plate- and strut-like structure to one that is mostly disconnected, dangling trabecular elements (a morphology not observed in non-OVX rats). Finally, a strong linear relationship was observed between connectivity and volume of trabecular bone.

Bonse et al. (1996) showed volume renderings of cancellous bone biopsies from a single patient over 15 years; the three-dimensional renderings clearly showed the large loss of bone as a result of chronic hemodialysis. Numerical values for the volume fraction of bone and the ratio of the bone surface area to bone volume for five biopsies suggested that the main change was in the number and connectivity of trabeculae and not in their thickness. Others have studied trabecular changes in pre- and postmenopausal women and numerically simulated changes in density and architecture beginning with the normal premenopausal structure (Müller and Hayes, 1997).

Images of microcallus formations within trabecular bone have been interpreted to show the healing process after microfracture (Bonse and Busch, 1996). Guldberg and coworkers inserted a small chamber into dogs' femora and tibia and observed tissue repair with microCT and other methods after the 7-mm internal diameter chambers were removed from the animals (Guldberg et al., 1997). Repair response with and without mechanical stimulation was studied, and after 12 weeks the bone volume fraction was 75 percent larger in the loaded group relative to the unloaded control group. Trabecular connectivity was increased for the loaded bone relative to the control groups; the trabeculae were much thicker in the loaded biopsies, but the trabecular spacing was the same, as was the number of trabeculae. The results support the hypothesis that newly formed bone is deposited in mechanically appropriate locations, and Guldberg et al. suggest that the limitation of strain magnitude is important in the process of tissue adaptation.

8.4.3 Cancellous Bone: Growth and Aging

Most studies of cancellous bone have been on human tissue or several popular animal models. Because high-definition microCT datasets covering trabecular networks can be extremely large, special approaches are required for connectivity analyses (Apostol and Peyrin, 2007). Although one might regard the need for such expedients as transitory, given the rapid increase in RAM and other aspects of computational power, and soon to go the way of foreground and background computing techniques, this is probably not the case: the push for ever-higher definition, that is, more voxels per slice, and the increasing data collection rates suggest pressure on analysis algorithms will continue.

Before addressing the main subject of this subsection, brief mention might be useful of microCT studies of other animals' trabecular bone, studies that do not fit into the flow of the other text. MicroCT of the trabecular portion of deer antler revealed a porosity profile along the antler diameter, with a granulometry-determined pore size distribution dependent on the sample's original site (Leonard et al., 2007). Skeletal pneumaticity, produced by the extension of pulmonary air sacs into marrow spaces, is a hallmark adaptation of some birds, and microCT was used to quantify differences between thoracic vertebrae in one species of duck that possesses a pneumatic structure and a second with apneumatic vertebrae (Fajardo et al., 2007). One expects differences in bone microarchitecture for animals that are phylogenically close relatives but have different modes of locomotion, and microCT of trabecular bone in the femoral head and neck for two such primate genera showed no difference in cancellous bone volume density but did show differences in trabecular orientation (MacLatchy and Müller, 2002).

Bone undergoes large changes during development and growth, and a few such microCT studies are mentioned in this paragraph and in scattered places throughout subsequent paragraphs. Synchrotron microCT and synchrotron x-ray diffraction were used to study the microarchitectural and physical changes of human vertebrae during fetal growth (Nuzzo et al., 2003); the trabecular bone network was denser than in adult vertebrae, perhaps in compensation for the immature cortical shell in the fetal bone. Tibia and spine from IGF-1 deficient and wild-type fetal mice (18th day of gestation) were imaged with synchrotron microCT; considerable variation in mineral level and bone structure was observed along the length of the tibiae, and, interestingly, von Kossa staining, a standard histological procedure for detecting mineral in tissues, revealed no mineral content in the IGF-1 −/− spinal ossification center, whereas synchrotron microCT clearly indicated the presence of highly attenuating components (Burghardt et al., 2007).

An important maturation stage in the long bones of many animals is the fusion of the growth plate, and, until microCT was applied to the problem, it was unclear whether epiphyseal growth plates remained open throughout life in rats, a popular skeletal (trabecular) model for aging humans. Rats are not good models for Haversian bone remodeling because this process does not occur, and, if their bones never ceased to extend, however slowly, this would further limit the usefulness of the rat model. The microCT data of Martin et al. (2003) suggest that growth plate fusion does take place in rats and the fraction closed follows a sigmoidal pattern.

Bone adapts during aging, and microCT has been employed to investigate gender- and site-specific differences in trabecular bone in humans. For the femoral neck, in vivo densitometry of nearly 700 women and direct microCT measurements of 13 pairs of postmortem specimens revealed that strength in old age was largely achieved during growth by differences (from individual to individual) in the distribution rather than the amount of bone in a given femoral neck cross-section (Zebaze et al., 2007). Transmenopausal changes in trabecular bone were assessed in 38 paired transilial biopsies (pre- and postmenopause specimens from the same individual), and variables characterizing bone structure significantly decreased from values before menopause (BV/TV, trabecular number) and an increased structure model index suggested that trabecular microarchitecture was transforming from a platelike to a rodlike structure (Akhter et al., 2007). In vitro microCT of bone structure in subjects older than 52 years revealed significant sex differences in microarchitectural parameters at the distal radius, femoral neck, and femoral trochanter but not the iliac crest, calcaneus, and lumbar vertebral body (Eckstein et al., 2007).

Baseline studies of gender and age effects on trabecular bone have also been the subject of studies in animal models. Paired biopsies of goat iliac crest (pre- and post-OVX) demonstrated significant deterioration of the bone microarchitecture six months after OVX (Siu et al., 2004). As rats and mice lack Haversian remodeling, marmosets, which do exhibit this type of remodeling, were investigated as a model of age-related changes to bone which was found to resemble human bone (Bagi et al., 2007). MicroCT was used to assess age-related changes in bone architecture of male and female C57BL/6J mice (eight time points between 1 and 20 months of age), and deterioration in vertebral and femoral microarchitecture began early, continued throughout life, was more pronounced at the femoral metaphysis than in the vertebrae, and was greater in females than males (Glatt et al., 2007). Gender and LRP-5 gene-related changes in trabecular bone were investigated using microCT, and significant differences were noted at multiple sites (Dubrow et al., 2007).

There have been many studies of response of bone in animal models to different challenges to bone integrity and to drugs administered to combat this degradation. The experimental details of one study on a mouse model of osteoporosis (but not its results) are described in the following paragraph as an exemplar of how microCT is being employed. Note because one animal is used per time point, the design differs from longitudinal studies involving in vivo observation.

Bouxsein et al. (2005) compared OVX-induced bone loss among five inbred mouse strains to establish whether genetic regulation of bone loss occurred under the influence of estrogen deficiency. Between six and nine animals for each strain and for OVX and sham-OVX (control) were studied, and lab microCT with 12-μm isotropic voxels was performed on dissected bones: right femur, right tibia, and fifth lumbar vertebra. Note that sham operations are necessary for the controls because the stress of the surgery could affect bone retention. For vertebrae, 250–300 transverse slices were collected to cover the bone; the volume from the growth plate to 1.8-mm distal was scanned in the tibia; diaphysis characteristics were measured in the femur. Binary segmentation was used along the scanner's software to calculate assumption-free bone characteristics such as bone volume fraction, trabecular thickness, and SMI. Analysis of variance was used to determine whether observed differences were statistically significant.

MicroCT has examined trabecular bone loss accompanying steroid treatment for inflammation (McLaughlin et al., 2002), sciatic neurectomy (Ito et al., 2002b), and gastrectomy (Stenstrom et al., 2000). Patterns of bone resorption and the effect of dose were examined for calvarial culture and IL-1 and PTH treatment (Stock et al., 2004b). Effects of disease processes on bone (infections, tumors, arthritis) have also been examined with microCT

(Balto et al., 2000; Kurth and Müller, 2001; Pettit et al., 2001; Boyd et al., 2002; Patel et al., 2003; Morenko et al., 2004; Sone et al., 2004). Degradation of the 3D microarchitecture of subchondral trabecular human bone has been studied in osteoarthritis and comparisons made of these areas without cartilage with cores from neighboring sites that retained their cartilage (Chappard et al., 2006).

Bone loss at menopause has been studied extensively, for example, using OVX in animal models to mimic estrogen loss (Lane et al., 1998; Ito et al., 2002b). A study of five inbred mouse strains found OVX-related skeletal response varied in a site- and compartment-specific fashion among the different strains, supporting the hypothesis that bone loss during and after menopause is partly genetically regulated (Bouxsein et al., 2005). Trabecular bone in senescent marmosets and its response to alendronate therapy resembled human bone (Bagi et al., 2007). Postmenopausal osteoporosis is often treated with bisphosphonates, and several studies of bisphosphonates' effects on bone loss have been reported for animal models (Nuzzo et al., 2002; Dufresne et al., 2003; Ito et al., 2003; Borah et al., 2005). Parathyroid hormone (PTH; Dempster et al. (2001), Lane et al. (2003b), Gittens et al. (2004), and Lotinun et al. (2004)) or fibroblast growth factor (FGF; Lane et al. (2003a,b)) treatment of osteoporosis models have also been investigated.

MicroCT has been employed to examine several other strategies for improving bone fragility. Ex vivo data on vertebrae of rats administered strontium ranelate revealed improved mechanical properties (increased plastic energy to failure) of the vertebrae, increased elastic modulus in individual trabeculae (via nanoindentation), and improved microarchitecture (via microCT) compared to controls (Ammann et al., 2007). Extremely low-level, high-frequency oscillatory motion was examined in chronically unloaded mouse limbs; compared to normal age-matched controls, oscillations attenuated the decline in trabecular microarchitecture-related mechanical properties as assessed by finite element modeling (Ozcivici et al., 2007).

Alveolar bone is highly porous and supports teeth through the connection of the periodontal ligaments. Dalstra and coworkers used synchrotron microCT to image human, simian, and porcine jaw segments and were able to visualize blood vessels and fibers of the periodontal ligaments (Dalstra et al., 2006, 2007). Differences in the distribution of linear attenuation coefficient values in the close vicinity of the ligaments suggested remodeling activity at those locations.

Bone structure is expected to change in response to in vivo loading. Changes in bone loading history are easily imposed in rat and mouse models and form the basis for a number of microCT studies (Ishijima et al., 2002; Amblard et al., 2003; David et al., 2003; Joo et al., 2003; Squire et al., 2004). Craniomandibular joints are expected to change in response to different forces encountered in eating (i.e., to exhibit adaptive plasticity) and to attachment of different-sized muscles. MicroCT was used to examine such changes in rabbits (animals eating a hard diet versus those on a soft diet), in myostatin-deficient mice versus normal mice (the larger-than-normal muscle mass of the former has led to the informal label "muscle mice"), and in subfossils of the extinct primate *Archaeolemur* (Nicholson et al., 2006; Ravosa et al. 2006, 2007a,b,c). Tanaka et al. (2007) used lab microCT and several calibration standards to compare mineral levels in the mandibles of rats fed hard versus soft diets; significant differences were observed between groups as well as at different locations.

Mulder and coworkers used microCT to study altering architecture and mineralization during development of trabecular bone in the mandibular condyle (Mulder et al., 2007b); the inhomogeneous finite element models (tissue moduli scaled to the local degree of mineralization) derived from the data identified important structural anisotropy. In one related paper, microCT of pig mandibular condyles revealed mean degree of mineralization and intratrabecular differences in mineralization between surfaces and cores of trabecular elements increased during development from 8 weeks prepartum to 108 weeks postpartum (Willems et al., 2007). Figure 8.3 shows 3D microCT renderings of the condyle of a 43-week-old pig; the more highly magnified images of several trabeculae show, via color, the intratrabecular distribution of mineral. In a related paper on pig mandibular condyles, tissue stiffness (via nanoindentation) and mineral level (micro-CT) were low at trabecular surfaces and higher in the cores, directly demonstrating the link between developing stiffness and degree of mineralization (Mulder et al., 2007a). A fourth paper in this series found that remodeling in human mandibular condylar trabecular bone was larger than in the adjacent cortical bone (Renders et al., 2007).

In many animals, mineral content varies considerably from older mature areas (highest mineral content) to newly remodeled osteons (lower mineral content). The sensitivity limits of microCT make it ill-suited for quantifying small differences in composition, but a number of studies have shown that the mineralization levels in bone can be mapped. Synchrotron microCT of low-mineralized versus high-mineralized volumes in trabecular as a function of bisphosphonate treatment was the subject of one study (Borah et al., 2005). Microradiography of thin sections of bone has long been used to show remodeling, and determination of the degree of bone mineralization via this method has been compared to that with synchrotron microCT (Nuzzo et al., 2002). Synchrotron microCT has also been used to study the degree of mineralization in human normal vertebra, osteoblastic metastases and sites with degenerative osteosclerosis (Sone et al., 2004), and different mouse genetic strains (Martín-Badosa et al., 2003; Bayat et al., 2005).

Several in vivo and in situ microCT characterization studies have appeared. Multiple observations of rat bone by synchrotron microCT

Cellular or Trabecular Solids 189

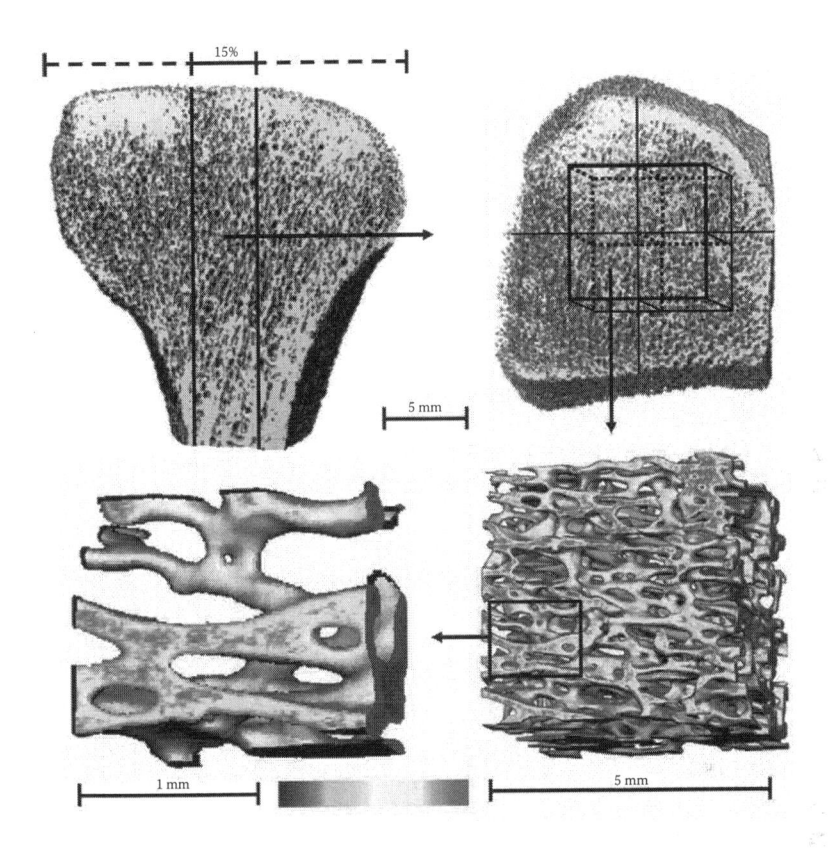

FIGURE 8.3 (SEE COLOR INSERT FOLLOWING PAGE 144.)
Rendering of the left condyle of a 43-week-old pig. The upper right image is a frontal cross-section. The upper left image is the medial view of the sagittal slice (posterior at the left) showing selection of four subvolumes. The lower left subvolume (posterioinferior) is shown magnified at the lower right, and a small number of trabeculae are shown still further magnified at the loft left. The magnified images at the bottom shows increasing degree of mineralization from white to gray (Willems et al., 2007).

are one example (Lane et al., 1998, 2003b), and this approach has been extended to mice recently (Bayat et al., 2005). The reader should consult the latter study for detailed discussion of the relationship between x-ray dose and signal-to-noise ratio in the reconstructions. Observation of changes in trabeculae in the murine hind-limb unloading model is another example (David et al., 2003).

Determination of longitudinal changes in trabeculae by direct comparison of the reconstructed volumes was a rather ambitious approach employed in an in vivo lab microCT comparison of OVX and control rats (Waarsing et al., 2004). In this study, image registration was difficult in the OVX animals because of the large changes, but clear patterns of bone resorption and apposition were documented for the OVX rats, and new bone formation was shown to correspond with areas of calcein labeling. It is also possible to administer Pb and Sr to living animals in order to label the bone in much the same way as fluorochromes such as tetracycline or calcein are used in optical microscopy of thin sections; imaging to either side of the absorption edge increases sensitivity in synchrotron microCT studies (Kinney and Ryaby, 2001).

Ruimerman and coworkers proposed a 3D numerical model of developing bone that takes into account different mechanobiological pathways affecting bone (Ruimerman et al., 2005): osteocyte stimulation of osteoblast bone formation, an effect of elevated strain in the bone matrix; microcrack; and disuse promotion of osteoclast resorption. The model appears to reproduce features seen in microCT scans of growing pigs and, if osteoclast resorption frequencies are increased, also produces structures such as those seen in osteoporotic bone.

A prominent feature of clinical radiographs is the texture produced by the (partially) overlapping trabeculae at the ends of long bones and within vertebrae. An obvious question is whether analysis of the texture can provide useful information about the trabecular microarchitecture, and this has, in fact, received attention. Luo et al. (1999) used synchrotron microCT to show that plain radiographs contain architectural information directly related to the underlying 3D structure; they suggested that a well-controlled sequence of radiographs might allow monitoring of trabecular changes in vivo and identifying individuals at increased risk of osteoporotic fracture. Later synchrotron microCT-based analysis by others examined which 2D texture parameters correlated best with the underlying 3D structure (Apostol et al., 2004). This approach may prove useful for analysis of other cellular solids, for example, metal foams undergoing high strain rate deformation (with complete tomographic reconstructions only possible of the initial and final states), but much more detailed descriptions of the radiograph's texture need to be developed (instead of one or a few global measures of texture, a dozen or more parameters might be appropriate) and a priori knowledge could be utilized (for this example, the known initial structure and corresponding radiograph, the final structure and radiograph, and the fact that the ith radiograph resulted from the $(i-1)$th structure and evolved into the $(i+1)$th structure). Other related papers have appeared (Showalter et al., 2006; Jennane et al., 2007).

In addition to absorption microCT, trabecular bone has been imaged with scattered x-radiation (Kleuker and Schulze, 1997). Recently, Stock and coworkers performed x-ray scattering-based reconstruction of the diaphyses of a rabbit femur (Stock et al., 2008). One hopes for more studies of this sort, perhaps coupled with use of x-ray scattering-based quantification of internal strain states accompanying loading (Almer and Stock, 2005, 2007).

8.4.4 Cancellous Bone: Deformation, Damage, and Modeling

The relationship between elastic modulus and microarchitecture of trabecular bone is the first subject of this subsection. Damage and fracture studies are described afterwards.

In one early study, numerous 8-mm cubes of human cancellous bone (four cadavers without known bone disorders, various bones, and cube orientations) were imaged with 50-µm isotropic voxels prior to mechanical determination of modulus and ultimate strength (Goldstein et al., 1993; Goulet et al., 1994). Bone volume fractions ranged from 6 to 36 percent, trabeculae thickness from 0.10 to 0.19 mm, and spacings from 0.32 to 1.67 mm. In normal bone, strong correlations were found among the independent structural measures of bone volume fraction, trabecular plate number, and connectivity, and strong relationships were found for modulus and ultimate strength, accounting for 80 to 90 percent of the variance in these properties. Other work on modeling elastic moduli, based on three-dimensional networks defined by microCT, appear elsewhere (Hildebrand et al., 1997; Kinney et al., 1997; Ladd and Kinney, 1997).

Elastic moduli from indentation measurements have been determined for PTH- and FGF-treated bones and structures compared with microCT (Lane et al., 2003b). Bone modeling and simulation was the subject of another study (Müller, 1999). MicroCT-derived bone microstructures are routinely imported into finite element models and the response to mechanical loading investigated numerically (Kinney et al., 2000; Niebur et al., 2000; Borah et al., 2001; Homminga et al., 2001; van Rietbergen, 2001; van Rietbergen and Huiskes, 2001; Boyd et al., 2002; Hara et al., 2002; Pistoia et al., 2003; Stölken and Kinney, 2003; van Eijden et al., 2004).

Smoothing methods have been examined for automatic finite element model generation from microCT volumetric data (Boyd and Müller, 2006); this is important in terms of model efficiency and accuracy. In another microCT study, effects of three types of thresholding techniques applied to trabecular bone did not change predictions of apparent mechanical properties and structural properties, both of which agreed well with experimental measurements (Kim et al., 2007).

Selective laser sintering of polymer powders has been used to reproduce trabecular bone structures derived from microCT datasets (Cosmi and Dreossi, 2007; Su et al., 2007). Numerical modeling of the moduli of observed structure was then compared to moduli measured experimentally for the fabricated specimens in both these studies.

Most mechanical tests of trabecular bone are on excised specimens where the peripheral trabeculae have lost some of their load-bearing capability due to loss of connectivity. This introduces a bias in the measured value of the elastic modulus that must be corrected if models are to correctly represent the actual load-carrying capacity of the structure. MicroCT-based finite element models of aged human vertebral trabecular revealed that the widths over which the peripheral trabeculae were mostly unloaded were between 0.19 and 0.58 mm in ~8-mm-diameter cores (Ün et al., 2006). Note that trabecular thickness averages about 0.14 mm in humans (McCreadie et al., 2001) and is slightly greater than 0.10 mm in mice (McLaughlin et al., 2002). Actual experiments were found to underestimate the true, "side"-artifact-free Young's modulus by 27 percent on the average and by more than 50 percent in certain experiments (Ün et al., 2006).

As mentioned above, trabecular bone specimens can have primarily platelike elements or mainly rodlike elements. Based on their synchrotron microCT data, Follet et al. (2007) found one finite element model worked better predicting the compressive Young's modulus for specimens containing rodlike trabecular arrays (those with very low bone volume fraction), and a second was superior with platelike trabecular samples.

A digital correlation technique was developed for strain measurement in trabecular bone; typical values for the standard deviation of strain approached 0.01 (Verhulp et al., 2004), indicating that, at that level of development, the technique was limited to strain measurements beyond the yield strain. Trabecular bone can be expected to differ from site to site and between different mammals; several different digital volume correlation methods and microCT were used to compare real and simulated displacement fields for the bovine distal femur, bovine proximal tibia, rabbit distal femur, rabbit proximal tibia, rabbit vertebra, and human vertebra; across all bone types, differences in strains ranged between 345 and 794 µε (Liu and Morgan, 2007). Regional variations in apparent and tissue-level mechanical parameters were studied in 90 trabecular bone samples from six human L4 vertebrae combining microfinite element analysis and microCT; significant differences were observed (Gong et al., 2007), a very interesting result indeed.

Imaging of entire vertebrae allowed, via importation into a finite element model, assessment of the contribution of cortical and trabecular structure to strength; this study also compared actual and numerical estimates of yield stress for OVX versus control versus OVX plus alfacalcidol rats (Ito et al., 2002a). In a similar study, Eswaran and coworkers used microCT-derived finite element models of human vertebrae to determine that the biomechanical role of the thin cortical shell can be substantial and vary with position: 45 percent of the load carried at the midtransverse section versus 15 percent close to the endplates (Eswaran et al., 2006). Strength in bisphosphonate-treated bone has been characterized along with bone microarchitecture in OVX rats (Ito et al., 2003). Healthy and osteoporotic human proximal femora were compared using microCT microstructure and 3D micro- and continuum-finite element models of loading during a fall (Verhulp et al., 2006). Figure 8.4 compares maximum principal stress on the midplane of the healthy and diseased bones at the three different finite element resolutions; the different resolutions produced comparable results. Importation into numerical models is particularly valuable

Cellular or Trabecular Solids 193

in highly variable biological systems because the same starting structure can be perturbed virtually and the response determined numerically.

Mulder and coworkers focused on the distribution of strains within individual trabeculae as they developed in fetal and newborn porcine mandibular condylar bone (Mulder et al., 2007b). MicroCT measurements of the spatial distribution of mineral level within the trabecula (higher in the center, lower toward the surfaces) was the input for the finite element model, and the inhomogeneous mineral distribution and finite element results suggested to the authors that the mineral distribution contributed to development of a structure that is able to resist increasing loads without an increase in average deformation.

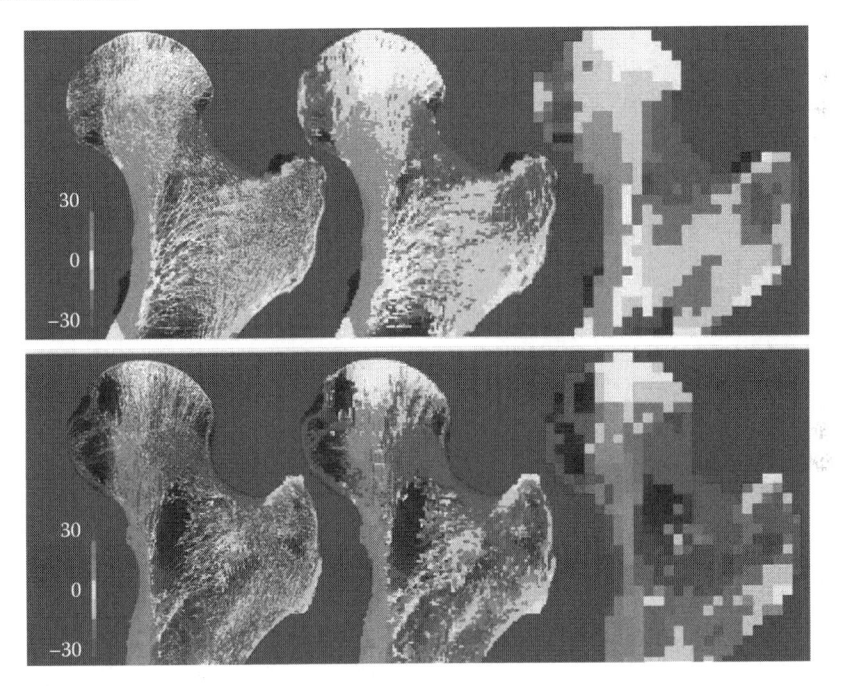

FIGURE 8.4 (SEE COLOR INSERT FOLLOWING PAGE 144.)
Finite element models at three resolutions for a healthy proximal femur (top row) and an osteoporotic femur (bottom row). Maximum principal stresses (MPa, in gray) from the 3D models are shown for the bones' midplanes; the stress values in the binned models (center and right) were divided by the local volume fraction of bone in order to estimate the average tissue-level stresses and to provide a better comparison with the highest-resolution model. The left column is the model based on 80-µm isotropic voxels, the elements in the middle column were enlarged 8 times (0.64-mm edges), and the elements in the right column were enlarged 38 times (3.04-mm edges). The head of the femur is at the upper left, the applied force of 1,000 N is distributed over the left side of the head, the nodes at the surface of the trochanter (far right side of the femur in the figure) were fixed in the horizontal direction, and the nodes at the distal end of the femur (bottom of the figure) were restricted to vertical motion. These loading conditions are those used in standard models of falling forces (Verhulp et al., 2006).

The position of teeth within the mandible can affect fractures at the mandibular angle; in particular, impacted third molars are associated clinically with increased risk of fracture at this site. MicroCT and finite element analysis were used to investigate the effect of partially impacted third molars on clinically observed fractures (Takada et al., 2006). The 3D microstructure did not differ between human mandibles with and without third molars, but the finite element analysis showed stress concentration around the root apex of the third molar and its transmission in a direction matching clinical findings of angle fractures.

Damage accumulation in cancellous bone specimens has been followed during in situ loading (Nazarian and Muller, 2004; Thurner et al., 2004; Nazarian et al., 2005). Yeh and Keaveny (2001) investigated the relative effects of trabecular microfracture and microdamage on reloading elastic modulus. In situ straining of bone was the subject of another study (Bleuet et al., 2004). Compressive fatigue testing of human vertebral trabecular bone led to the proposal of a simple power law relationship among volume fraction of bone, fabric eigenvalue, applied stress, and number of cycles to failure and the suggestion that fatigue life of such trabecular systems can be predicted for this loading mode (Rapillard et al., 2006).

8.5 Implants and Tissue Scaffolds

This section describes another very active research area with numerous microCT studies. The first subsection describes solid implants and bone formation; although the implants are not cellular materials, the bone formed fits into this category. Section 8.5.2 describes some of the scaffold processing studies. Section 8.5.3 covers bone ingrowth into scaffolds.

8.5.1 Implants

Bone formation around implants and the resulting structural integrity (or lack thereof) is a biomedical engineering topic of increasing importance as the number of hip (and knee) replacements increases. Examples of microCT studies include the 3D analysis of bone formation around Ti implants (Bernhardt et al., 2006), characterization of Mg bone implant degradation (Witte et al., 2006a,b), study of bone around dental implants (Cattaneo et al., 2004), and investigation of human tooth–alveolar bone complex (Dalstra et al., 2006). One complication in many such imaging studies is that the high absorptivity of many implants such as stainless steel or Ti washes out contrast in bone. In the author's experience, this can cripple analysis if contrast is confined to linear, 256-level grayscale; such an 8-bit approach seems to be favored in many different analyses, so this is not an academic concern. Use of nonlinear contrast scales or high-end clipping can be effective; contrast enhancements for bone structures adjacent to implants are described elsewhere (Tesei et al., 2005; DaPonte et al., 2006). In synchrotron microCT, phase contrast at edges of different low-absorption tissue types may also aid segmentation in the presence of a very absorbing material.

Itokawa and coworkers examined a novel composite material, PMMA containing hydroxyapatite particles, for closing cranial defects (Itokawa et al., 2007). After 12 months implantation in beagles, microCT revealed that bone was preferentially attached to hydroxyapatite particles intersecting the composite's surface.

Four-millimeter mid-diaphyseal defects in murine femoral were grafted with live autografts or processed allografts and allowed to heal for 6, 9, 12, or 18 weeks (Reynolds et al., 2007). Significant statistical correlations were found between combinations of microCT-derived structural parameters (graft and callus volume and cross-sectional polar moment of inertia) with measured ultimate torque and torsional rigidity.

Measurement of pre-existing bone geometry by CT and design of matching replacement implants via rapid prototyping techniques such as selective laser sintering is a developing area of clinical interest. MicroCT of a minipig model found significant bone growth interior and exterior to an implanted polycaprolactone condylar ramus unit at one- and three-month time points (Smith et al., 2007).

8.5.2 Scaffold Structures and Processing

Scaffolds for bone replacement via osteointegration have been actively researched (Hollister et al., 2001), and microCT is a natural tool for studying these structures that generally mimic trabecular bone. Biocompatible materials including scaffolds were discussed in Müller et al. (2001), and other reports of scaffold structures and function include Thurner et al. (2001, 2003), Donath et al. (2004), Thurner et al. (2004), and Irsen et al. (2006). In another study, porosity characteristics were quantified in six very different graft materials; an order of magnitude spread in specific internal surface areas was observed and the porosity was interconnecting in all but one material (Vanis et al., 2006).

Starch/polycaprolactone biodegradable scaffolds were produced by a 3D-plotting technology, and coverage of the scaffolds by bonelike apatite layers was investigated by microCT (Oliveira et al., 2007). Deposition under static, agitation, and circulating flow perfusion conditions was compared, and the last was the most effective at producing well-defined apatite layers in the inner parts of the scaffolds.

Partap and coworkers described controlled porosity alginate hydrogels synthesized by simultaneous micelle templating and internal gelation (Partap et al., 2007). MicroCT revealed relatively monodisperse pore sizes, high total pore volume, and high degrees of porosity. Higher surfactant concentrations produced smaller pores with lower pore volume, and the surfactant could be completely washed from the scaffolds, something that is essential for use in tissue engineering.

Fourier phase segmentation methods were used to study porosity in biodegradable poly (propylene fumarate) scaffolds formed by the solvent casting particulate (sodium chloride) leaching process (Rajagopalan et al., 2005). A similar NaCl dissolution processing route was used with a composite of polylactic acid and soluble calcium phosphate glass; a three-factor (NaCl particle size, glass particle size, glass volume fraction), two-level design of experiments approach was employed, and microCT characterization and finite element modeling of the structure were compared with stiffness measurements from compression tests (Charles-Harris et al., 2007).

Jones and coworkers examined foaming sol–gel-derived bioactive glass scaffolds, focusing on the pores and interconnects in the structure (Jones et al., 2007b). They extracted small volumes containing single pores, rendered only the pores, and labeled the portions of the surface comprising interconnects to adjacent pores. Their maps of predicted flow paths (Figure 8.5) and analyses of predicted permeability as a function of the size of the representative elemental volume are of particular interest.

8.5.3 Bone Growth into Scaffolds

Cancedda and coauthors recently reviewed bulk and surface studies on trabecular bone, scaffolds, and tissue-engineered bones (Cancedda et al., 2007). The combination of microCT imaging and microbeam x-ray scattering provides very useful information, and this paper is covered in more detail in Chapter 12.

Bone ingrowth into scaffolds occurs via several steps. Before bone mineral can be formed, the matrix material must be in place, and cell seeding must precede deposition of the matrix. Hofmann et al. (2007) used the NaCl particle dissolution route to fabricate porous silk scaffolds with designed gradients of pores (mimicking what is seen in vertebrae) and compared static versus dynamic mesenchymal stem cell seeding in the scaffolds. The fluid-dynamic microenvironment in tissue construct was examined in another microCT study (Cioffi et al., 2006).

Quantification of bone on hydroxyapatite scaffolds (Mastrogiacomo et al., 2004) and 3D study of bone ingrowth into calcium phosphate biomaterials (Weiss et al., 2003) are studies of interest. Komlev et al. (2006) examined highly porous hydroxyapatite scaffolds before implantation and after they were seeded in vitro with bone marrow stromal cells and implanted for 8, 16, or 24 weeks in immunodeficient mice and measured volume fraction, average thickness, and distribution of newly formed bone. Komlev and coworkers found new bone thickness increased from week 8 to week 16, new bone thickness did not increase from week 16 to week 24, but mineralization of the bone matrix continued during this time.

Cellular or Trabecular Solids

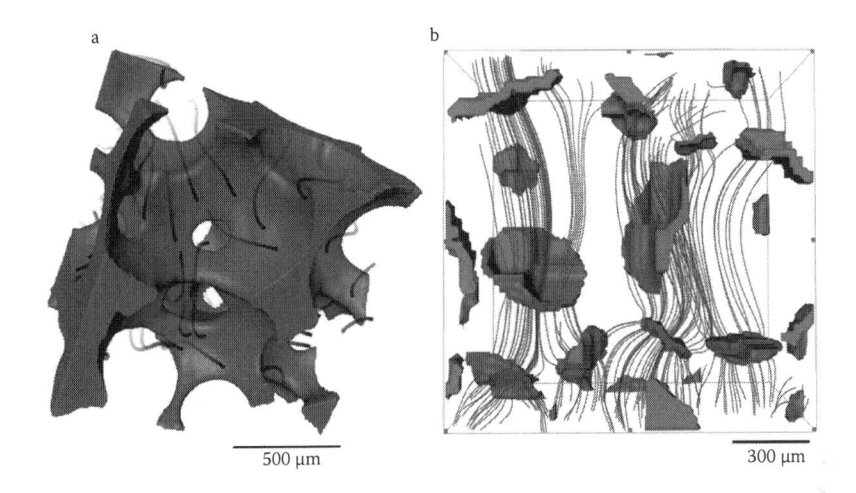

FIGURE 8.5
(a) Small subvolume of a scaffold rendered as a solid object, with flow illustrated by streak lines, ribbons representing the path a massless particle would follow. (b) A larger volume, with the solid removed and only the pore interconnects shown (i.e., the opposite of how the interconnects are shown in (a)). The streak lines converge at the interconnects and illustrate the importance of interconnect size on flow (Jones et al. 2007b).

Bone formation in polymeric scaffolds has been evaluated by proton magnetic resonance microscopy and microCT (Washburn et al., 2004). Other studies include pore interconnectivity in bioactive glass foams (Atwood et al., 2004), structure and properties of clinical coralline implants (Knackstedt et al., 2006), scaffolds of carbonated apatite–collagen sponges (Itoh et al., 2004), and polymer composite scaffolds (Mathieu et al., 2006).

Jones et al. (2007a) compared bone ingrowth in hydroxyapatite scaffolds with a cellular structure with those with a more strutlike structure. Porosity, pore size distribution, pore constriction sizes, and pore topology were measured for the original scaffold structure, and the distribution of bone ingrowth after 4- or 12-week implantation was quantified. For the early time period, growth was from the periphery of the scaffold, with a constant decrease in bone mineralization into the scaffold volume. Bone ingrowth was strongly enhanced for pore diameters >100 µm.

The efficacy of engineered scaffolds for bone formation is typically evaluated in nonweight-bearing locations or in the presence of stress-shielding devices (bone plates or external fixation), primarily because these scaffolds are not designed to carry weight. Chu and coworkers designed a weight-carrying, biodegradable, tubular porous scaffold for repair of segmental bone defects in rat femora (Chu et al., 2007). Scaffolds containing BMP-2 maintained bone length throughout and allowed bone bridging; those without collapsed after 15 weeks and failed to induce bridging.

References

Adrien, J., E. Maire, N. Gimenez, and V. Sauvant-Moynot (2007). Experimental study of the compression behaviour of syntactic foams by in situ X-ray tomography. *Acta Mater* **55**: 1667–1679.

Akhter, M.P., J.M. Lappe, K.M. Davies, and R.R. Recker (2007). Transmenopausal changes in the trabecular bone structure. *Bone* **41**: 111–116.

Almer, J.D. and S.R. Stock (2005). Internal strains and stresses measured in cortical bone via high-energy x-ray diffraction. *J Struct Biol* **152**: 14–27.

Almer, J.D. and S.R. Stock (2007). Micromechanical response of mineral and collagen phases in bone. *J Struct Biol* **157**: 365–370.

Amblard, D., M.H. Lafage-Proust, A. Laib, T. Thomas, P. Rüegsegger, C. Alexandre, and L. Vico (2003). Tail suspension induces bone loss in skeletally mature mice in the C57BL/6J strain but not in the C3H/HeJ strain. *J Bone Miner Res* **18**: 561–569.

Ammann, P., I. Badoud, S. Barraud, R. Dayer, and R. Rizzoli (2007). Strontium ranelate treatment improves trabecular and cortical intrinsic bone tissue quality, a determinant of bone strength. *J Bone Miner Res* **22**: 1419–1425.

Apostol, L. and F. Peyrin (2007). Connectivity analysis in very large 3D microtomographic images. *IEEE Trans Nucl Sci* **54**: 167–172.

Apostol, L., F. Peyrin, S. Yot, O. Basset, C. Odet, J. Tabary, J.M. Dinten, E. Boller, V. Boudousq, and P.O. Kotzki (2004). A procedure for the evaluation of 2D radiographic texture analysis to assess 3D bone micro-architecture. *Medical Imaging 2004: Image Processing*. J. M. Fitzpatrick and M. Sonka (Eds.). Bellingham, WA, SPIE. *SPIE Proc Vol* **5370**: 195–206.

Atwood, R.C., J.R. Jones, P.D. Lee, and L.L. Hench (2004). Analysis of pore interconnectivity in bioactive glass foams using x-ray microtomography. *Scripta Mater* **51**: 1029–1033.

Babin, P., G.D. Valle, H. Chiron, P. Cloetens, J. Hoszowska, P. Pernot, A.L. Réguerre, L. Salvo, and R. Dendievel (2006). Fast x-ray tomography analysis of bubble growth and foam setting during breadmaking. *J Cereal Sci* **43**: 393–397.

Badel, E., J.M. Letang, G. Peix, and D. Babot (2003). Quantitative microtomography: Measurement of density distribution in glass wool and local evolution during a one-dimensional compressive load. *Meas Sci Technol* **14**: 410–420.

Bagi, C.M., M. Volberg, M. Moalli, V. Shen, E. Olson, N. Hanson, E. Berryman, and C.J. Andresen (2007). Age-related changes in marmoset trabecular and cortical bone and response to alendronate therapy resemble human bone physiology and architecture. *Anat Rec* **290**: 1005–1016.

Balch, D.K., J.G. O'Dwyer, G.R. Davis, C.M. Cady, G.T.G. III and D. C. Dunand (2005). Plasticity and damage in aluminum syntactic foams deformed under dynamic and quasi-static conditions. *Mater Sci Eng A* **391**: 408–417.

Balto, K., R. Müller, D.C. Carrington, J. Dobeck, and P. Stashenko (2000). Quantification of periapical bone destruction in mice by microcomputed tomography. *J Dent Res* **79**: 35–40.

Bart-Smith, N., A.F. Bastawros, D.R. Mumm, A.G. Evans, D.J. Sypeck, and H.N.G. Wadley (1998). Compressive deformation and yielding mechanisms in cellular Al alloys determined using x-ray tomography and surface strain mapping. *Acta Mater* **46**: 3583–3592.

Bayat, S., L. Apostol, E. Boller, T. Borchard, and F. Peyrin (2005). In vivo imaging of bone microarchitecture in mice with 3D synchrotron radiation microtomography. *Nucl Instrum Meth A* **548**: 247–252.

Benaouli, A.H., L. Froyen, and M. Wevers (2000). Micro focus computed tomography of aluminium foams. *X-Ray Tomography in Materials Science*. J. Baruchel, J.Y. Buffière, E. Maire, P. Merle, and G. Peix (Eds.). Paris, Hermes Science: 139–154.

Benouali, A.H., L. Froyen, T. Dillard, S. Forest, and F. N'Guyen (2005). Investigation on the influence of cell shape anisotropy on the mechanical performance of closed cell aluminum foams using micro-computed tomography. *J Mater Sci* **40**: 5801–5811.

Bernhardt, R., D. Scharnweber, B. Müller, F. Beckmann, J. Goebbels, J. Jansen, H. Schliephake, and H. Worch (2006). 3D analysis of bone formation around Ti implants using microcomputed tomography (µCT). *Developments in X-Ray Tomography V*. U. Bonse (Ed.). Bellingham, WA, SPIE. *SPIE Proc Vol* **6318**: 631807-1 to –10.

Blacher, S., A. Léonard, B. Heinrichs, N. Tcherkassova, F. Ferauche, M. Crine, P. Marchot, E. Loukine, and J. P. Pirard (2004). Image analysis of x-ray microtomograms of Pd–Ag/SiO2 xerogels catalysts supported on Al2O3 foams. *Colloids Surf A* **241**: 201–206.

Bleuet, P., J.P. Roux, Y. Dabin, and G. Boivin (2004). In situ microtomography study of human bones under strain with synchrotron radiation. *Developments in X-ray Tomography IV*. U. Bonse (Ed.). Bellingham, WA, SPIE. *SPIE Proc Vol* **5535**: 129–136.

Bonse, U. and F. Busch (1996). X-ray computed microtomography (µCT) using synchrotron radiation. *Prog Biophys Molec Biol* **65**: 133–169.

Bonse, U., F. Busch, O. Gunnewig, F. Beckmann, G. Delling, M. Hahn, and A. Kvick (1996). Microtomography (µCT) applied to structure analysis of human bone biopsies. *ESRF Newsletter* (March): 21–23.

Bonse, U., F. Busch, O. Gunnewig, F. Beckmann, R. Pahl, G. Delling, M. Hahn, and W. Graeff (1994). 3-D computed x-ray tomography of human cancellous bone at 8 microns spatial and 10-4 energy resolution. *Bone Miner* **25**: 25–38.

Borah, B., G.J. Gross, T.E. Dufresne, T.S. Smith, M.D. Cockman, P.A. Chmielewski, M.W. Lundy, J.R. Hartke, and E.W. Sod (2001). Three-dimensional microimaging (MRµI and µCT), finite element modeling, and rapid prototyping provide unique insights into bone architecture in osteoporosis. *Anat Rec* **265**: 101–110.

Borah, B., E.L. Ritman, T.E. Dufresne, S.M. Jorgensen, S. Liu, J. Sacha, R.J. Phipps, and R.T. Turner (2005). The effect of risedronate on bone mineralization as measured by microcomputed tomography with synchrotron radiation: Correlation to histomorphometric indices of turnover. *Bone* **37**: 1–9.

Bourne, N.K., K. Bennett, A.M. Milne, S.A. MacDonald, J.J. Harrigan, and J.C.F. Millett (2008). The shock response of aluminium foams. *Scripta Mater* **58**: 154–157.

Bouxsein, M.L., K.S. Myers, K.L. Schultz, L.R. Donahue, C.J. Rosen, and W.G. Beamer (2005). Ovariectomy-induced bone loss varies among inbred strains of mice. *J Bone Min Res* **20**: 1085–1092.

Boyd, S.K. and R. Müller (2006). Smooth surface meshing for automated finite element model generation from 3D image data. *J Biomech* **39**: 1287–1295.

Boyd, S.K., R. Müller and R.F. Zernicke (2002). Mechanical and architectural bone adaptation in early stage experimental osteoarthritis. *J Bone Miner Res* **17**: 687–694.

Breunig, T.M., M.C. Nichols, J.S. Gruver, J.H. Kinney, and D.L. Haupt (1994). Servo-mechanical load frame for in situ, non-invasive, imaging of damage development. *Ceram Eng Sci Proc* **15**: 410–417.

Breunig, T.M., S.R. Stock, and R.C. Brown (1993). Simple load frame for in situ computed tomography and x-ray tomographic microscopy. *Mater Eval* **51**: 596–600.

Breunig, T.M., S.R. Stock, S.D. Antolovich, J.H. Kinney, W.N. Massey, and M.C. Nichols (1992). A framework relating macroscopic measures and physical processes of crack closure of Al–Li Alloy 2090. *Fracture Mechanics: Twenty-Second Symposium* (Vol. 1). H.A. Ernst, A. Saxena, and D.L. McDowell (Eds.). Philadelphia, ASTM. *ASTM STP* **1131**: 749–761.

Brunke, O., S. Oldenbach, and F. Beckmann (2004). Structural characterization of aluminium foams by means microcomputed tomography. *Developments in X-ray Tomography IV*. U. Bonse (Ed.). Bellingham, WA, SPIE. *SPIE Proc Vol* **5535**: 453–463.

Brunke, O., S. Oldenbach, and F. Beckmann (2005). Quantitative methods for the analysis of synchrotron-μCT datasets of metallic foams, Eur. *Eur J Appl Phys* **29**: 73–81.

Brydon, A.D., S.G. Bardenhagen, E.A. Miller, and G.T. Seidler (2005). Simulation of the densification of real open-celled foam microstructures. *J Mech Phys Sol* **53**: 2638–2660.

Burghardt, A.J., Y. Wang, H. Elalieh, X. Thibault, D. Bikle, F. Peyrin, and S. Majumdar (2007). Evaluation of fetal bone structure and mineralization in IGF-I deficient mice using synchrotron radiation microtomography and Fourier transform infrared spectroscopy. *Bone* **40**: 160–168.

Cancedda, R., A. Cedola, A. Giuliani, V. Komlev, S. Lagomarsino, M. Mastrogiacomo, F. Peyrin, and F. Rustichelli (2007). Bulk and interface investigations of scaffolds and tissue-engineered bones by x-ray microtomography and x-ray microdiffraction. *Biomater* **28**: 2505–2524.

Cattaneo, P.M., M. Dalstra, F. Beckmann, T. Donath, and B. Melsen (2004). Comparison of conventional and synchrotron-radiation-based microtomography of bone around dental implants. *Developments in X-ray Tomography IV*. U. Bonse (Ed.). Bellingham, WA, SPIE. *SPIE Proc Vol* **5535**: 757–764.

Chappard, C., F. Peyrin, A. Bonnassie, G. Lemineur, B. Brunet-Imbault, E. Lespessailles, and C.L. Benhamou (2006). Subchondral bone micro-architectural alterations in osteoarthritis: a synchrotron micro-computed tomography study. *Osteoarth Cartilage* **14**: 215–223.

Charles-Harris, M., S. del Valle, E. Hentges, P. Bleuet, D. Lacroix, and J.A. Planell (2007). Mechanical and structural characterisation of completely degradable polylactic acid/calcium phosphate glass scaffolds. *Biomater* **28**: 4429–4438.

Chu, T.M.G., S.J. Warden, C.H. Turner, and R.L. Stewart (2007). Segmental bone regeneration using a load-bearing biodegradable carrier of bone morphogenetic protein-2. *Biomater* **28**: 459–467.

Cioffi, M., F. Boschetti, M.T. Raimondi, and G. Dubini (2006). Modeling evaluation of the fluid-dynamic microenvironment in tissue-engineered construct: A microCT based model. *Biotech Bioeng* **93**: 500–510.

Cody, D.D., D.A. McCubbrey, G.W. Divine, G.J. Gross, and S.A. Goldstein (1996). Predictive value of proximal femoral bone densitometry in determining local orthogonal material properties. *J Biomech* **29**: 753–761.

Cosmi, F. and D. Dreossi (2007). Numerical and experimental structural analysis of trabecular architectures. *Meccan* **42**: 85–93.

Cowin, S.C. (1985). The relationship between the elasticity tensor and the fabric tensor. *Mech Mater* **4**: 137–147.

Cowin, S.C. (Ed.) (2001). *Bone Mechanics Handbook*. Boca Raton, FL: CRC Press.

Currey, J.D. (2002). *Bones — Structure and Mechanics*. Princeton, NJ: Princeton University Press.

Dalstra, M., P.M. Cattaneo, and F. Beckmann (2007). Synchrotron radiation-based microtomography of alveolar support tissues. *Orthod Craniofacial Res* **9**: 199–205.

Dalstra, M., P.M. Cattaneo, F. Beckmann, M.T. Sakima, G. Lemor, M.G. Laursen, and B. Melsen (2006). Microtomography of the human tooth–alveolar bone complex. *Developments in X-Ray Tomography V*. U. Bonse (Ed.). Bellingham, WA, SPIE. *SPIE Proc Vol* **6318**: 631804–1 to –9.

DaPonte, J.S., M. Clark, P. Nelson, T. Sadowski, and E. Wood (2006). Quantitative confirmation of visual improvements to microCT bone density images. *Visual Information Processing XV*. Z. Rahman, S.E. Reichenbach, and M.A. Neifeld (Eds.). Bellingham,WA, SPIE. *SPIE Proc Vol* **6246**: 62460D–1 to –9.

David, V., N. Laroche, B. Boudignon, M.H. Lafage-Proust, C. Alexandre, P. Rüegsegger, and L. Vico (2003). Noninvasive in vivo monitoring of bone architecture alterations in hindlimb-unloaded female rats using novel three-dimensional microcomputed tomography. *J Bone Miner Res* **18**: 1622–1631.

Dedrick, D.K., S.A. Goldstein, K.D. Brandt, B.L. O'Connor, R.W. Goulet, and M. Albrecht (1993). A longitudinal study of subchondral plate and trabecular bone in cruciate-deficient dogs with osteoarthritis followed up for 54 months. *Arthritis Rheum* **36**: 1460–1467.

Dedrick, D.K., R. Goulet, L. Huston, S.A. Goldstein, and G.G. Bole (1991). Early bone changes in experimental osteoarthritis using microscopic computed tomography. *J Rheumatol Suppl* **27**: 44–45.

Demetriou, M.D., J.C. Hanan, C. Veazey, M. Di Michiel, N. Lenoir, E. Üstündag, and W.L. Johnson (2007). Yielding of metallic glass foam by percolation of an elastic buckling instability. *Adv Mater* **19**: 1957–1962.

Dempster, D.W., F. Cosman, E. Kurland, H. Zhou, J. Nieves, L. Woelfert, E. Shane, K. Plaveti, M. Müller, J. Bilezikian, and R. Lindsay (2001). Effects of daily treatment with parathyroid hormone on bone microarchitecture and turnover in patients with osteoporosis: A paired biopsy study. *J Bone Miner Res* **16**: 1846–1853.

Desischer, H.P., A. Kottar, and B. Foroughi (2000). Determination of local mass density distribution. *X-Ray Tomography in Materials Science*. J. Baruchel, J.Y. Buffière, E. Maire, P. Merle, and G. Peix (Eds.). Paris, Hermes Science: 165–176.

Di Prima, M.A., M. Lesniewski, K. Gall, D.L. McDowell, T. Sanderson, and D. Campbell (2007). Thermomechanical behavior of epoxy shape memory polymer foams. *Smart Mater Struct* **16**: 2330–2340.

Dillard, T., F. N'Guyen, E. Maire, L. Salvo, S. Forest, Y. Bienvenu, J.D. Bartout, M. Croset, R. Dendievel, and P. Cloetens (2005). 3D quantitative image analysis of open-cell nickel foams under tension and compression loading using x-ray microtomography. *Phil Mag* **85**: 2147–2175.

Donath, T., F. Beckmann, R.G.J.C. Heijkants, O. Brunke, and A. Schreyer (2004). Characterization of polyurethane scaffolds using synchrotron-radiation-based computed microtomography. *Developments in X-ray Tomography IV*. U. Bonse (Ed.). Bellingham, WA, SPIE. *SPIE Proc Vol* **5535**: 775–782.

Dubrow, S.A., P.M. Hruby, and M.P. Akhter (2007). Gender specific LRP5 influences on trabecular bone structure and strength. *J Musculoskelet Neuronal Interact* **7**: 166–173.

Dufresne, T.E., P.A. Chmielewski, M.D. Manhart, T.D. Johnson, and B. Borah (2003). Risedronate preserves bone architecture in early postmenopausal women in 1 year as measured by three-dimensional microcomputed tomography. *Calcif Tiss Int* **73**: 423–432.

Eckstein, F., M. Matsuura, V. Kuhn, M. Priemel, R. Müller, T.M. Link, and E.M. Lochmüller (2007). Sex differences of human trabecular bone microstructure in aging are site-dependent. *J Bone Min Res* **22**: 817–824.

Elliott, J.A., A.H. Windle, J.R. Hobdell, G. Eeckhaut, R.J. Oldman, W. Ludwig, E. Boller, P. Cloetens, and J. Baruchel (2002). In-situ deformation of an open-cell flexible polyurethane foam characterised by 3D computed microtomography. *J Mater Sci* **37**: 1547–1555.

Elliott, J.C., P. Anderson, X.J. Gao, F.S.L. Wong, G.R. Davis, and S.E.P. Dowker (1994). Application of scanning microradiography and x-ray microtomography to studies of bone and teeth. *J X-ray Sci Technol* **4**: 102–117.

Elmoutaouakkil, A., G. Fuchs, P. Bergounhon, R. Péres, and F. Peyrin (2003). Three-dimensional quantitative analysis of polymer foams from synchrotron radiation x-ray microtomography. *J Phys D* **36**: A37–A43.

Engelke, K., G. Umgießer, S. Prevrhal, and W. Kalendar (1997). Three-dimensional analysis of trabecular bone structure: The need for spongiosa standard models. *Developments in X-ray Tomography*. U. Bonse (Ed.). Bellingham, WA, SPIE. *SPIE Proc Vol* **3149**: 53–61.

Eswaran, S.K., A. Gupta, M.F. Adams, and T.M. Keaveny (2006). Cortical and trabecular load sharing in the human vertebral body. *J Bone Min Res* **21**: 307–314.

Fajardo, R.J., E. Hernandez, and P.M. O'Connor (2007). Post cranial skeletal pneumaticity: A case study in the use of quantitative microCT to assess vertebral structure in birds. *J Anat* **211**: 138–147.

Fan, L. and C.W. Chen (1999). Integrated approach to 3D warping and registration from lung images. *Developments in X-ray Tomography II*. U. Bonse (Ed.). Bellingham, WA, SPIE. *SPIE Proc Vol* **3772**: 24–35.

Feldkamp, L.A., S.A. Goldstein, A.M. Parfitt, G. Jesion, and M. Kleerekopes (1989). The direct examination of three-dimensional bone architecture in vitro by computed tomography. *J Bone Miner Res* **4**: 3–11.

Follet, H., F. Peyrin, E. Vidal-Salle, A. Bonnassie, C. Rumelhart, and P.J. Meunier (2007). Intrinsic mechanical properties of trabecular calcaneus determined by finite-element models using 3D synchrotron microtomography. *J Biomech* **40**: 2174–2183.

Friis, E.M., P.R. Crane, K.R. Pedersen, S. Bengtson, P.C.J. Donoghue, G.W. Grimm, and M. Stampanoni (2007). Phase contrast x-ray microtomography links Cretaceous seeds with Gnetales and Bennettitales. *Nature* **450**: 549–552.

Gibson, L.J. and M.F. Ashby (1999). *Cellular Solids: Structure and Properties*, 2nd ed. Cambridge: Cambridge University Press.

Gittens, S.A., G.R. Wohl, R.F. Zernicke, J.R. Matyas, P. Morley, and H. Uludag (2004). Systemic bone formation with weekly PTH administration in ovariectomized rats. *J Pharm Pharmaceut Sci* **7**: 27–37.

Glatt, V., E. Canalis, L. Stadmeyer, and M.L. Bouxsein (2007). Age-related changes in trabecular architecture differ in female and male C57BL/6J mice. *J Bone Min Res* **22**: 1197–1207.

Goldstein, S.A., R. Goulet, and D. McCubbrey (1993). Measurement and significance of three-dimensional architecture to the mechanical integrity of trabecular bone. *Calcif Tissue Int* **53**(Suppl 1): S127–S133.

Gong, H., M. Zhang, L. Qin, and Y. Hou (2007). Regional variations in the apparent and tissue-level mechanical parameters of vertebral trabecular bone with aging using micro-finite element analysis. *Annal Biomed Eng* **35**: 1622–1631.

Goodall, R., A. Marmottant, L. Salvo, and A. Mortensen (2007). Spherical pore replicated microcellular aluminium: Processing and influence on properties. *Mater Sci Eng A* **465**: 124–135.

Goulet, R.W., S.A. Goldstein, M.J. Ciarelli, J.L. Kuhn, M.B. Brown, and L.A. Feldkamp (1994). The relationship between the structural and orthogonal compressive properties of trabecular bone. *J Biomech* **27**: 375–389.

Guldberg, R.E., N.J. Caldwell, X.E. Guo, R.W. Goulet, S.J. Hollister, and S.A. Goldstein (1997). Mechanical stimulation of tissue repair in the repair in the hydraulic bone chamber *J Bone Miner Res* **12**: 1295–1302.

Hara, T., E. Tanck, J. Homminga, and R. Huiskes (2002). The influence of micro-computed tomography threshold variations on the assessment of structural and mechanical trabecular bone properties. *Bone* **31**: 107–109.

Helfen, L., T. Baumbach, P. Pernot, P. Cloetens, H. Stanzick, K. Schladitz, and J. Banhart (2005). Investigation of pore initiation in metal foams by synchrotron-radiation tomography. *Appl Phys Lett* **86**: 231907–1 to –3.

Hildebrand, T. and P. Rüegsegger (1997). Quantification of bone microarchitecture with the structure model index. *Comp Meth Biomed Eng* **1**: 15–23.

Hildebrand, T., A. Laib, D. Ulrich, A. Kohlbrenner, and P. Rüegsegger (1997). Bone structure as revealed by microtomography. *Developments in X-ray Tomography*. U. Bonse (Ed.). Bellingham, WA, SPIE. *SPIE Proc Vol* **3149**: 34–43.

Hirano, T., K. Usami, Y. Tanaka, and C. Masuda (1995). In situ x-ray CT under tensile loading using synchrotron radiation. *J Mater Res* **10**: 381–385.

Hofmann, S., H. Hagenmüller, A.M. Koch, R. Müller, G. Vunjak-Novakovic, D.L. Kaplan, H.P. Merkle, and L. Meinel (2007). Control of in vitro tissue-engineered bone-like structures using human mesenchymal stem cells and porous silk scaffolds. *Biomater* **28**: 1152–1162.

Hollister, S.J., T.M.G. Chu, J.W. Halloran, and S.E. Feinberg (2001). Design and manufacture of bone replacement scaffolds. *Bone Mechanics Handbook*, 2nd ed. S.C. Cowin (Ed.). Boca Raton, FL, CRC Press: 14–1 to –19.

Homminga, J., R. Huiskes, B. v. Rietbergen, P. Rüegsegger, and H. Weinans (2001). Introduction and evaluation of a gray-value voxel conversion technique. *J Biomech* **34**: 513–517.

Illman, B. and B. Dowd (1999). High resolution microtomography for density and spatial information about wood structures. *Developments in X-ray Tomography II*. U. Bonse (Ed.). Bellingham, WA, SPIE. *SPIE Proc Vol* **3772**: 198–204.

Irsen, S.H., B. Leukers, B. Bruckschen, C. Tille, H. Seitz, F. Beckmann, and B. Müller (2006). Image-based analysis of the internal microstructure of bone replacement scaffolds fabricated by 3D printing. *Developments in X-Ray Tomography V*. U. Bonse (Ed.). Bellingham, WA, SPIE. *SPIE Proc Vol* **6318**: 631809–1 to –10.

Ishijima, M., K. Tsuji, S.R. Rittling, T. Yamashita, H. Kurosawa, D.T. Denhardt, A. Nifuji, and M. Noda (2002). Resistance to unloading-induced three-dimensional bone loss in osteopontin-deficient mice. *J Bone Miner Res* **17**: 661–667.

Ito, M., Y. Azuma, H. Takagi, T. Kamimura, K. Komoriya, T. Ohta, and H. Kawaguchi (2003). Preventive effects of sequential treatment with alendronate and 1 alpha-hydroxyvitamin D3 on bone mass and strength in ovariectomized rats. *Bone* **33**: 90–99.

Ito, M., A. Nishida, A. Koga, S. Ikeda, A. Shiraishi, M. Uetani, K. Hayashi, and T. Nakamura (2002a). Contribution of trabecular and cortical components to the mechanical properties of bone and their regulating parameters. *Bone* **31**: 351–358.

Ito, M., A. Nishida, T. Nakamura, M. Uetani, and K. Hayashi (2002b). Differences of three-dimensional trabecular microstructure in osteopenic rat models caused by ovariectomy and neurectomy. *Bone* **30**: 594–598.

Itoh, M., A. Shimazu, I. Hirata, Y. Yoshida, H. Shintani, and M. Okazaki (2004). Characterization of CO_3Ap-collagen sponges using x-ray high resolution microtomography. *Biomater* **25**: 2577–2583.

Itokawa, H., T. Hiraide, M. Moriya, M. Fujimoto, G. Nagashima, R. Suzuki, and T. Fujimoto (2007). A 12 month in vivo study on the response of bone to a hydroxyapatite–polymethylmethacrylate cranioplasty composite. *Biomater* **28**: 4922–4927.

Jennane, R., R. Harba, G. Lemineur, S. Bretteil, A. Estrade, and C.L. Benhamou (2007). Estimation of the 3D self-similarity parameter of trabecular bone from its 2D projection. *Med Image Anal* **11**: 91–98.

Jones, A.C., C.H. Arns, A.P. Sheppard, D.W. Hutmacher, B.K. Milthorpe, and M.A. Knackstedt (2007a). Assessment of bone ingrowth into porous biomaterials using microCT. *Biomater* **28**: 2491–2504.

Jones, J.R., G. Poologasundarampillai, R.C. Atwood, D. Bernard, and P.D. Lee (2007b). Non-destructive quantitative 3D analysis for the optimisation of tissue scaffolds. *Biomater* **28**: 1404–1413.

Joo, Y.I., T. Sone, M. Fukunaga, S.G. Lim, and S. Onodera (2003). Effects of endurance exercise on three-dimensional trabecular bone microarchitecture in young growing rats. *Bone* **33**: 485–493.

Kádár, C., E. Maire, A. Borbély, G. Peix, J. Lendvai, and Z. Rajkovits (2004). X-ray tomography and finite element simulation of the indentation behavior of metal foams. *Mater Sci Eng A* **387–389**: 321–325.

Keaveny, T.M. (2001). Strength of trabecular bone. *Bone Mechanics Handbook*, 2nd ed. S.C. Cowin (Ed.). Boca Raton, FL, CRC Press: 16–1 to –42.

Kim, C.H., H. Zhang, G. Mikhail, D. von Stechow, R. Müller, H.S. Kim, and X.E. Guo (2007). Effects of thresholding techniques on μCT-based finite element models of trabecular bone. *J Biomed Eng* **129**: 481–486.

Kinney, J.H. and J.T. Ryaby (2001). Resonant markers for noninvasive, three-dimensional dynamic bone histomorphometry with x-ray microtomography. *Rev Sci Instrum* **72**: 1921–1923.

Kinney, J.H., D.L. Haupt, and A.J.C. Ladd (1997). Applications of synchrotron microtomography in osteoporosis research. *Developments in X-ray Tomography*. U. Bonse (Ed.). Bellingham, WA, SPIE. *SPIE Proc Vol* **3149**: 64–68.

Kinney, J.H., D.L. Haupt, M. Balooch, A.J.C. Ladd, J.T. Ryaby, and N.E. Lane (2000). Three-dimensional morphometry of the L6 vertebra in the ovariectomized rat model of osteoporosis: Biomechanical implications. *J Bone Miner Res* **15**: 1981–1991.

Kinney, J.H., N.E. Lane, and D.L. Haupt (1995). In vivo, three-dimensional microscopy of trabecular bone. *J Bone Min Res* **10**: 264–270.

Kinney, J.H., N. Lane, S. Majumdar, S.J. Marshall, and G.W.J. Marshall (1992). Noninvasive three-dimensional histomorphology using x-ray tomographic microscopy. *J Bone Miner Res* **7**(Suppl 1): S136.

Kinney, J.H., G.W. Marshall, S.J. Marshall, and D.L. Haupt (2001). Three-dimensional imaging of large compressive deformations in elastomeric foams. *J Appl Polym Sci* **80**: 1746–1755.

Kirby, B.J., J.R. Davis, J.A. Grant, and M.J. Morgan (1997). Monochromatic microtomographic imaging of osteoporotic bone. *Phys Med Biol* **42**: 1375–1385.

Kleuker, U. and C. Schulze (1997). A novel approach to the Rayleigh-to-Compton method: Wavelength dispersive tomography. *Developments in X-ray Tomography*. U. Bonse (Ed.). Bellingham, WA, SPIE. *SPIE Proc Vol* **3149**: 177–185.

Knackstedt, M.A., C.H. Arns, T.J. Senden, and K. Gross (2006). Structure and properties of clinical coralline implants measured via 3D imaging and analysis. *Biomater* **27**: 2776–2786.

Komlev, V.S., F. Peyrin, M. Mastrogiacomo, A. Cedola, A. Papadimitropoulos, F. Rustichelli, and R. Cancedda (2006). Kinetics of in vivo bone deposition by bone marrow stromal cells into porous calcium phosphate scaffolds: An X-ray computed microtomography study. *Tiss Eng* **12**: 3449–3458.

Kuhn, J.L., R.W. Goulet, M. Pappas, and S.A. Goldstein (1990). Morphometric and anisotropic symmetries of the canine distal femur. *J Orthop Res* **8**: 776–780.

Kurth, A.A. and R. Müller (2001). The effect of an osteolytic tumor on the three-dimensional trabecular bone morphology in an animal model. *Skeletal Radiol* **30**: 94–98.

Ladd, A.J.C. and J.H. Kinney (1997). Elastic constants of cellular structures. *Physica A* **240**: 349–360.

Lambert, J., I. Cantat, R. Delannay, A. Renault, F. Graner, J.A. Glazier, I. Veretennikov, and P. Cloetens (2005). Extraction of relevant physical parameters from 3D images of foams obtained by x-ray tomography. *Colloids Surf A* **263**: 295–302.

Lane, N.E., J. Kumer, W. Yao, T. Breunig, T. Wronski, G. Modin, and J.H. Kinney (2003a). Basic fibroblast growth factor forms new trabeculae that physically connect with pre-existing trabeculae, and this new bone is maintained with an anti-resorptive agent and enhanced with an anabolic agent in an osteopenic rat model. *Osteoporos Int* **14**: 374–382.

Lane, N.E., J.M. Thompson, D. Haupt, D.B. Kimmel, G. Modin, and J.H. Kinney (1998). Acute changes in trabecular bone connectivity and osteoclast activity in the ovariectomized rat in vivo. *J Bone Miner Res* **13**: 229–236.

Lane, N.E., W. Yao, J.H. Kinney, G. Modin, M. Balooch, and T.J. Wronski (2003b). Both hPTH(1–34) and bFGF increase trabecular bone mass in osteopenic rats but they have different effects on trabecular bone architecture. *J Bone Miner Res* **18**: 2105–2115.

Layton, M.W., S.A. Goldstein, R.W. Goulet, L.A. Feldkamp, D.J. Kubinski, and G.G. Bole (1988). Examination of subchondral bone architecture in experimental osteoarthritis by microscopic computed axial tomography. *Arthritis Rheum* **31**: 1400–1405.

Lee, S.B. (1993). Nondestructive examination of chemical vapor infiltration of 0°/90° SiC/Nicalon composites, Ph.D. Thesis. Atlanta, Georgia Institute of Technology.

Léonard, A., S. Blacher, C. Nimmol, and S. Devahastin (2008). Effect of far-infrared radiation assisted drying on microstructure of banana slices: An illustrative use of X-ray microtomography in microstructural evaluation of a food product. *J Food Eng* **85**: 154–162.

Leonard, A., L.P. Guiot, J.P. Pirard, M. Crine, M. Balligand, and S. Blacher (2007). Non-destructive characterization of deer (*Cervus Elaphus*) antlers by X-ray microtomography coupled with image analysis. *J Microsc* **225**: 258–263.

Liu, L. and E.F. Morgan (2007). Accuracy and precision of digital volume correlation in quantifying displacements and strains in trabecular bone. *J Biomech* **40**: 3516–3520.

Lotinun, S., G.L. Evans, J.T. Bronk, M.E. Bolander, T.J. Wronski, E.L. Ritman, and R.T. Turner (2004). Continuous parathyroid hormone induces cortical porosity in the rat: Effects on bone turnover and mechanical properties. *J Bone Miner Res* **19**: 1165–1171.

Luo, G., J.H. Kinney, J.L. Kaufman, D. Haupt, A. Chiabrera, and R.S. Siffert (1999). Relationship between plain radiographic patterns and three-dimensional trabecular architecture in the human calcaneus. *Osteoporos Int* **9**: 339–345.

MacLatchy, L. and R. Müller (2002). A comparison of the femoral head and neck trabecular architecture of Galago and Perodicticus using microcomputed tomography. *J Hum Evol* **43**: 89–105.

Madi, K., S. Forest, M. Boussuge, S. Gailliègue, E. Lataste, J.Y. Buffière, D. Bernard, and D. Jeulin (2007). Finite element simulations of the deformation of fused-cast refractories based on X-ray computed tomography. *Compu Mater Sci* **39**: 224–229.

Maire, E. and J. Buffière (2000). X-ray tomography of aluminium foams and Ti/SiC composites. *X-Ray Tomography in Materials Science*. J. Baruchel, J.Y. Buffière, E. Maire, P. Merle, and G. Peix (Eds.). Paris, Hermes Science: 115–126.

Maire, E., A. Elmoutaouakkil, A. Fazekas, and L. Salvo (2003a). In situ x-ray tomography measurements of deformation in cellular solids. *MRS Bull* **28**: 284–289.

Maire, E., A. Fazekas, L. Salvo, R. Dendievel, S. Youssef, P. Cloetens, and J.M. Letang (2003b). X-ray tomography applied to the characterization of cellular materials. Related finite element modeling problems. *Compos Sci Technol* **63**: 2431–2443.

Majumdar, S. and B. K. Bay (Eds.) (2001). *Hierarchical Structure of Bone and Micro-Computed Tomography. Noninvasive Assessment of Trabecular Bone Architecture and the Competence of Bone*. New York: Kluwer.

Martin, E.A., E.L. Ritman, and R.T. Turner (2003). Time course of epiphyseal growth plate fusion in rat tibiae. *Bone* **32**: 261–267.

Martin, R.B., D.B. Burr, and N.A. Sharkey (1998). *Skeletal Tissue Mechanics*. New York: Springer.

Martín-Badosa, E., D. Amblard, S. Nuzzo, A. Elmoutaouakkilo, L. Vico, and F. Peyrin (2003). Excised bone structures in mice: Imaging at three-dimensional synchrotron microCT. *Radiol* **229**: 921–928.

Mastrogiacomo, M., V.S. Komlev, M. Hausard, F. Peyrin, F. Turquier, S. Casari, A. Cedola, F. Rustichelli, and R. Cancedda (2004). Synchrotron radiation microtomography of bone engineered from bone marrow stromal cells. *Tiss Eng* **10**: 1767–1774.

Mathieu, L.M., T.L. Mueller, P.E. Bourban, D.P. Pioletti, R. Müller, and J.A.E. Månson (2006). Architecture and properties of anisotropic polymer composite scaffolds for bone tissue engineering. *Biomater* **27**: 905–916.

Matsumoto, Y., A.H. Brothers, S.R. Stock, and D.C. Dunand (2007). Uniform and graded chemical milling of aluminum foams. *Mater Sci Eng A* **447**: 150–157.

McCreadie, B.R., R.W. Goulet, L.A. Feldkamp, and S.A. Goldstein (2001). Hierarchical structure of bone and micro-computed tomography. *Noninvasive Assessment of Trabecular Bone Architecture and the Competence of Bone*. S. Majumdar and B.K. Bay (Eds.). New York, Kluwer: 67–83.

McLaughlin, F., J. Mackintosh, B.P. Hayes, A. McLaren, I.J. Uings, P. Salmon, J. Humphreys, E. Meldrum, and S.N. Farrow (2002). Glucocorticoid-induced osteopenia in the mouse as assessed by histomorphometry, microcomputed tomography and biochemical markers. *Bone* **30**: 924–930.

Mendoza, F., P. Verboven, H.K. Mebatsion, G. Kerckhofs, M. Wevers, and B. Nicolai (2007). Three-dimensional pore space quantification of apple tissue using x-ray computed tomography. *Planta* **226**: 559–570.

Montanini, R. (2005). Measurement of strain rate sensitivity of aluminium foams for energy dissipation. *Int J Mech Sci* **47**: 26–42.

Morenko, B.J., S.E. Bove, L. Chen, R.E. Guzman, P. Juneau, T.M.A. Bocan, G.K. Peter, R. Arora, and K.S. Kilgore (2004). In vivo micro computed tomography of subchondral bone in the rat after intra-articular administration of monosodium iodoacetate. *Contemp Top Lab Anim Sci* **43**: 39–43.

Mulder, L., J.H. Koolstra, J.M.J. den Toonder, and T.M.G.J. van Eijden (2007a). Intratrabecular distribution of tissue stiffness and mineralization in developing trabecular bone. *Bone* **41**: 256–265.

Mulder, L., L.J. van Ruijven, J.H. Koolstra, and T.M.G.J. van Eijden (2007b). Biomechanical consequences of developmental changes in trabecular architecture and mineralization of the pig mandibular condyle. *J Biomech* **40**: 1575–1582.

Mulder, L., L.J. van Ruijven, J.H. Koolstra, and T.M.G.J. van Eijden (2007c). The influence of mineralization on intrabecular stress and strain distribution in developing trabecular bone. *Annal Biomed Eng* **35**: 1668–1677.

Müller, B., P. Thurnier, F. Beckmann, T. Weitkamp, C. Rau, R. Bernhardt, E. Karamuk, L. Eckert, J. Brandt, S. Buchloh, E. Wintermantel, D. Scharnweber, and H. Worch (2001). Nondestructive three-dimensional evaluation of biocompatible materials by microtomography using synchrotron radiation. *Developments in X-ray Tomography III*. U. Bonse (Ed.). Bellingham, WA, SPIE. *SPIE Proc Vol* **4503**: 178–188.

Müller, R. (1999). Microtomographic imaging in the process of bone modeling and simulation. *Developments in X-ray Tomography II*. U. Bonse (Ed.). Bellingham, WA, SPIE. *SPIE Proc Vol* **3772**: 63–76.

Müller, R. and W.C. Hayes (1997). Biomechanical competence of microstructural bone in the progress of adaptive bone remodeling. *Developments in X-ray Tomography*. U. Bonse (Ed.). Bellingham, WA, SPIE. *SPIE Proc Vol* **3149**: 69–81.

Nazarian, A. and R. Muller (2004). Time-lapsed microstructural imaging of bone failure behavior. *J Biomech* **37**: 55–65.

Nazarian, A., M. Stauber, and R. Müller (2005). Design and implementation of a novel mechanical testing system for cellular solids. *J Biomed Mater Res* **73B**: 400–411.

Nicholson, E.K., S.R. Stock, M.W. Hamrick, and M.J. Ravosa (2006). Biomineralization and adaptive plasticity of the temporomandibular joint in myostatin knock-out mice. *Arch Oral Biol* **51**: 37–49.

Niebur, G.L., M.J. Feldstein, J.C. Yuen, T.J. Chen, and T.M. Keaveny (2000). High-resolution finite element models with tissue strength asymmetry accurately predict failure of trabecular bone. *J Biomech* **33**: 1575–1583.

Nuzzo, S., M.H. Lafage-Proust, E. Martin-Badosa, G. Boivin, T. Thomas, C. Alexandre, and F. Peyrin (2002). Synchrotron radiation microtomography allows the analysis of three-dimensional microarchitecture and degree of mineralization of human iliac crest biopsy specimens: Effects of etidronate treatment. *J Bone Miner Res* **17**: 1372–1382.

Nuzzo, S., C. Meneghini, P. Braillon, R. Bouvier, S. Mobilio, and F. Peyrin (2003). Microarchitectural and physical changes during fetal growth in human vertebral bone. *J Bone Miner Res* **18**: 760–768.

Odgard, A. and H.J.G. Gunderson (1993). Quantification of connectivity in cancellous bone, with special emphasis on 3-D reconstructions. *Bone* **14**: 173–182.

Odgard, A. (1997). Three-dimensional methods for quantification of cancellous bone architecture. *Bone* **20**: 315–328.

Odgard, A. (2001). Quantification of cancellous bone architecture. *Bone Mechanics Handbook*, 2nd ed. S.C. Cowin (Ed.). Boca Raton, FL, CRC Press: 14–1 to –19.

Ohgaki, T., H. Toda, M. Kobayashi, K. Uesugi, M. Niinom, T. Akahori, T. Kobayashi, K. Makii, and Y. Aruga (2006). In situ observations of compressive behaviour of aluminium foams by local tomography using high-resolution tomography. *Phil Mag* **86**: 4417–4438.

Oliveira, A.L., P.B. Malafaya, S.A. Costa, R.A. Sousa, and R.L. Reis (2007). Microcomputed tomography as a potential tool to assess the effect of dynamic coating routes on the formation of biomimetic apatite layers on 3D-plotted biodegradable polymeric scaffolds. *J Mater Sci Mater Med* **18**: 211–223.

Ozcivici, E., R. Garman, and S. Judex (2007). High-frequency oscillatory motions enhance the simulated mechanical properties of non-weight bearing trabecular bone. *J Biomech* **40**: 3404–3411.

Partap, S., A. Muthutantri, I.U. Rehman, G.R. Davis, and J.A. Darr (2007). Preparation and characterisation of controlled porosity alginate hydrogels made via a simultaneous micelle templating and internal gelation process. *J Mater Sci* **42**: 3502–3507.

Patel, V., A.S. Issever, A. Burghardt, A. Laib, M. Ries, and S. Majumdar (2003). MicroCT evaluation of normal and osteoarthritic bone structure in human knee specimens. *J Orthop Res* **21**: 6–13.

Pettit, A.R., H. Ji, D. v. Stechow, R. Müller, S.R. Goldring, Y. Choi, C. Benoist, and E.M. Gravallese (2001). TRANCE/RANKL knockout mice are protected from bone erosion in a serum transfer model of arthritis. *Am J Pathol* **159**: 1689–1699.

Peyrin, F., M. Salome, P. Cloetens, A.M. Laval-Jeantet, E. Ritman, and P. Rüegsegger (1998). MicroCT examinations of trabecular bone samples at different resolutions: 14, 7 and 2 micron level. *Technol Health Care* **6**: 391–401.

Peyrin, F., M. Salomé, F. Denis, P. Braillon, A.M. Laval-Jeanlet, and P. Cloetens (1997). 3D imaging of fetus vertebra by synchrotron radiation microtomography. *Developments in X-ray tomography*. U. Bonse (Ed.). Bellingham, WA, SPIE. *SPIE Proc Vol* **3149**: 44–52.

Pistoia, W., B. v. Rietbergen, and P. Rüegsegger (2003). Mechanical consequences of different scenarios for simulated bone atrophy and recovery in the distal radius. *Bone* **33**: 937–945.

Plougonven, E., D. Bernard, and P. Viot (2006). Quantitative analysis of the deformation of polypropylene foam under dynamic loading. *Developments in X-Ray Tomography V*. U. Bonse (Ed.). Bellingham, WA, SPIE. *SPIE Proc Vol* **6318**: 631813–1 to –10.

Rajagopalan, S., L. Lu, M.J. Yaszemski, and R.A. Robb (2005). Optimal segmentation of microcomputed tomographic images of porous tissue-engineering scaffolds. *J Biomed Mater Res* **75A**: 877–887.

Rapillard, L., M. Charlebois, and P.K. Zysset (2006). Compressive fatigue behavior of human vertebral trabecular bone. *J Biomech* **39**: 2133–2139.

Ravosa, M.J., E.P. Kloop, J. Pinchoff, S.R. Stock, and M. Hamrick (2007a). Plasticity of mandibular biomineralization in myostatin-deficient mice. *J Morphol*: **268**: 275–282.

Ravosa, M.J., R. Kunwar, E.K. Nicholson, E.B. Klopp, J. Pinchoff, S.R. Stock, and M.S. Stack (2006). Adaptive plasticity in mammalian masticatory joints. *Developments in X-ray Tomography V*. U. Bonse (Ed.). Bellingham, WA, SPIE. *SPIE Proc Vol* **6318**: 63180D–1 to –9.

Ravosa, M.J., R. Kunwar, S.R. Stock, and M.S. Stack (2007b). Pushing the limit: Masticatory stress and adaptive plasticity in mammalian craniomandibular joints. *J Exp Biol* **210**: 628–641.

Ravosa, M.J., S.R. Stock, E.L. Simons, and R. Kunwar (2007c). MicroCT analysis of symphyseal ontogeny in a subfossil lemur (Archaeolemur). *Int J Primatology* **28**: 1385–1396.

Remigy, J.C. and M. Meireles (2006). Assessment of pore geometry and 3-D architecture of filtration membranes by synchrotron radiation computed microtomography. *Desalination* **199**: 501–503.

Remigy, J.C., M. Meireles, and X. Thibault (2007). Morphological characterization of a polymeric microfiltration membrane by synchrotron radiation computed microtomography. *J Membrane Sci* **305**: 27–35.

Renders, G.A.P., L. Mulder, L.J. van Ruijven, and T.M.G.J. van Eijden (2007). Porosity of human mandibular condylar bone. *J Anat* **210**: 239–248.

Reynolds, D.G., C. Hock, S. Shaikh, J. Jacobson, X. Zhang, P.T. Rubery, C.A. Beck, R.J. O'Keefe, A.L. Lerner, E.M. Schwarz, and H.A. Awad (2007). Microcomputed tomography prediction of biomechanical strength in murine structural bone grafts. *J Biomech* **40**: 3178–3186.

Robert, G., D.R. Baker, M.L. Rivers, E. Allard, and J. Larocque (2004). Comparison of the bubble size distribution in silicate foams using 2-dimensional images and 3-dimensional x-ray microtomography. *Developments in X-ray Tomography IV*. U. Bonse (Ed.). Bellingham, WA, SPIE. *SPIE Proc Vol* **5535**: 505–513.

Ruimerman, R., P. Hilbers, B. van Rietbergen, and R. Huiskes (2005). A theoretical framework for strain-related trabecular bone maintanence and adaptation. *J Biomech* **38**: 931–941.

Salvo, L., P. Belestin, E. Maire, M. Jacquesson, C. Vecchionacci, E. Boller, M. Bornert, and P. Doumalin (2004). Structure and mechanical properties of AFS sandwiches studied by in-situ compression tests in x-ray microtomography. *Adv Eng Mater* **6**: 411–415.

Scheffler, F., R. Herrmann, W. Schwieger, and M. Scheffler (2004). Preparation and properties of an electrically heatable aluminium foam/zeolite composite. *Micropor Mesopor Mater* **67**: 53–59.

Showalter, C., B.D. Clymer, B. Richmond, and K. Powell (2006). Three-dimensional texture analysis of cancellous bone cores evaluated at clinical CT resolutions. *Osteopor Int* **17**: 259–266.

Siu, W.S., L. Qin, W.H. Cheung, and K.S. Leung (2004). A study of trabecular bones in ovariectomized goats with micro-computed tomography and peripheral quantitative computed tomography. *Bone* **35**: 21–26.

Smith, A.B. (1980). *Stereom Microstructure of the Echinoid Test. Special Papers in Palaeontology* No. 25. London: The Palaeontology Association.

Smith, M.H., C.L. Flanagan, J.M. Kemppainen, J.A. Sack, H. Chung, S. Das, S.J. Hollister, and S.E. Feinberg (2007). Computed tomography-based tissue-engineered scaffolds in craniomaxillofacial surgery. *Int J Med Robotics Compu Assist Surg* **3**: 207–216.

Sone, T., T. Tamada, Y. Jo, H. Miyoshi, and M. Fukunaga (2004). Analysis of three-dimensional microarchitecture and degree of mineralization in bone metastases from prostate cancer using synchrotron microcomputed tomography. *Bone* **35**: 432–438.

Squire, M., L.R. Donahue, C. Rubin, and S. Judex (2004). Genetic variations that regulate bone morphology in the male mouse skeleton do not define its susceptibility to mechanical unloading. *Bone* **35**: 1353–1360.

Stenstrom, M., B. Olander, D. Lehto-Axtelius, J.E. Madsen, L. Nordsletten, and G.A. Carlsson (2000). Bone mineral density and bone structure parameters as predictors of bone strength: An analysis using computerized microtomography and gastrectomy-induced osteopenia in the rat. *J Biomech* **33**: 289–297.

Steppe, K., V. Cnudde, C. Girard, R. Lemeur, J.P. Cnudde, and P. Jacobs (2004). Use of x-ray computed microtomography for non-invasive determination of wood anatomical characteristics. *J Struct Biol* **148**: 11–21.

Stock, S.R., F. De Carlo, and J.D. Almer (2008). High energy x-ray scattering tomography applied to bone. *J Struct Biol* **161**: 144–150.

Stock, S.R., K. Ignatiev, and F.D. Carlo (2004a). Very high resolution synchrotron microCT of sea urchin ossicle structure. *Echinoderms: München.* T. Heinzeller and J.H. Nebelsick (Eds.). London, Taylor & Francis: 353–358.

Stock, S.R., K.I. Ignatiev, S.A. Foster, L.A. Forman, and P.H. Stern (2004b). MicroCT quantification of in vitro bone resorption of neonatal murine calvaria exposed to IL-1 or PTH. *J Struct Biol* **147**: 185–199.

Stölken, J.S. and J.H. Kinney (2003). On the importance of geometric nonlinearity in finite element simulations of trabecular bone failure. *Bone* **33**: 494–504.

Su, R., G. M. Campbell, and S. K. Boyd (2007). Establishment of an architecture-specific experimental validation approach for finite element modeling of bone by rapid prototyping and high resolution computed tomography. *Med Eng Phys* **29**: 480–490.

Takada, H., S. Abe, Y. Tamatsu, S. Mitarashi, H. Saka, and Y. Ide (2006). Three-dimensional bone microstructures of the mandibular angle using micro-CT and finite element analysis: Relationship between partially impacted mandibular thrid molars and angle fractures. *Dent Traumatol* **22**: 18–24.

Tanaka, E., R. Sano, N. Kawai, G.E.J. Langenbach, P. Brugman, K. Tanne, and T.M.G.J. van Eijden (2007). Effect of food consistency on the degree of mineralization in the rat mandible. *Annal Biomed Eng* **35**: 1617–1621.

Tesei, L., F. Casseler, D. Dreossi, L. Mancini, G. Tromba, and F. Zanini (2005). Contrast-enhanced x-ray microtomography of the bone structure adjacent to oral implants. *Nucl Instrum Meth A* **548**: 257–263.

Thurner, P., E. Karamuk, and B. Müller (2001). 3D characterization of fibroblast cultures on PRT textiles. *Euro Cells Mater* **2** Suppl 1: 57–58.

Thurner, P., B. Müller, F. Beckmann, T. Weitkamp, C. Rau, R. Müller, J.A. Hubbell, and U. Sennhauser (2003). Tomography studies of human foreskin fibroblasts on polymer yarns. *Nucl Instrum Meth B* **200**: 397–405.

Thurner, P.J., P. Wyss, R. Voide, M. Stauber, B. Müller, M. Stampanoni, J.A. Hubbell, R. Müller, and U. Sennhauser (2004). Functional microimaging of soft and hard tissue using synchrotron light. *Developments in X-ray Tomography IV.* U. Bonse (Ed.). Bellingham, WA, SPIE. *SPIE Proc Vol* **5535**: 112–128.

Toda, H., T. Ohgaki, K. Uesugi, M. Kobayashi, N. Kuroda, T. Kobayashi, M. Niinomi, T. Akahori, K. Makii, and Y. Aruga (2006a). Quantitative assessment of microstructure and its effects on compression behavior of aluminum foams via high-resolution synchrotron x-ray tomography. *Metall Mater Trans A* **37**: 1211–1219.

Toda, H., M. Takata, T. Ohgaki, M. Kobayashi, T. Kobayashi, K. Uesugi, K. Makii, and Y. Aruga (2006b). 3-D image-based mechanical simulation of aluminium foams: Effects of internal microstructure. *Adv Eng Mater* **8**: 459–467.

Trtik, P., J. Dual, D. Keunecke, D. Mannes, P. Niemz, P. Stahli, A. Kaestner, A. Groso, and M. Stampanoni (2007). 3D imaging of microstructure of spruce wood. *J Struct Biol* **159**: 46–55.

Ün, K., G. Bevill, and T.M. Keaveny (2006). The effects of side-artifacts on the elastic modulus of trabecular bone. *J Biomech* **39**: 1955–1963.

van Eijden, T.M.G J., L.J. Ruijven, and E.B.W. Giesen (2004). Bone tissue stiffness in the mandibular condyle is dependent on the direction and density of the cancellous structure. *Calcif Tiss Int* **75**: 502–508.

van Rietbergen, B. (2001). Micro-FE analyses of bone — State of the art. *Noninvasive Assessment of Trabecular Bone Architecture and the Competence of Bone*. B.K.B.S Majumdar (Ed.). New York: Kluwer/Plenum. *Adv Exp Med Biol Vol* **496**: 21–30.

van Rietbergen, B. and R. Huiskes (2001). Elastic constants of cancellous bone. *Bone Mechanics Handbook*, 2nd ed. S.C. Cowin (Ed.). Boca Raton, FL, CRC Press: 15–1 to –24.

Vanis, S., O. Rheinbach, A. Klawonn, O. Prymak, and M. Epple (2006). Numerical computation of the porosity of bone substitution materials from synchrotron micro computer tomographic data *Mat-wiss Werkstofftech* **37**: 469–473.

Verhulp, E., B. van Rietbergen, and R. Huiskes (2004). A three-dimensional digital image correlation technique for strain measurements in microstructures. *J Biomech* **37**: 1313–1320.

Verhulp, E., B. van Rietbergen, and R. Huiskes (2006). Comparison of micro-level and continuum-level voxel models of the proximal femur. *J Biomech* **39**: 2951–2957.

Vetter, L.D., V. Cnudde, B. Masschaele, P.J.S. Jacobs, and J.V. Acker (2006). Detection and distribution analysis of organosilicon compounds in wood by means of SEM-EDX and micro-CT. *Mater Char* **56**: 39–48.

Viot, P., D. Bernard, and E. Plougonven (2007). Polymeric foam deformation under dynamic loading by the use of the microtomographic technique. *J Mater Sci* **42**: 7202–7213.

Waarsing, J.H., J.S. Day, J.C. v. d. Linden, A.G. Ederveen, C. Spanjers, N D. Clerck, A. Sasov, J.A.N. Verhaar, and H. Weinans (2004). Detecting and tracking local changes in the tibiae of individual rats: A novel method to analyse longitudinal in vivo microCT data. *Bone* **34**: 163–169.

Wang, M., X.F. Hu, and X.P. Wu (2006). Internal microstructure evolution of aluminum foams under compression. *Mater Res Bull* **41**: 1949–1958.

Washburn, N. R., M. Weir, P. Anderson, and K. Potter (2004). Bone formation in polymeric scaffolds evaluated by proton magnetic resonance microscopy and x-ray microtomography. *J Biomed Mater Res Pt A* **69**: 738–747.

Weiss, P., L. Obadia, D. Magne, X. Bourges, C. Rau, T. Weitkamp, I. Khairoun, J.M. Bouler, D. Chappard, O. Gauthier, and G. Daculsi (2003). Synchrotron X-ray microtomography (on a micron scale) provides three-dimensional imaging representation of bone ingrowth in calcium phosphate biomaterials. *Biomater* **24**: 4591–4601.

Wilkes, T. E. (2008). Metal/ceramic composites via infiltration of an interconnected wood-derived ceramic, Ph.D. Thesis. Northwestern University.

Willems, N.M.B.K., L. Mulder, G.E.J. Langenbach, T. Grünheid, A. Zentner, and T.M.G.J. van Eijden (2007). Age-related changes in microarchitecture and mineralization of cancellous bone in the porcine mandibular condyle. *J Struct Biol* **158**: 421–427.

Witte, F., J. Fischer, J. Nellesen, and F. Beckmann (2006a). Microtomography of magnesium implants in bone and their degradation. *Developments in X-Ray Tomography V*. U. Bonse (Ed.). Bellingham, WA, SPIE. *SPIE Proc Vol* **6318**: 631806–1 to –9.

Witte, F., J. Fischer, J. Nellesen, H.A. Crostack, V. Kaese, A. Pisch, F. Beckmann, and H. Windhagen (2006b). In vitro and in vivo corrosion measurements of magnesium alloys. *Biomater* **27**: 1013–1018.

Wu, J.P., A.R. Boccaccini, and P.D. Lee (2007). Thermal and mechanical properties of a foamed glass–ceramic material produced from silicate wastes. *Glass Technol A* **48**: 133–141.

Wu, J.P., A.R. Boccaccini, P.D. Lee, M.J. Kershaw, and R.D. Rawlings (2006). Glass ceramic foams from coal ash and waste glass: Production and characterization. *Adv Appl Ceram* **105**: 32–39.

Yeh, O.C. and T.M. Keaveny (2001). Relative roles of microdamage and microfracture in the mechanical behavior of trabecular bone. *J Orthop Res* **19**: 1001–1007.

Youssef, S., E. Maire, and R. Gaertner (2005). Finite element modeling of the actual structure of cellular materials determined by x-ray tomography. *Acta Mater* **53**: 719–730.

Zebaze, R.M.D., A. Jones, M. Knackstedt, G. Maalouf, and E. Seeman (2007). Construction of the femoral neck during growth determines its strength in old age. *J Bone Miner Res* **22**: 1055–1061.

9
Networks

The subject of this chapter is microCT studies of network. The first section concerns engineered network solids, materials that consist of many fibers consolidated into paper or fiberboard. Networks of pores and fluid flow and localization therein are the subject of the second section. Coverage flows into microCT studies of animal circulatory systems in the third section. The chapter concludes with a review of respiratory system studies.

9.1 Engineered Network Solids

Fibrous network solids differ somewhat from most of the cellular materials discussed in Chapter 8. They are collections of long, thin (nominally 1D) objects. Whereas trabecular bone often consists of struts, much of this bone also has a platelike character and so differs from the network solids described in this section.

Cellulosic fibrous networks are encountered in papers and in engineered, low-density, wood-based fiberboards. Synchrotron microCT of low-, medium-, and high-density fiberboards was used to determine fiber network characteristics and to generate a realistic, ABAQUS™-based model for calculating thermal conductivity, a property important in construction materials; a design of experiments (DOE) analysis of the model parameters revealed that fiber density was responsible for 60 percent of local conductivity of the network and that fiber orientation and tortuosity represented 25 percent of the influence of network conductivity (Faessel et al., 2005).

Walther et al. (2006) examined various medium-density fiberboards with synchrotron microCT and provided a particularly clear account of the challenging image analysis needed to differentiate the fiber volume from the surrounding air. With a voxel size of ~2.3 µm, the lignocellulose fiber walls and the hollow lumens were clearly resolved; both needed to be included in the fiber volume, but real and apparent (noise-related) breaks in the walls were a complication that was overcome. The maximum fiber thickness was ~14 µm (6 voxels), so the analysis (quantification of fiber diameters and related quantities) should be regarded as quite robust. Quantities such as volume fraction (fiber walls, lumen volume,

FIGURE 9.1
3D renderings of a $(256)^3$ subvolume of a medium density fiberboard. (Left) The complete volume. (Middle) A selection of individually labeled fibers. (Right) Individually labeled fiber fragments with volume less than 10^3 voxels. The edges of the display cube equal 588 μm (Walther et al., 2006).

and interfiber air) and total surface area per unit volume (fiber–lumen and fiber–outer air) were measured as a function of specimen density. Individual fibers and fiber fragments and their orientations were quantified, and fiber bundles (groups of fibers each with contact areas >10^4 μm^2) were mapped. Figure 9.1 shows 3D renderings of a subvolume of a medium-density fiberboard; it compares a segmented image of all of the fibers, a selection of individual fibers identified from the entire volume, and a set of very small fiber fragments. Lux et al. (2006) provide analysis similar to that of Walther et al. (but with different image analysis tools) of synchrotron microCT data for a very low density fiberboard and found the resulting density agreed with that measured macroscopically

Synchrotron microCT studies of paper have appeared in which paper's extreme sensitivity to changes in humidity was overcome: shrinkage of even 1 percent produces displacements of a few micrometers, leading to sample motion artifacts (Antoine et al., 2002; Rolland Du Roscoat et al., 2005). Yang and Lindquist (2000) applied automatic 3D image analysis to synchrotron microCT data from a fibrous polymer mat. Eberhardt and Clarke (2002) advanced another algorithm for analyzing fiber characteristics in cloth.

Porosity and specific surface area of four types of industrial paper materials were quantified (Rolland Du Roscoat et al., 2007). Two papers contained fillers and two did not. In all four papers, synchrotron microCT revealed two boundary layers (surfaces, totaling about 50 percent of the volume) with strong gradients in porosity and a central bulk layer with nearly constant porosity. The microstructure of the four papers was transversely isotropic, and the anisotropy of the filler-containing papers was less pronounced than in the other two materials. Representative elemental volumes were also determined for the material.

In four different 3D studies of textiles, lab microCT was used to quantify yarn diameter and spacing and the variability in these quantities

Networks 217

(Desplentere et al., 2005). Incorporating these data into stochastic models of mechanical properties produced good agreement with experiments, not only for the value of the in-plane Young's modulus but also for the scatter in results. MicroCT analysis of the structure of polymer fabric used in paper drying has been incorporated into lattice Boltzmann simulations of fluid flow and Monte Carlo simulations of vapor diffusion in the structure (Ramaswamy et al., 2004), but the report is unclear (at least to the author) as to whether 2D (possibly very inaccurate) or 3D assessments of pore tortuosity were used in the models. Contaminant classification in cotton fabrics is another area where microCT was applied (Pavani et al., 2004).

The relatively good environmental stability of metals makes metal fibrous networks attractive for heat transfer, filtration, and catalyst support applications. Lab microCT and skeletonization analysis were used to study two bonded stainless steel fiber assemblies (Tan et al., 2006). The distribution of fiber segment lengths between the two specimens differed, as did the distribution of fiber orientations (shown on stereographic projections), and these data could certainly be imported into numerical models of stiffness (or other quantities) as in Clyne et al. (2005). One example is the microCT determination of short-fiber angular and spatial distribution that was used as input for a multiscale model of composite properties (Pyrz and Schjødt-Thomsen, 2006). Determination of fiber orientation and fiber length distributions is also important to the behavior of short-fiber reinforced foams, and lab microCT of a phenolic foam incorporating short glass fibers revealed fiber preferred orientation attributed to shear generated during foaming (Shen et al., 2004).

Plant roots can be regarded as a fibrous network and can be analyzed with approaches similar to those described above. For example, Kaestner et al. (2006) studied root networks in two alder specimens.

9.2 Networks of Pores

Channel structures are simply structures complementary to those of cellular solids: once again, the volume of interest occupies the smaller fraction of the total, but here the empty space (or other phase filling the channels) is of interest. Many analyses used for cellular solids can be and are used for channel structures with the solid and empty space subvolumes inverted. Although fluid transport and partition within porous systems could be covered equally appropriately in the chapter on structure evolution (Chapter 10), porosity characterization studies are included in this section.

Understanding the role of porosity, in particular, the pore connectivity, on material transport is crucial in fields as diverse as oil production, filter operation, and composite material densification. A priori microstructural knowledge, for example, could aid in the design of a drilling fluid that would cause minimal formation damage from particle invasion (Jasti et al., 1993).

Low-resolution as well as high-resolution tomography systems have been applied to porous materials; the lower-resolution, wider-field-of-view systems can obtain representative data on larger spatial scales, something that can be very important in highly heterogeneous geological systems. Fluid content and drainage were studied in a wide range of porous specimen sizes using an industrial CT system (76-mm diameter with (368 µm)^3 voxels), a medical CT scanner (76-mm diameter with (150 µm)^2 × 2 mm voxels) and synchrotron microCT (27-mm diameter with (76-µm)^3 voxels down to 1.5-mm diameter with (6.7-µm)^3 voxels) (Wildenschild et al., 2002). Note that similar comparisons were performed for trabecular bone (Peyrin et al., 1998) and for bone containing Ti implants (Bernhardt et al., 2004), and this subject was addressed in more detail in Chapter 5. The industrial and medical systems detected heterogeneous drainage patterns but not the pore spaces. In coarse sand, synchrotron microCT partitioned the specimen volume into KI-doped water, air, and solid and reliably detected different interfaces; changes in fluid distribution after drainage were also imaged clearly (Wildenschild et al., 2002). Contrast sensitivity for water in pores (undoped and doped with tracers such as KI) was established in other work, for example, Altman et al. (2005), and this subject was discussed in Chapter 5.

Disordered packing of idealized particles (e.g., spheres, equilateral cylinders) were examined with tomographic techniques (Aste et al., 2004; Zhang et al., 2006; Vasic et al., 2007). Packing in various multiphase systems was also studied, and properties of pore networks (relatively large open volumes connected by much narrower throats) received attention (Willson et al., 2004; Al-Raoush and Willson, 2005; Videla et al. 2006, 2007). Interfacial areas in fluid containing porous materials and water saturation were studied (Culligan et al., 2004; Brusseau et al., 2006).

The influence of flow rate and porous media microstructure on macroscopic relationships such as capillary pressure versus saturation was examined effectively with synchrotron microCT (Wildenschild et al., 2004), and the 3D rendering in Figure 9.2 shows pendular rings of water in the network, rings that provide a significant contribution to saturation and specific interfacial area on secondary drainage. Detailed examination of aqueous and nonaqueous fluid interactions on quantities such as interfacial area suggests that this quantity may be a primary determinant of nonaqueous liquid phase (e.g., oil) removal efficiency (Culligan et al., 2006). One idealized packing system (alumina cylinders, 1-mm diameter, 3–4-mm length), pore-scale modeling, and microCT visualization were used to study the spatiotemporal evolution of liquid phase clusters; behavior in untreated (hydrophilic) and silanized (neutral) packings was compared (Kohout et al., 2006). Observations of the evolution of water distribution

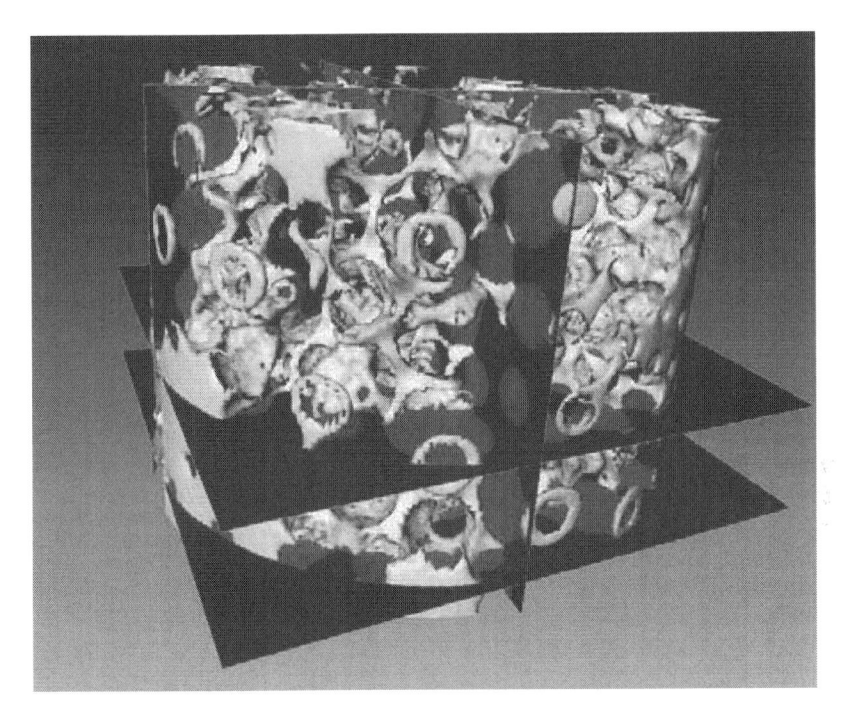

FIGURE 9.2
3D rendering of packed glass beads showing the KI-containing water as the white phase and orthogonal sectioning planes with empty space in black and glass beads in dark gray. The specimen diameter is 7.0 mm, and the imaging voxel size is 17 µm (Culligan et al. (2004); © 2004 American Geophysical Union).

with drying were compared with numerical simulations derived from the initial microstructure, and this sort of analysis typifies the direction many microCT investigations are taking. Numerical modeling of equilibrium distributions of water within partially saturated rock has been based on microCT-derived structures and appears to be a very valuable approach (Berkowitz and Hansen, 2001).

Sheppard et al. (2006) characterized a number of materials with porosities between 18.5 percent and 54 percent using microCT and either compared these determinations to results from other techniques or used the data as input to fluid transport or other models. In addition to quantification of grain size, these authors looked at grain coordination number in different types of specimens, pore connectivity, and local flow paths. One specific approach was to view the porosity as a system of throats and pores, using the microCT data to define the characteristics of the network (throat diameter, throat length, mean throat orientation, etc.); Figure 9.3 is a 3D rendering of the throats and pores in several very different specimens and shows them to scale and in the proper orientation. Similar analyses were performed by Knackstedt et al. (2005) on eight different industrial foams

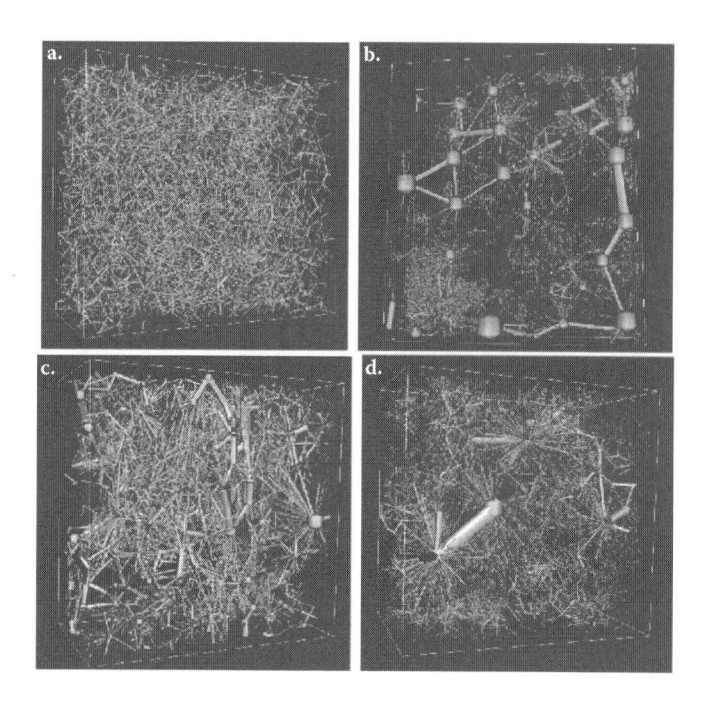

FIGURE 9.3
3D renderings of the throats and pores in subsets of four very different specimens. The variation in structure across the specimens is dramatic. Note that the size of the pores and throats reflects their actual size in the segmented tomographic dataset (Sheppard et al., 2006).

(microcellular polyurethane formed by four different processes), including thermal conductivity, permeability, and elastic properties (Young's modulus). The same group described a vertically integrated center for the analysis of transport (and mechanical) properties in a wide class of solids (Sakellariou et al., 2004).

A Fontainebleau sandstone (porosity fraction = 0.197 and experimentally measured permeability K_{exp} = 1860 mD, where 1 Darcy = 10^{-12} m^2 and the permeability relates the volumetric flow rate divided by the cross-sectional area to the pressure gradient divided by the Newtonian viscosity) was studied (Jasti et al., 1993), and the calculated permeability for the imaged sample agreed with that measured directly in larger samples. A second study of Fontainebleau sandstone (Auzerais et al., 1996) examined 3.5-mm-diameter samples, about seven grains on a side, which was close to the minimum size to estimate permeability with accuracy, and calculated permeabilities from the microstructure of 1,000–1,300 mD, in good agreement with experimental values determined with several other techniques. Because Fontainebleau sandstones are reported to be remarkably homogeneous for geological materials, the agreement between these two studies was not surprising. Modeling of fluid permeability in both studies showed large variability in fluid flow, depending on which subset of the reconstructed volumes was used to define the three-dimensional channel structure. Because several different numerical methods were used, it appears unlikely that the variability was due to anything other than changes in channel architecture. This demonstrates how powerful microCT imaging can be in determining local transport conditions and, on the other hand, how much care must be taken to ensure representative volumes are imaged and to avoid overinterpretation of limited data.

MicroCT results from three Berea sandstones, a Texas creme chalk, and a glass bead sphere pack were reported (Jasti et al., 1993); the air permeabilities of the geological samples ranged over nearly three orders of magnitude (from 1,000 to 16 mD), and the results confirmed that pore structures could be followed in materials representative of those found in petroleum reservoirs. When the glass bead pack was infiltrated with a mixture of oil and water, the three fluid phases were resolved. The ability to resolve oil from water or other phases is important in understanding how, during secondary recovery, some oil being swept by the driving fluid becomes immobilized by capillary forces and forms isolated unextractable blobs.

In natural systems such as sandstone, the extent to which pores are connected dictates not only reservoir characteristics (see Bernard et al. (2000) and the material in Chapter 7) but also the rate of contaminant transport. The distribution of pore diameters and local volume fraction of pores for a Botucatu sandstone were studied with microCT (Appoloni et al., 2007). Many different analysis approaches were outlined for different classes of porous media in a description of a "toolbox" (Levitz, 2007). Cluster labeling analysis of synchrotron microCT data of a 14 percent porous sandstone suggested that isolated pores constitute only 11 percent of the total porosity and unresolved connections between pores contribute significantly to fluid transport (Nakashima et al., 2004).

Multiscale modeling (lattice Boltzmann method simulating single and multicomponent fluid flow) of real microCT-derived 3D structures of sandstone revealed good agreement with permeability measurements on similar rocks (Martys et al., 1999; Martys and Hagedorn, 2002; Verma et al., 2007); 2D sampling of 3D volumes of small fragments of sandstone, however, did not yield the porosity and permeability of the sample but rather relative trends requiring calibration by independent data (Kameda et al., 2006). MicroCT-derived oil and water distributions and pore structures (pore volume, pore surface area, throat surface area, principal direction diameters for pores and throats) were used with lattice Boltzmann computations to consider the effect of water-based gels on the relative permeabilities of water and oil (Prodanović et al., 2007).

The presence of organic, water-immiscible liquids, often as blobs, is a key complicating factor in the remediation of hazardous waste sites. The pore-scale morphology of such liquids residing in porous media (sand) was investigated with synchrotron microCT, and the distribution of blob sizes and morphologies was quantified (Schnaar and Brusseau, 2005). A follow-on study tracked blob dissolution after each of three column rinsing cycles; changes in blob number, volume, and surface area determined with microCT correlated well with effluent concentrations and validated a first-order mass transfer expression for effluent concentration (Schnaar and Brusseau, 2006). Note that these studies used a contrast agent to enhance blob visibility, and a more recent report from these investigators has appeared (Brusseau et al., 2007). Digital laminography with an x-ray tube source was used to map residual oil and water saturation in porous media (Palchikov et al., 2006). X-ray imaging in a lab microCT system visualized the dynamic adsorption of organic vapor and water vapor on activated carbon (Lodewyckx et al., 2006), but the report left unclear whether microCT or simple radiography was the modality used.

Natural gas hydrate is a crystalline solid composed of water and natural gas that is stable at high-pressure and low-temperature conditions, and it occurs as part of a highly dynamic fluid–sediment system. Interest in natural gas hydrate stems from its potential impact on energy resources, its significance as a drilling hazard, and so on. The porosity of methane hydrate host sediments was studied with microCT as a first step in determining whether the hydrate is free-floating in the sediment, whether it cements and stiffens the bulk sediments, or whether it contacts but does not cement the sediment grains (Jones et al., 2007b).

Transport during filtration has aspects related to flushing of porous materials, and one microCT study determined filter structure and related it to droplet formation for different fluids (Vladisavljević et al., 2007). Other filtration studies focused on efficient elimination of water from the remaining solid. Flocculation is one method of extracting suspended solids, and characteristics of the resulting aggregates affect downstream solids recovery. MicroCT was combined with numerical modeling to predict permeability in a model flocculated system (Selomulya et al., 2005, 2006). Drying of mechanically dewatered wastewater sludges and of filtered particulates in mineral processing (filter cake) provides two examples where microCT has suggested how to improve energy efficiency of these processes. MicroCT identified microstructures resulting from drying sludges and their correlation with drying rates and energy consumption (Leonard et al., 2002, 2003, 2004). Lin and Miller (2000a,b; 2004) used numerical modeling to examine flow through well-defined simulated filter cake, and they also simulated flow based on structures directly derived from microCT.

Polymer foams can be used to absorb oil from spills, and the ingestion of fluids into cellular materials such as polyurethane foams has, therefore, considerable practical importance but has received relatively little attention. Oil uptake (two densities of oil) of polyurethane foams (three densities) were examined for two temperatures (simulating winter and summer seawater temperatures) using lab microCT (Duong and Burford, 2006). Weight uptake as a function of time was the main measure of polyurethane foam performance, and a few reconstructions and 3D renderings of the as-received foam were provided. Unfortunately, the foam structure was not analyzed, structure was not correlated with performance, and reconstructions were not presented of the foam with absorbed oil.

Colloid transport in porous media is often treated as a filtration problem, and Li and coworkers observed (with lab microCT) the deposition of 36-µm gold-coated microspheres in two porous media (glass beads and quartz sand, both ~780-µm mean particle size; Li et al., 2006). In the absence of an energy barrier (native particle surfaces), the logarithm of the deposited microsphere concentration decreased linearly with increasing transport distance, a result consistent with filtration theory. In the presence of an energy barrier (treatment of the columns with polyoxyethylene laurel ether), colloid deposition was strongly influenced by local geometry (particularly grain–grain contacts) and did not vary monotonically with transport distance.

As was mentioned in Chapter 6, texture analysis has been applied to 3D microCT datasets of five porous specimens (mineral carbon forms from different geographical locations with similar topological structure that differed mainly in textural quality; Jones et al., 2007a). This approach appears quite promising.

9.3 Circulatory System

Although most of the microCT studies of the circulatory system have been on mammals, there has been some work done on arthropods. Corrosion casts of the circulatory system in the shrimplike mysidaceans were studied with microCT, and the results were used for phylogenic analyses of a complex mix of characters (Wirkner and Richter, 2007). Corrosion casting and microCT were also applied to study the hemolymph vascular system in the major lineages of scorpions (Wirkner and Prendini, 2007).

Blood vessel structure in organs is an important topic in medicine and biomedical engineering, and microCT-based analysis of these systems of channels is an efficient quantification method. Ritman and coworkers described blood vessel quantification for different organ systems, including Haversian systems in bone, coronary arteries, hind limb vascular trees, kidney glomerular microvasculature, and biliary trees (Ritman et al., 1997; Jorgensen et al., 1998; Kwon et al., 1998; Wilson et al., 2002; Fortepiani et al., 2003; Masyuk et al., 2004). The hundreds of branches present in a typical organ are usually filled with a contrast agent, but Haversian canals can be imaged directly. Figure 9.4 shows two views through a volume of bone; each interconnected Haversian canal system is a different color in this inverse 3D rendering; that is, only the low-absorption voxels are shown (Ritman

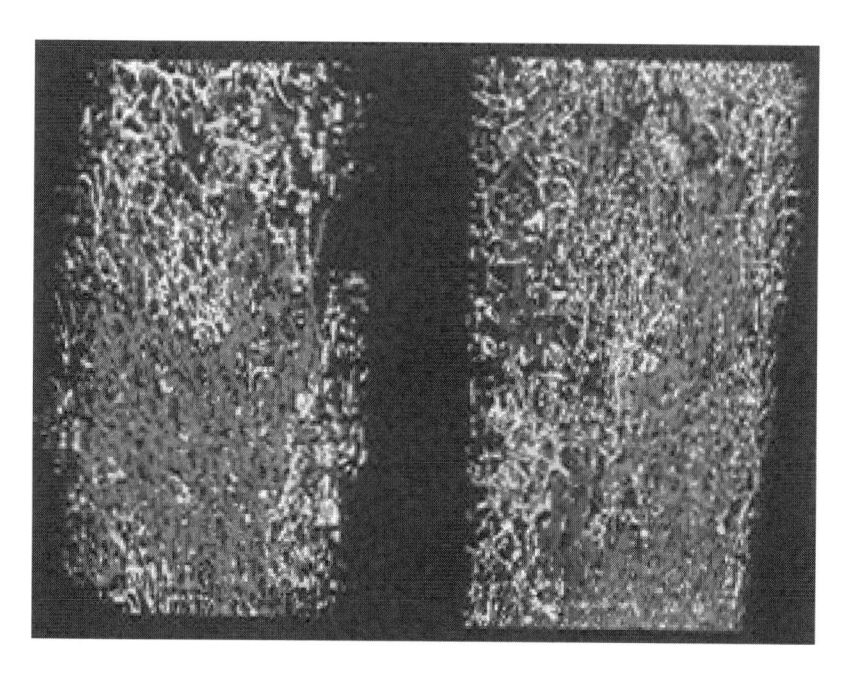

FIGURE 9.4 (SEE COLOR INSERT FOLLOWING PAGE 144.)
Haversian canals in a 3D rendering showing only the low-absorption voxels in 1 mm × 1 mm × 3 mm volume. The low-absorption voxels that are interconnected are shown in the same shade (Ritman et al., 1997).

et al., 1997). Phase imaging can also be used to image blood vessels in soft tissue without contrast agents (Momose et al., 2000; Hwu et al., 2004).

Regardless of the contrast mechanism, automated analysis routines are required if an adequate number of replicates are to be analyzed, and this is particularly important in biological studies where interindividual variability is very large. In common with analyses of cellular structures, the first step in the analysis is extraction of the vessels from the image and suppression of noise. It is important to note the treelike nature of the vessel system and branches, because this differs from the situation pertaining to cellular materials and conditions analysis algorithms. After extraction of the vessel tree, quantitative data can be computed and numerical analysis of the tree characteristics can be performed (e.g., partition into mother/child/sibling relationships for the different branches that focus analysis of functionality and dimensional changes onto equivalent portions of the network).

Wan et al. (2002) studied the coronary arterial tree of a rat and focused quantification on the arterial lumen cross-sectional area, interbranch segment length, branch surface area at equivalent generation, and interbranch and intrabranch levels. In coronary circulation, arterial blood inflow increases during diastole and venous outflow increases during systole; direct comparison of microvessels in the myocardium for diastolic-arrested and systolic-arrested hearts allowed investigators to identify the characteristics of the capacitance vessels (Toyota et al., 2001). In vivo microCT with retrospective gating quantified ejection fraction and cardiac output for free-breathing mice (Drangova et al., 2007).

Anomalies in patterning of coronary arteries are associated with congenital heart disease and can have a profound impact on the outcome of surgical palliation. Patterning of coronary arteries (coronary artery insertion and branching patterns of the proximal coronary stems) was compared in mice (wild type vs. a knockout strain) in order to develop a patterning defect model (Clauss et al., 2006). Right ventricular dysfunction in patients with congenital heart disease may be due to different structural response of microvessels to increased pressure load in the right ventricle, and, using microCT, Ohuchi and collaborators found a significant difference in volume of porcine myocardium perfused per vessel cross-sectional area between right and left ventricle walls at 5 months of age but not at 1 month or neonatally (Ohuchi et al., 2007).

Myocardial microvasculature is difficult to image in microCT, and visualizing vessels at the capillary level (~5-µm diameters) led to the examination of novel contrast agents, including the clinically relevant lyophilic salts $CaSO_4$, $SrSO_4$, and $BaSO_4$ (Müller et al., 2006). Incomplete filling remains a problem, and the question of whether it has occurred should be considered carefully in every analysis. Early morphological changes in coronary arteries of hypercholesterolemic pigs were followed using OsO_4, a novel microCT contrast agent: specifically, coronary artery wall structure, early lesion formation, and changes in microvascular permeability (Zhu et al., 2007a). Microvascular permeability accompanying lipopolysaccharide-induced sepsis was investigated using 70-nm particles as contrast agents in rats, and the endothelial defect size was estimated to be larger than 70 nm and smaller than 1 µm (Langheinrich and Ritman, 2006). In another ex vivo microCT study, Zhu et al. (2007b) found simvastin prevented microvascular remodeling and hypertrophy in swine renovascular hypertension. Myocardial fatty acid metabolism was measured using a contrast agent and fluorescence microCT in hamsters, and the observed significant decrease in contrast agent uptake in the cardiomyopathic heart compared to the normal heart was not due to cardiomyopathy-related fibrosis but, rather, to abnormal fatty acid metabolism (Lwin et al., 2007).

Pulmonary arterial wall distensibility, decreases of which are important in various lung pathologies, was examined with a liquid contrast agent injected into pulmonary arteries and imaged with lab microCT at different arterial pressures spanning the physiological range (Johnson et al., 1999b, 2000); main arterial trunk diameter versus distance from the inlet was determined for each of four pressures, and a linear relationship between diameter and pressure was found for a single point on one vessel. This same group used the self-similarity of the arterial trees to improve analysis efficiency (Johnson et al., 1999a) and employed these tools to compare vascular remodeling for hypoxia-treated rats with rats living under normal oxygen partial pressures (Molthen et al., 2004).

Bentley et al. (2002) reviewed the use of microCT to study alterations in renal microvasculature caused by development of cirrhosis in a rat model; they found changes in microvascular volume fractions in different portions of the kidney that may contribute to changes in salt and water retention that accompany cirrhosis. Sled and coworkers applied semi-automatic analysis to microvasculature in mouse kidneys (simple threshold, distance transformation, special routines for local contact between vessels; Sled et al., 2004); Toyota et al. (2004) studied heterogeneity of glomerular volume distribution in a rat model of early diabetic nephropathy and found statistically significant (greater) coefficient of variation within individuals of the model compared to controls.

Montet and coworkers investigated iodated liposomes as a contrast agent for microvessels in the murine liver, compared controls with liver-micrometastatase containing animals, and found they could detect 250-μm liver tumors after injection of iodated liposomes at 2 gI/kg body weight (Montet et al., 2007). Op Den Buijs et al. (2006) hypothesized that hepatitc vasculature could more closely approach optimal branching geometry than the vasculature of the lung or myocardium because the liver has fewer anatomical constraints in branching. Using microCT to study rat livers, the investigators found the contrary to be true: the liver showed variation in branching morphology (e.g., branching angles oscillating between that predicted by the optimality principle of minimum power loss and volume and that of minimum shear stress and surface) that was similar to that of other organs.

Quantitative microCT analysis of collateral vessel development after ischemic injury in a mouse model revealed the vascular volume was reconstituted as a series of highly connected, small-diameter, closely spaced, and isotropically oriented vessels as soon as three days after surgical ligation of the femoral artery (Duvall et al., 2004). Simple thresholding and vessel diameter measurement based on distance transformation were used.

Corrosion casting of the vasculature system in the brain (perfusion of a polymer followed by masceration of the soft tissue and decalcification of the bone) and a two-resolution approach (lab microCT, 16-μm voxels, to identify volumes of interest for subsequent local tomography reconstruction with synchrotron microCT, 1.4-μm voxels) have been used to compare an Alzheimer's disease model mouse with the wild type using metrics described above (Heinzer et al., 2004). Figure 9.5 reproduces a comparison of matching SEM and microCT views of the network of vessels, a comparison that highlights several very minor artifacts in the rendering of the microCT data (Heinzer et al., 2006).

The nutrient canal system (blood vessels) within cortical bone has received extensive attention, but the focus of microCT studies has been on

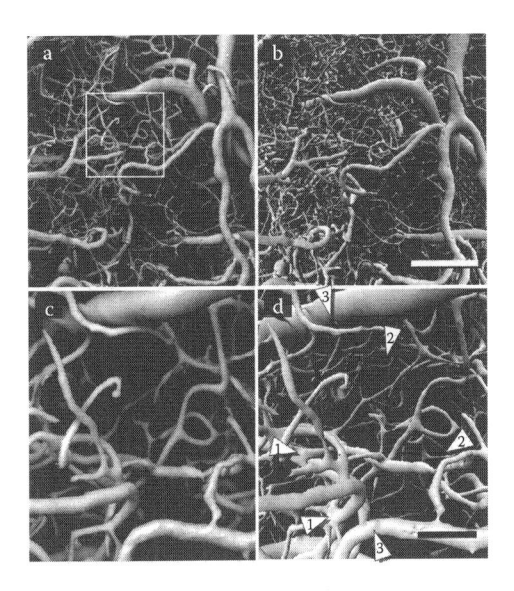

Networks

FIGURE 9.5
Validation of microCT rendering of murine brain vascular corrosion castings by comparison to SEM images. (a) SEM image of vasculature at the cortical surface. (b) Surface rendered microCT data of the same region and with the same viewing perspective as in (a). The scale bar equals 200 µm. (c) SEM detail of the subregion outlined in (a). (d) MicroCT rendering of the same subregion as in (c). At this magnification, three types of minor imaging artifacts become visible (arrows): vessels in close proximity fuse ("1"), some capillaries are too thin or even disappear ("2"), and indentations or contusions appear in the walls of the larger vessels ("3") (Heinzer et al., 2006).

the effect on mechanical properties and on bone remodeling. Figure 9.4, described earlier in this section, shows interconnected Haversian systems in cortical bone, and these systems can be quite complex (Ritman et al., 1997; Cooper et al., 2007). Further discussion of nutrient canal networks in bone is postponed to Chapter 10.

9.4 Respiratory System

Corrosion casts of a canine lung and a mouse lung were imaged with conventional CT and microCT, respectively, and, due to effects such as bubbles in the polymer, a small amount of manual segmentation was required (Chaturvedi and Lee, 2005). Analysis of the airway tree structure in terms of generations and so on was similar to that reported by others (Wan et al., 2002); interestingly, the approach was accurate and efficient for up to six generations for the canine cast and ten generations for the murine cast, presumably because of instrumentation differences.

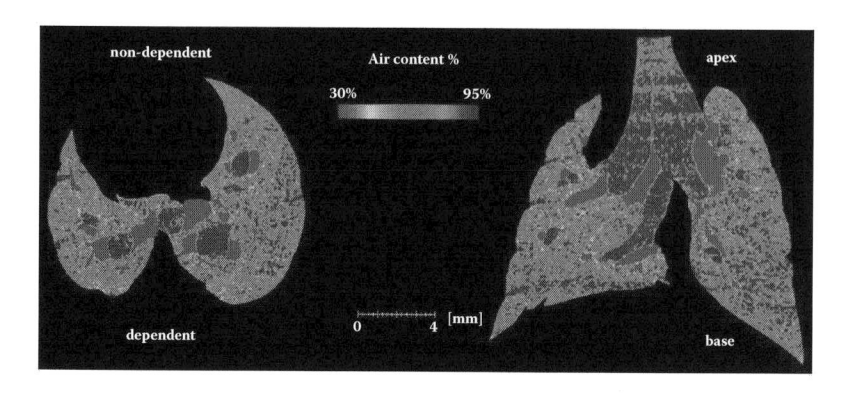

FIGURE 9.6 (SEE COLOR INSERT FOLLOWING PAGE 144.)
Axial (left) and coronal (right) 2D sections of a typical mouse lung showing regional air content differences, represented by the shades shown in the gray scale bar (Namati et al., 2006).

Resolution-related errors in high-resolution (clinical) CT imaging of inflation-fixed porcine lung cubes were evaluated using microCT, and systematic, size-dependent underestimation of the lumen area and overestimation of the wall area were found for the clinical system (Dame Carroll et al., 2006). Respiratory gated microCT of anesthetized free-breathing mice was used to quantify lung tidal volume and functional residual capacity (Ford et al., 2007). Several versions of respiratory gating were compared with computer-controlled intermittent isopressure breath hold technique in in vivo microCT mouse studies, and Namati et al. (2006) reported that the last technique yielded the most reproducible lung volume and air content measurements. Figure 9.6, from Namati et al., shows axial and coronal sections showing air content in a typical mouse lung.

In vivo quantification of regional lung gas volumes in rabbits was reported using synchrotron microCT, mechanical ventilation with Xe-O gas, and K-edge subtraction imaging (Bayat et al., 2001; Monfrais et al., 2005). Bayat et al. (2006) used Xe-enhanced synchrotron microCT imaging to examine in vivo response to histamine administration in healthy anesthesized and mechanically ventilated rabbits; proximal airway cross-sectional area decreased by 57 percent by 20 min and recovered gradually but incompletely within 60 min, whereas ventilated alveolar area decreased by 55 percent immediately after histamine inhalation and recovered rapidly thereafter. Hyperresponsiveness in allergically inflamed mouse lungs was the subject of another in vivo microCT study (Lundblad et al., 2007). MicroCT compared in vivo pulmonary compliance of healthy mice with a common pulmonary fibrosis model, damage induced by bleomycin exposure (Cavanaugh et al., 2006; Shofer et al., 2007). In vivo imaging with respiratory gating of laboratory rodents was also used to examine lung damage from tumors and from chemotherapy (Cody et al., 2004).

References

Al-Raoush, R.I. and C.S. Willson (2005). Extraction of physically realistic pore network properties from three-dimensional synchrotron x-ray microtomography images of unconsolidated porous media systems. *J Hydrol* **300**: 44–64.

Altman, S.J., W.J. Peplinski, and M.L. Rivers (2005). Evaluation of synchrotron x-ray computerized microtomography for the visualization of transport processes in low porosity materials. *J Contam Hydrol* **78**: 167–183.

Antoine, C., P. Nygard, O.W. Gregersen, R. Holmstad, T. Weitkamp, and C. Rau (2002). 3D images of paper obtained by phase-contrast x-ray microtomography: Image quality and binarisation. *Nucl Instrum Meth A* **490**: 392–402.

Appoloni, C.R., C.P. Fernandes, and C.R.O. Rodrigues (2007). X-ray microtomography study of a sandstone reservoir rock. *Nucl Instrum Meth A* **580**: 629–632.

Aste, T., M. Saadatfar, A. Sakellariou, and T. Senden (2004). Investigating the geometrical structure of disordered sphere packings. *Physica A* **339**: 16–23.

Auzerais, F.M., J. Dunsmuir, B.B. Ferreol, N. Martys, J. Olson, T.S. Ramakrishan, D.H. Rothman and L.M. Schwartz (1996). Transport in sandstones: A study based on three dimensional microtomography. *Geophys Res Lett* **23**: 705–708.

Bayat, S., G. Le Duc, L. Porra, G. Berruyer, C. Nemoz, S. Monfraix, S. Fiedler, W. Thomlinson, P. Suortti, C.G. Standertskjold-Nordenstam, and A.R.A. Sovijarvi (2001). Quantitative functional lung imaging with synchrotron radiation using inhaled xenon as contrast agent. *Phys Med Biol* **46**: 3287–3299.

Bayat, S., L. Porra, H. Suhonen, C. Nemoz, P. Suortti, and A.R.A. Sovijärvi (2006). Differences in the time course of proximal and distal airway response to inhaled histamine studied by synchrotron radiation CT. *J Appl Physiol* **100**: 1964–1973.

Bentley, M.D., M.C. Ortiz, E.L. Ritman, and J.C. Romero (2002). The use of micro-computed tomography to study microvasculature in small rodents. *Am J Physiol Reg Integ Comp Physiol* **282**: R1267–R1279.

Berkowitz, B. and D.P. Hansen (2001). A numerical study of the distribution of water in partially saturated porous rock. *Transport Porous Media* **45**: 303–319.

Bernard, D., G.L. Vignoles, and J.M. Heintz (2000). Modelling porous materials evolution. *X-Ray Tomography in Materials Science*. J. Baruchel, J.Y. Buffière, E. Maire, P. Merle, and G. Peix (Eds.). Paris, Hermes Science: 177–192.

Bernhardt, R., D. Scharnweber, B. Müller, P. Thurnier, H. Schliephake, P. Wyss, F. Beckmann, J. Goebbels, and H. Worch (2004). Comparison of microfocus and synchrotron x-ray tomography for the analysis of osteointegration around Ti6AlV4 implants. *Euro Cells Mater* **7**: 42–51.

Brusseau, M.L., S. Peng, G. Schnaar, and M.S. Costanza-Robinson (2006). Relationships among air-water interfacial area, capillary pressure, and water saturation for a sandy porous medium. *Water Resour Res* **42**: W03501.

Brusseau, M.L., S. Peng, G. Schnaar, and A. Murao (2007). Measuring air-water interfacial areas with x-ray microtomography and interfacial partitioning tracer tests. *Env Sci Technol* **41**: 1956–1961.

Cavanaugh, D., E.L. Travis, R.E. Price, G. Gladish, R.A. White, M. Wang, and D.D. Cody (2006). Quantification of bleomycin-induced murine lung damage in vivo with micro-computed tomography. *Acad Radiol* **13**: 1505–1512.

Chaturvedi, A. and A. Lee (2005). Three-dimensional segmentation and skeletonization to build an airway tree data structure for small animals. *Phys Med Biol* **50**: 1405–1419.

Clauss, S.B., D.L. Walker, M.L. Kirby, D. Schimel, and C.W. Lo (2006). Patterning of coronary arteries in wildtype and connexin43 knockout mice. *Dev Dynamics* **235**: 2786–2794.

Clyne, T.W., A.E. Markaki, and J.C. Tan (2005). Mechanical and magnetic properties of metal fibre networks with and without a polymeric matrix. *Compos Sci Technol* **65**: 2492–2499.

Cody, D., D. Cavanaugh, R.E. Price, B. Rivera, G. Gladish, and E. Travis (2004). Lung imaging of laboratory rodents in vivo. *Developments in X-ray Tomography IV*. U. Bonse (Ed.). Bellingham, WA, SPIE. *SPIE Proc Vol* **5535**: 43–52.

Cooper, D., A. Turinsky, C. Sensen, and B. Hallgrimsson (2007). Effect of voxel size on 3D microCT analysis of cortical bone porosity. *Calcif Tiss Int* **80**: 211–219.

Culligan, K.A., D. Wildenschild, B.S.B. Christensen, W.G. Gray, and M.L. Rivers (2006). Pore-scale characteristics of multiphase flow in porous media: A comparison of air–water and oil–water experiments. *Adv Water Res* **29**: 227–238.

Culligan, K.A., D. Wildenschild, B.S.B. Christensen, W.G. Gray, M.L. Rivers, and A.F.B. Tompson (2004). Interfacial area measurements for unsaturated flow through a porous medium. *Water Resour Res* **40**: W12413/1–/12.

Dame Carroll, J.R., A. Chandra, A.S. Jones, N. Berend, J.S. Magnussen, and G.G. King (2006). Airway dimensions measured from micro-computed tomography and high-resolution computed tomography. *Eur Respir J* **28**: 712–720.

Desplentere, F., S.V. Lomov, D.L. Woerdeman, I. Verpoest, M. Wevers, and A. Bogdanovich (2005). Micro-CT characterization of variability in 3D textile architecture. *Compos Sci Technol* **65**: 1920–1930.

Drangova, M., N.L. Ford, S.A. Detombe, A.R. Wheatley, and D.W. Holdsworth (2007). Fast retrospectively gated quantitative four-dimensional (4D) cardiac micro computed tomography imaging of free-breathing mice. *Invest Radiol* **42**: 85–94.

Duong, H.T.T. and R.P. Burford (2006). Effect of foam density, oil viscosity and temperature on oil sorption behavior of polyurethane. *J Appl Polymer Sci* **99**: 360–367.

Duvall, C.L., W.R. Taylor, D. Weiss, and R.E. Guldberg (2004). Quantitative micro-computed tomography analysis of collateral vessel development after ischemic injury. *Am J Physiol Heart Circ Physiol* **287**: H302–H310.

Eberhardt, C.N. and A.R. Clarke (2002). Automated reconstruction of curvilinear fibres from 3D datasets acquired by x-ray microtomography. *J Microsc* **206**: 41–53.

Faessel, M., C. Delisée, F. Bos, and P. Castéra (2005). 3D modeling of random cellulosic fibrous networks based on x-ray tomography and image analysis. *Compos Sci Technol* **65**: 1931–1940.

Ford, N.L., E.L. Martin, J.F. Lewis, R.A.W. Veldhuizen, M. Drangova, and D.W. Holdsworth (2007). In vivo characterization of lung morphology and function in anesthetized free-breathing mice using micro-computed tomography. *J Appl Physiol* **102**: 2046–2055.

Fortepiani, L.A., M.C.O. Ruiz, F. Passardi, M.D. Bentley, J. Garcia-Estan, E.L. Ritman, and J.C. Romero (2003). Effect of losartan on renal microvasculature during chronic inhibition of nitric oxide visualized by microCT. *Am J Physiol Renal Physiol* **285**: F852–F860.

Heinzer, S., T. Krucker, M. Stampanoni, R. Abela, E.P. Meyer, A. Schuler, P. Schneider, and R. Müller (2004). Hierarchical bioimaging and quantification of vasculature in disease models using corrosion casts and microcomputed tomography. *Developments in X-ray Tomography IV*. U. Bonse (Ed.). Bellingham, WA, SPIE. *SPIE Proc Vol* **5535**: 65–76.

Heinzer, S., T. Krucker, M. Stampanoni, R. Abela, E.P. Meyer, A. Schuler, P. Schneider, and R. Müller (2006). Hierarchical microimaging for multiscale analysis of large vascular networks. *Neuroimage* **32**: 626–636.

Hwu, Y., W.L. Tsai, J.H. Je, S.K. Seol, B. Kim, A. Groso, G. Margaritondo, K.H. Lee, and J.K. Seong (2004). Synchrotron microangiography with no contrast agent. *Phys Med Biol* **49**: 501–508.

Jasti, J.K., G.J. Jesion, and L. Feldkamp (1993). Microscopic imaging of porous media with x-ray computer tomography. *SPE Formation Eval* **8**: 189–193.

Johnson, R.H., K.L. Karau, R.C. Molthen, and C.A. Dawson (1999a). Exploiting self-similarity of arterial trees to reduce the complexity of image analysis. *Medical Imaging 2000: Physiology and Function from Multidimensional Images*. C.T. Chen and A.V. Clough (Eds.). Bellingham, WA, SPIE. *SPIE Proc Vol* **3660**: 351–361.

Johnson, R.H., K.L. Karau, R.C. Molthen, and C.A. Dawson (1999b). Quantification of pulmonary arterial wall distensibility using parameters extracted from volumetric microCT images. *Developments in X-ray Tomography II*. U. Bonse (Ed.). Bellingham, WA, SPIE. *SPIE Proc Vol* **3772**: 15–23.

Johnson, R.H., K.L. Karau, R.C. Molthen, S.T. Haworth, and C.A. Dawson (2000). Micro-CT image-derived metrics quantify arterial wall distensibility reduction in a rat model of pulmonary hypertension. *Medical Imaging 2000: Physiology and Function from Multidimensional Images*. C.T. Chen and A.V. Clough (Eds.). Bellingham, WA, SPIE. *SPIE Proc Vol* **3978**: 320–330.

Jones, A.S., A. Reztsov, and C.E. Loo (2007a). Application of invariant grey scale features for analysis of porous minerals. *Micron* **38**: 40–48.

Jones, K.W., H. Feng, S. Tomov, W.J. Winters, M. Prodanović, and D. Mahajan (2007b). Characterization of methane hydrate host sediments using synchrotron-computed microtomography (CMT). *J Petrol Sci Eng* **56**: 136–145.

Jorgensen, S.M., O. Demirkaya, and E.L. Ritman (1998). Three-dimensional imaging of vasculature and parenchyma in intact rodent organs with x-ray microCT. *Am J Physiol 275 (Heart Circ Physiol 44)* **275**: H1103–H1114.

Kaestner, A., M. Schneebeli, and F. Graf (2006). Visualizing three-dimensional root networks using computed tomography. *Geoderma* **136**: 459–469.

Kameda, A., J. Dvorkin, Y. Keehm, A. Nur, and W. Bosl (2006). Permeability-porosity transforms from small sandstone fragments. *Geophys* **71**: N11–N19.

Knackstedt, M., C. Arns, M. Saadatfar, T. Senden, A. Sakellariou, A. Sheppard, R. Sok, W. Schrof, and H. Steininger (2005). Virtual materials design: Properties of cellular solids derived from 3D tomographic images. *Adv Eng Mater* **7**: 238–243.

Kohout, M., Z. Grof, and F. Stepanek (2006). Pore-scale modelling and tomographic visualisation of drying in granular media. *J Colloid Interface Sci* **299**: 342–351.

Kwon, H.M., G. Sangiori, E.L. Ritman, A. Lerman, C. McKenna, R. Virmani, W.D. Edwards, D.R. Holmes, and R.S. Schwartz (1998). Adventitial vasa vasorum in balloon-injuried coronary arteries — Visualization and quantitation by microscopic three-dimensional computed tomography technique. *J Am Coll Cardiol* **32**: 2072–2079.

Langheinrich, A.C. and E.L. Ritman (2006). Quantitative imaging of microvascular permeability in a rat model of lipopolysaccharide-induced sepsis: Evaluation using cryostatic microcomputed tomography. *Invest Radiol* **41**: 645–650.

Leonard, A., S. Blacher, P. Marchot, and M. Crine (2002). Use of x-ray microtomography to follow the convective heat drying of wastewater sludges. *Drying Technol* **20**: 1053–1069.

Leonard, A., S. Blacher, P. Marchot, J. Pirard, and M. Crine (2003). Image analysis of x-ray tomograms of soft materials during convective drying. *J Microsc* **212**: 197–204.

Leonard, A., S. Blacher, P. Marchot, J. Pirard, and M. Crine (2004). Measurement of shrinkage and cracks associated to convective drying of soft materials by x-ray microtomography. *Drying Technol* **22**: 1695–1708.

Levitz, P. (2007). Toolbox for 3D imaging and modeling of porous media: Relationship with transport properties. *Cement Concr Res* **37**: 351–359.

Li, X., C.L. Lin, I.D. Miller, and W.P. Johnson (2006). Pore-scale observation of microsphere deposition at grain-to-grain contacts over assemblage-scale porous media domains using x-ray microtomography. *Env Sci Technol* **40**: 3762–3768.

Lin, C.L. and J.D. Miller (2000a). Network analysis of filter cake pore structure by high resolution x-ray microtomography. *Chem Eng J* **77**: 79–86.

Lin, C.L. and J.D. Miller (2000b). Pore structure and network analysis of filter cake. *Chem Eng J* **80**: 221–231.

Lin, C.L. and J.D. Miller (2004). Pore structure analysis of particle beds for fluid transport simulation during filtration. *Int J Miner Process* **73**: 281–294.

Lodewyckx, P., S. Blacher, and A. Leonard (2006). Use of x-ray microtomography to visualise dynamic adsorption of organic vapour and water vapour on activated carbon. *Adsorp* **12**: 19–26.

Lundblad, L.K.A., J. Thompson-Figueroa, G.B. Allen, L. Rinaldi, R.J. Norton, C.G. Irvin, and J.H.T. Bates (2007). Airway hyperresponsiveness in allergically inflamed mice: The role of airway closure. *Am J Respir Crit Care Med* **175**: 768–774.

Lux, J., C. Delisée, and X. Thibault (2006). 3D characterization of wood based fibrous materials: An application. *Image Anal Stereol* **25**: 25–35.

Lwin, T.T., T. Takeda, J. Win, N. Sunaguchi, T. Murakami, S. Mouri, S. Nasukawa, Q. Huo, T. Yuasa, K. Hyodo, and T. Akatsuka (2007). Preliminary quantitative analysis of myocardial fatty acid metabolism from fluoresecent x-ray computed tomography imaging. *J Synchrotron Rad* **14**: 158–162.

Martys, N.S. and J.G. Hagedorn (2002). Multiscale modeling of fluid transport in heterogeneous materials using discrete Boltzmann methods. *Mater Struct* **35**: 650–658.

Martys, N.S., J.G. Hagedorn, D. Goujon, and J.E. Devaney (1999). Large scale simulations of single and multi-component flow in porous media. *Developments in X-ray Tomography II*. U. Bonse (Ed.). Bellingham, WA, SPIE. *SPIE Proc Vol* **3772**: 205–213.

Masyuk, T.V., B.G. Huang, A.I. Masyuk, E.L. Ritman, V.E. Torres, X. Wang, P.C. Harris, and N.F. LaRusso (2004). Biliary dysgenesis in the PCK rat, an orthologous model of autosomal recessive polycystic kidney disease. *Am J Pathol* **165**: 1719–1730.

Molthen, R.C., K.L. Karau, and C.A. Dawson (2004). Quantitative models of the rat pulmonary arterial tree morphometry applied to hypoxia-induced arterial remodeling. *J Appl Physiol* **97**: 2372–2384.

Momose, A., T. Takeda, and Y. Itai (2000). Blood vessels: Depiction at phase contrast x-ray imaging without contrast agents in the mouse and rat — feasibility study. *Radiol* **217**: 593–596.

Monfrais, S., S. Bayat, L. Porra, G. Berruyer, C. Nemoz, W. Tomlinson, P. Suortti, and A.R.A. Sovijärvi (2005). Quantitative measurement of regional lung gas volume by synchrotron radiation computed tomography. *Phys Med Biol* **50**: 1–11.

Montet, X., C.M. Pastor, J.P. Vallée, C.D. Becker, A. Geissbuhler, D.R. Morel, and P. Meda (2007). Improved visualization of vessels and hepatic tumors by micro-computed tomography (CT) using iodinated liposomes. *Invest Radiol* **42**: 652–658.

Müller, B., J. Fischer, U. Dietz, P.J. Thurner, and F. Beckmann (2006). Blood vessel staining in the myocardium for 3D visualization down to the smallest capillaries. *Nucl Instrum Meth B* **246**: 254–261.

Nakashima, Y., T. Nakano, K. Nakamura, K. Uesugi, A. Tsuchiyama, and S. Ikeda (2004). Three-dimensional diffusion of non-sorbing species in porous sandstone: Computer simulation based on x-ray microtomography using synchrotron radiation. *J Contam Hydrol* **74**: 253–264.

Namati, E., D. Chon, J. Thiesse, E.A. Hoffman, J. de Ryk, A. Ross, and G. McLennan (2006). In vivo micro-CT lung imaging via a computer-controlled intermittent iso-pressure breath hold (IIBH) technique. *Phys Med Biol* **51**: 6061–6075.

Ohuchi, H., P.E. Beighley, Y. Dong, M. Zamir, and E.L. Ritman (2007). Microvascular development in porcine right and left ventricular walls. *Ped Res* **61**: 676–680.

Op Den Buijs, J., Ž. Bajzer, and E.L. Ritman (2006). Branching morphology of the rat hepatic portal vein tree: A micro-CT study. *Annal Biomed Eng* **34**: 1420–1428.

Palchikov, E.I., Y.A. Schemelinin, A.G. Skripkin, D.Y. Mekhontsev, and N.A. Kondratiev (2006). Mapping of residual oil and water saturation in porous media by means of digital x-ray laminography. *Part Part Syst Charact* **23**: 254–259.

Pavani, S.K., M.S. Dogan, H. Sari-Sarraf, and E.F. Hequet (2004). Segmentation and classification of four common cotton contaminants in x-ray microtomographic images. *Machine Vision Applications in Industrial Inspection XII*. J.R. Price and F. Meriaudeau (Eds.). Bellingham,WA, SPIE. *SPIE Proc Vol* **5303**: 1–13.

Peyrin, F., M. Salome, P. Cloetens, A.M. Laval-Jeantet, E. Ritman, and P. Rüegsegger (1998). MicroCT examinations of trabecular bone samples at different resolutions: 14, 7 and 2 micron level. *Technol Health Care* **6**: 391–401.

Prodanović, M., W.B. Lindquist, and R.S. Seright (2007). 3D image-based characterization of fluid displacement in a Berea core. *Adv Water Res* **30**: 214–226.

Pyrz, R. and J. Schjødt-Thomsen (2006). Bridging the length-scale gap—Short fibre composite material as an example. *J Mater Sci* **41**: 6737–6750.

Ramaswamy, S., M. Gupta, A. Goel, U. Aaltosalmi, M. Kataja, A. Koponen, and B.V. Ramarao (2004). The 3D structure of fabric and its relationship to liquid and vapor transport. *Colloids Surf A* **241**: 323–333.

Ritman, E.L., S.M. Jorgensen, P.E. Lund, P.J. Thomas, J.H. Dunsmuir, J.C. Romero, R.T. Turner, and M.E. Bolander (1997). Synchrotron-based micro-CT of in situ biological basic functional units and their integration. *Developments in X-ray Tomography*. U. Bonse (Ed.). Bellingham, WA, SPIE. *SPIE Proc Vol* **3149**: 13–24.

Rolland Du Roscoat, S., J.F. Bloch, and X. Thibault (2005). Synchrotron radiation microtomography applied to investigation of paper. *J Phys D* **38**: A78–A84.

Rolland Du Roscoat, S., M. Decain, X. Thibault, C. Geindreau, and J.F. Bloch (2007). Estimation of microstructural properties from synchrotron x-ray microtomography and determination of the REV in paper materials. *Acta Mater* **55**: 2841–2850.

Sakellariou, A., T.J. Senden, T.J. Sawkins, M.A. Knackstedt, M.L. Turner, A.C. Jones, M. Saadatfar, R.J. Roberts, A. Limaye, C.H. Arns, A.P. Sheppard, and R.M. Sok (2004). An x-ray tomography facility for quantitative prediction of mechanical and transport properties in geological, biological and synthetic systems. *Developments in X-ray Tomography IV*. U. Bonse (Ed.). Bellingham, WA, SPIE. *SPIE Proc Vol* **5535**: 473–484.

Schnaar, G. and M.L. Brusseau (2005). Pore-scale characterization of organic immiscible-liquid morphology in natural porous media using synchrotron x-ray microtomography. *Env Sci Technol* **39**: 8403–8410.

Schnaar, G. and M.L. Brusseau (2006). Characterizing pore-scale dissolution of organic immiscible liquid in natural porous media using synchrotron x-ray microtomography. *Env Sci Technol* **40**: 6622–6629.

Selomulya, C., X. Jia, and R.A. Williams (2005). Direct prediction of structure and permeability of flocculated structures and sediments using 3D tomographic imaging. *Chem Eng Res Des* **83**: 844–852.

Selomulya, C., T.M. Tran, X. Jia, and R.A. Williams (2006). An integrated methodology to evaluate permeability from measured microstructures. *AIChE J* **52**: 3394–3400.

Shen, H., S. Nutt, and D. Hull (2004). Direct observation and measurement of fiber architecture in short fiber-polymer composite foam through micro-CT imaging. *Compos Sci Technol* **64**: 2113–2120.

Sheppard, A.P., C.H. Arns, A. Sakellariou, T.J. Senden, R.M. Sok, H. Averdunk, M. Saadatfar, A. Limaye, and M.A. Knackstedt (2006). Quantitative properties of complex porous materials calculated from x-ray μCT images. *Developments in X-ray Tomography V*. U. Bonse (Ed.). Bellingham, WA, SPIE. *SPIE Proc Vol* **6318**: 631811-1 to –15.

Shofer, S., C. Badea, S. Auerbach, D.A. Schwartz, and G.A. Johnson (2007). A micro-computed tomography-based method for the measurement of pulmonary compliance in healthy and bleomycin-exposed mice. *Exp Lung Res* **33**: 169–183.

Sled, J.G., M. Marxen, and R.M. Henkelman (2004). Analysis of microvasculature in whole kidney specimens using microCT. *Developments in X-ray Tomography IV*. U. Bonse (Ed.). Bellingham, WA, SPIE. *SPIE Proc Vol* **5535**: 53–64.

Tan, J.C., J.A. Elliott, and T.W. Clyne (2006). Analysis of tomography images of bonded fibre networks to measure distributions of fiber segment length and fiber orientation. *Adv Eng Mater* **8**: 495–500.

Toyota, E., K. Fujimoto, Y. Ogasawara, T. Kajita, F. Shigeto, T. Matsumoto, M. Goto, and F. Kajiya (2001). Dynamic changes in three-dimensional architecture and vascular volume of transmural coronary microvasculature between diastolic- and systolic-arrested rat hearts. *Circulation* **105**: 621–626.

Toyota, E., Y. Ogasawara, K. Fujimoto, T. Kajita, F. Shigeto, T. Asano, N. Watanabe, and F. Kajiya (2004). Global heterogeneity of glomerular volume distribution in early diabetic nephropathy. *Kidney Int* **66**: 855–861.

Vasic, S., B. Grobéty, J. Kuebler, T. Graule, and L. Baumgartner (2007). X-ray computed micro tomography as complementary method for the characterization of activated porous ceramic preforms. *J Mater Res* **22**: 1414–1424.

Verma, N., K. Salem, and D. Mewes (2007). Simulation of micro- and macro-transport in a packed bed of porous adsorbents by lattice Boltzmann methods. *Chem Eng Sci* **62**: 3685–3698.

Videla, A., C.L. Lin, and J.D. Miller (2006). Watershed functions applied to a 3D image segmentation problem for the analysis of packed particle beds. *Part Part Syst Charact* **23**: 237–245.

Videla, A.R., C.L. Lin, and J.D. Miller (2007). 3D characterization of individual multiphase particles in packed particle beds by x-ray microtomography (XMT). *Int J Miner Process* **84**: 321–326.

Vladisavljević, G.T., I. Kobayashi, M. Nakajima, R.A. Williams, M. Shimizu, and T. Nakashima (2007). Shirasu porous glass membrane emulsification: Characterisation of membrane structure by high-resolution X-ray microtomography and microscopic observation of droplet formation in real time. *J Membrane Sci* **302**: 243–253.

Walther, T., K. Terzic, T. Donath, H. Meine, F. Beckmann and H. Thoemen (2006). Microstructural analysis of lignocellulosic fiber networks. *Developments in X-ray Tomography V*. U. Bonse (Ed.). Bellingham, WA, SPIE. *SPIE Proc Vol* **6318**: 631812-1 to –10.

Wan, S.Y., E.L. Ritman, and W.E. Higgins (2002). Multi-generational analysis and visualization of the vascular tree in 3D microCT images. *Computers Biol Med* **32**: 55–71.

Wildenschild, D., K.A. Culligan, and B.S.B. Christensen (2004). Application of x-ray microtomography to environmental fluid flow problems. *Developments in X-ray Tomography IV*. U. Bonse (Ed.). Bellingham, WA, SPIE. *SPIE Proc Vol* **5535**: 432–441.

Wildenschild, D., J.W. Hopmans, C.M.P. Vaz, M.L. Rivers, D. Rickard, and B.S.B. Christensen (2002). Using x-ray tomography in hydrology: Systems, resolutions and limitations. *J Hydrol* **267**: 285–297.

Willson, C.S., R.W. Stacey, K. Ham, and K.E. Thompson (2004). Investigating the correlation between residual nonwetting phase liquids and pore-scale geometry and topology using synchrotron x-ray tomography. *Developments in X-ray Tomography IV*. U. Bonse (Ed.). Bellingham, WA, SPIE. *SPIE Proc Vol* **5535**: 101–111.

Wilson, S.H., J. Herrmann, L.O. Lerman, D.R.H. Jr., C. Napoli, E.L. Ritman, and A. Lerman (2002). Simvastatin preserves the structure of coronary adventitial vasa vasorum in experimental hypercholesterolemia independent of lipid lowering. *Circulation* **105**: 415–418.

Wirkner, C.S. and L. Prendini (2007). Comparative morphology of the hemolymph vascular system in scorpions — A survey using corrosion casting, MicroCT, and 3D-reconstruction. *J Morphol* **268**: 401–413.

Wirkner, C.S. and S. Richter (2007). The circulatory system in Mysidacea — Implications for the phylogenetic position of Lophogastrida and Mysida (Malacostraca, Crustacea). *J Morphol* **268**: 311–328.

Yang, H. and B.W. Lindquist (2000). Three-dimensional image analysis of fibrous materials. *Applications of Digital Image Processing XXIII*. A.G. Tesher (Ed.). Bellingham, WA, SPIE. *SPIE Proc Vol* **4115**: 275–282.

Zhang, W., K.E. Thompson, A.H. Reed, and L. Beenken (2006). Relationship between packing structure and porosity in fixed beds of equilateral cylindrical particles. *Chem Eng Sci* **61**: 8060–8074.

Zhu, X.Y., M.D. Bentley, A.R. Chade, E.L. Ritman, A. Lerman, and L.O. Lerman (2007a). Early changes in coronary artery wall structure detected by microcomputed tomography in experimental hypercholesterolemia. *Am J Physiol Heart Circ Physiol* **293**: H1997–H2003.

Zhu, X.Y., E. Daghini, A.R. Chade, C. Napoli, E.L. Ritman, A. Lerman, and L.O. Lerman (2007b). Simvastatin prevents coronary microvascular remodeling in renovascular hypertensive pigs. *J Am Soc Nephrol* **18**: 1209–1217.

10

Evolution of Structures

This chapter covers microCT studies in the broad area of evolution of microstructures. Being able to follow the 3D structure of a specimen as it changes is an enormous benefit of microCT. To be sure, great value also exists for single observations of specimens taken from within a sequence of processing steps or of time points, especially for (a) the class of specimens whose preparation for observation by other means is impractical (too laborious or fraught with artifacts) or (b) samples requiring volumes to be interrogated in order to avoid sampling bias from unanticipated microstructural anisotropy (e.g., Figure 6.1). The greatest impact occurs, however, when the sample itself can act as its own control during longitudinal studies.

Evolving microstructure studies covered in other chapters include changes in cellular materials (Chapter 8), changes in networks (Chapter 9), and changes in deformed or cracked solid materials (Chapter 11). Some papers that might have been covered in Chapters 8, 9, and 11 have found their way into this chapter. There are plenty of topics, nonetheless, to fill the current chapter, and this material is grouped into three sections: materials processing (Section 10.1), environmental interactions (Section 10.2), and hard and soft tissue adaptation (Section 10.3). In Section 10.1, microCT studies of solidification, vapor phase processing, plastic forming, and particle packing and sintering are reviewed. Section 10.2 covers geological applications, construction materials, degradation of biological structures, and corrosion of metals. Section 10.3 organizes the material into one subsection on mineralized tissue and implants, bone healing, bone mineral levels, and remodeling and a second subsection on degradation of tissue, primarily soft tissue, including studies on arthritis, on pathological calcification characterization, and tumor generation and growth.

10.1 Materials Processing

Difficulty in exactly reproducing conditions and a plethora of adjustable variables bedevil studies aimed at understanding materials processing. For metals, for example, not only processing temperature and time at temperature but also the rate of cooling and rate of heating figure into the microstructure produced. MicroCT is an extremely valuable tool, therefore, in this area of engineering/science, especially as longitudinal comparative data can improve predictive numerical models and, if accurate, enable virtual interrogation of many combinations of variables, thereby focusing attention on the most promising avenues.

10.1.1 Solidification

Solidification is a profitable processing application for microCT. One application that can be studied readily is the inhomogeneous distribution of particles in a discontinuously reinforced composite. Clustering of reinforcement particles can be deleterious from a fracture resistance perspective or can be used to great effect to provide a component with a hard, wear-resistant (albeit low-toughness) outer surface and a tough internal volume.

There are a number of situations such as restorative dentistry or orthopedics where implants or other structures must be fabricated for an exact match to the remaining hard tissue. MicroCT can be used to provide the numerical coordinates for computer-aided machining. If the structure is to be cast, then microCT can also provide the geometrical input for numerical models predicting how to avoid porosity and validate the success of each specific casting (Atwood et al., 2007).

In an in situ composite of Al and TiB_2, Watson and coworkers used a novel sampling procedure to withdraw material from the melt and observed boride particle clustering with synchrotron microCT as a function of melt hold time (Watson et al., 2005). Small amounts of melt were drawn off at times up to 2.6×10^5 s and quickly solidified. MicroCT revealed the maximum cluster size decreased from an initial value of 50 μm to 10 μm at the end of the experiment. Even though the sampling technique might have biased the clustering results somewhat, use of the same method throughout means that the changes observed almost surely reflected changes occurring in the melt vessel. In an Al/SiC_p functionally graded composite fabricated by centrifugal casting, Velhinho et al. (2003) observed slight gradients in particle volume fraction away from the SiC-rich surface. Because their mass attenuation coefficients are very similar, SiC and Al are difficult to distinguish based on absorption alone, and phase-enhanced interface contrast in synchrotron microCT can help segmentation considerably.

Solidification with segregation of atoms with different absorptivities is another area where microCT has been applied. This segregation, and the accompanying range of solidification temperatures, can lead to undesirable excess porosity or cracks (hot-tearing) appearing at the end of solidification. Ludwig et al. (2005) followed in situ solidification of Al-4 wt.% Cu in 3D with ultrafast synchrotron microCT. A complete 512 (perpendicular to the rotation axis) × 256 scan (500 projections over 180°) was recorded every ten seconds using polychromatic wiggler radiation; this was a case where contrast sensitivity was sacrificed for temporal resolution. A cooling rate of 0.1°C/s was used for these in situ experiments; despite the reconstructions encompassing structures averaged over a ~1°C temperature range, the slices were quite clear, showed growth and linkage of the solid-phase particles, and showed increasing Cu content in the liquid phase.

Evolution of the experimentally determined solid volume fraction with temperature was compared to different solidification models; shrinkage versus solid fraction was linear; S_V evolved as expected and mean wt.% Cu in the solid and liquid phases (determined from analysis of the linear attenuation coefficients) agreed reasonably well with that predicted by the liquidus and solidus temperatures of the equilibrium phase diagram (Ludwig et al., 2005). In an Al-15.8 wt.% Cu alloy, the investigators were also able to show that rapid quenching produces a higher volume fraction of solid phase than was present at the starting temperature (Baruchel et al., 2006); therefore, models based on quenching data may contain a bias that must be corrected. An earlier report of microCT materials quenched from the solid state appeared elsewhere (Verrier et al., 2000).

The grain size distribution and number of faces per grain were measured in synchrotron microCT-reconstructed volumes of solidified Al–Sn with 1, 2, or 3 at.% Sn (Krill et al., 2001; Dobrich et al., 2004); Sn is immiscible in Al and segregates to the Al–Al grain boundaries. Some areas of the grain boundaries appeared to be free of Sn, a possible effect of the sensitivity limit, and a special algorithm was derived to fill in the missing boundaries. The authors concluded that size distribution agreed with and the number of faces per grain differed from metallography data in the literature (Dobrich et al., 2004). In situ microCT of coarsening of a model binary semi-solid alloy (Al-32 wt.% Ge) revealed the particle network followed theoretical predictions (Zabler et al., 2007b).

Retained porosity in cast Al–Si was studied as a function of H_2/Ar gas ratio introduced during stirring of the melt (Buffière et al., 2000). Two populations of voids were observed. The smaller voids were associated with microshrinkage when the metal solidified, and the population characteristics (equivalent size vs. sphericity) did not vary with gas composition; the volume fraction of microshrinkage pores did not, as well. The larger voids were from artificial incorporation of gas bubbles, and the volume fraction of gas pores increased exponentially with H_2 content.

Evolution of micropores during homogenization of direct chill cast Al (2, 4, and 6 wt.% Mg) was studied with microCT (Chaijaruwanich et al., 2007); the specimens were machined from 2.5-mm-diameter rods of as-cast and heat-treated ingots (0, 1, 10, and 100 hr at 530°C). These investigators found that the tortuosity of the pore networks was very complex and that there was no increase in maximum pore length. Intrapore Ostwald ripening of the networks produced coarsening of both the asperities and interconnects. Hot tear damage was studied in four specimens of a direct chill cast commercial Al–Mg alloy (AA5182) tested to four tensile loads at 528°C (fraction solid ~0.98); growth of preexisting voids and formation of damage-related voids were observed and interpreted quantitatively (Phillion et al., 2006).

Porosity formation in epoxy-based composites used in filling cavities is one concern in restorative dentistry, and a second is marginal debonding of the light-cured material from dentin cavity walls. MicroCT was used to study volume fraction of micropores and interfacial gap in a model system (8-mm-diameter human dentin disks in which a 2-mm-diameter, 1-mm-deep cavity was filled with two types of composites; Kakaboura et al., 2007).

Highly nonequilibrium solidification occurs in spray deposition of thermal barrier coatings, and synchrotron microCT is invaluable for determining pore shapes and for quantifying pore volume fraction in the coating as a function of distance from the substrate (Kulkarni et al., 2000, 2004a,b, 2005). Different processes produced different pore shapes that can be directly visualized and compared with the results of techniques such as SANS (small angle neutron scattering) and nanoindentation for elastic moduli determination (Kulkarni et al., 2000, 2005). Gradients in porosity were also correlated with indentation-derived moduli and SAXS (Kulkarni et al., 2004a,b).

10.1.2 Vapor Phase Processing

Similar pore topology exists in ceramic matrix composites partially densified by CVI, that is, chemical vapor infiltration (Butts, 1993; Kinney et al., 1993; Lee, 1993; Kinney et al., 1994a; Kinney and Haupt, 1997; Lee et al., 1998). The goal with ceramic composites is to maximize the density produced by processing so that the toughening effect of the reinforcement is fully realized; an evolving structure must be understood in this case. Production of complex, near-net-shape parts results for materials that are very difficult to machine; cost savings for SiC/SiC composites can be enormous. The preforms for these composites are often constructed by laying layers of cloth to form the required thickness or by wrapping multiple layers of cloth around a mandrel. In CVI, the preform is infiltrated by a vapor precursor that decomposes and deposits the desired phase within the composite. Another approach, reaction sintering, fills the preform with a very fine powder and forms the matrix by reacting the precursor with a gas. Although reaction-sintered samples were studied with microCT for different processing conditions (Stock et al., 1989), the principal work has been in CVI of SiC/SiC composites based on cloths of Nicalon amorphous SiC fibers.

Use of a portable reaction chamber allowed Nicalon/SiC composites to be transported to a synchrotron radiation source between increments of CVI and to be imaged multiple times with microCT (Butts, 1993; Kinney et al., 1993; Lee, 1993; Lee et al., 1998). The graphite chambers were cylindrical with a 10-mm outside diameter, and the Nicalon cloth preforms, each 6.2 mm in diameter and 6.0 mm high, consisted of 20 layers stacked within a chamber. Each cloth was woven of bundles or tows of ~500, 10–20-μm diameter Nicalon fibers, and there were approximately 6.3 tows per cm. Two architectures were examined: 0°/90° layups where the tows in adjacent cloth layers were parallel and 0°/45° where the tows in adjacent layers were rotated 45°. The preforms were infiltrated with methyltrichlorosilane (MTS) at 975°C, and the MTS flow was adjusted so that after 3, 6, and 9 hours densification would be 33, 67, and 100 percent complete, respectively.

Samples were imaged prior to and after each increment of CVI, 400 contiguous slices were recorded, and the reconstructions consisted of isotropic 15.8-μm voxels. Data could not be collected at higher resolution because of the need to keep the entire reaction chamber in the field of view, that is, as a direct consequence of the number of elements in the CCD detector. If the experiment were repeated now, one expects that local tomography would be performed including only SiC-containing volume; voxel sizes might be ~3 μm.

In the 0°/90° preform, the structure of the channels between cloths depended on the relative displacement of the holes in the cloths bounding the channel (Butts, 1993; Kinney et al., 1993; Lee, 1993; Lee et al., 1998), and this relationship is illustrated in Figure 10.1. Offsets 45° to the tow axes resulted in very closed channels (Figure 10.1, top); hole displacements along the tow directions produced an array of parallel, one-dimensional open pipes (Figure 10.1, middle); and when the holes were aligned, a two-dimensional network of openings was observed in the channels (Figure 10.1, bottom). Figure 10.2 shows how the 2D pipe network closes after infiltration. Circle cuts, numerical sections along the surface of a cylinder unwrapped to provide a flat representation, are used to show where SiC was deposited within or on the SiC tows after three increments of infiltration: difference maps of the same cylinder through the composite are shown in Figure 10.3. The gas inlet is at the bottom of each image of Figure 10.3, and this is where the greatest change appears (the color bar shows increasing linear attenuation coefficient from left to right).

A consequence of the voxel size was that individual micropores within the tows could not be resolved, even in the preforms. Nonetheless, accurate volume fractions of Nicalon, deposited SiC, and micropores were obtained by taking averages over several subvolumes within tows for preform and for 3 and 6 hrs infiltration (Kinney et al., 1993; Lee, 1993). Comparison of a microCT section through a densified Nicalon/SiC composite with an optical micrograph of nearly the same section demonstrated the accuracy of the reconstruction (Kinney and Nichols, 1992). Measurement of surface area per unit volume as a function of fractional density (Kinney et al., 1993; Lee, 1993) agreed well with the uniform deposition model used

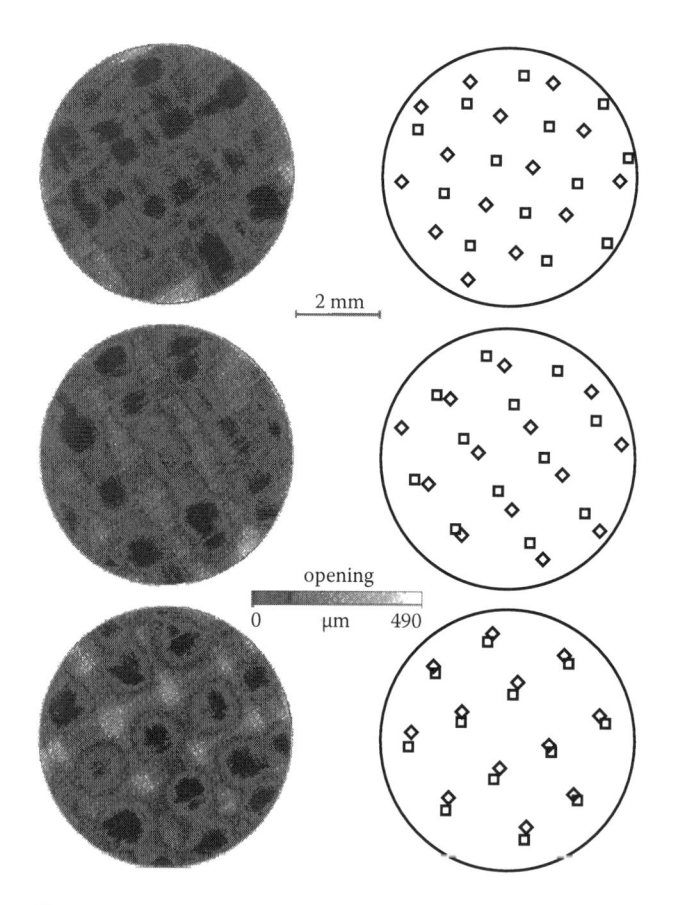

FIGURE 10.1 (SEE COLOR INSERT FOLLOWING PAGE 144.)
Maps of three patterns of channel width (left column) as a function of relative positions of the holes (squares and diamonds) in plane weave SiC cloths on either side of each channel (right column). The total channel width is projected onto a plane. Black pixels show position with no opening, with different shades showing increasing opening. (Top) Poorly defined, relatively closed channel due to hole displacements 45° to the tow (fiber bundle) direction. (Middle) Hole displacements along the tow direction and the resulting set of 1D channels. (Bottom) Closely aligned holes and the accompanying 2D network of opening (Lee (1993) and Lee et al. (1998); © Materials Research Society).

to describe the CVI process (Starr, 1992); the slight differences could be attributed to the fact that the surface area per unit volume of macroporosity (channels and holes through cloth layers) could be measured and the measurements could not include microporosity.

Bernard and coworkers reported some results of isobaric chemical vapor infiltration of a C/C composite (Bernard et al., 2000). Kang et al. (2004) showed some synchrotron microCT images of cracks in C fibers grown from the vapor phase. Other composites processed via a vapor phase route are mentioned in passing elsewhere in this volume.

Evolution of Structures   243

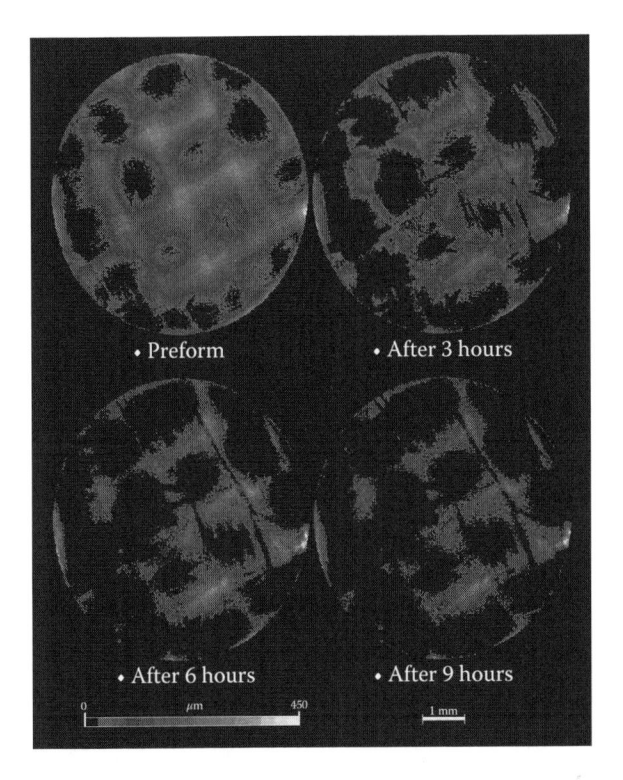

FIGURE 10.2 (SEE COLOR INSERT FOLLOWING PAGE 144.)
Variation of channel width in the SiC cloth composite described in Figure 10.1 as a function of SiC CVI infiltration time. The holes on either side of the channel are well aligned (2D array of pipes), the shades have the same meaning as in Figure 10.1, and the infiltration times are given below each map (Lee, 1993).

10.1.3 Plastic Forming

Superplastic forming (very high strains in certain alloys without rupture of the starting material) is finding application in aerospace and automotive fields. Superplastic deformation is generally limited by strain-induced cavitation leading to fracture. Using synchrotron microCT, Martin et al. (2000a,b) found that the number of cavities per unit volume versus strain ($\sim 1 < \varepsilon < \sim 1.7$) in AA5083 followed model predictions and observed developing cavity linkage. Pore evolution from rolling of an Al–6Mg alloy was studied with lab microCT, and the authors demonstrated that the highly tortuous pores would be difficult to detect in polished sections (Youssef et al., 2006). The tortuous pores spheroidized during homogenization, and accelerated centerline intrapore coarsening observed during initial, low-reduction-ratio rolling passes was attributed (through finite element modeling) to local tensile conditions, a counterintuitive but not unreasonable result (Youssef et al., 2006).

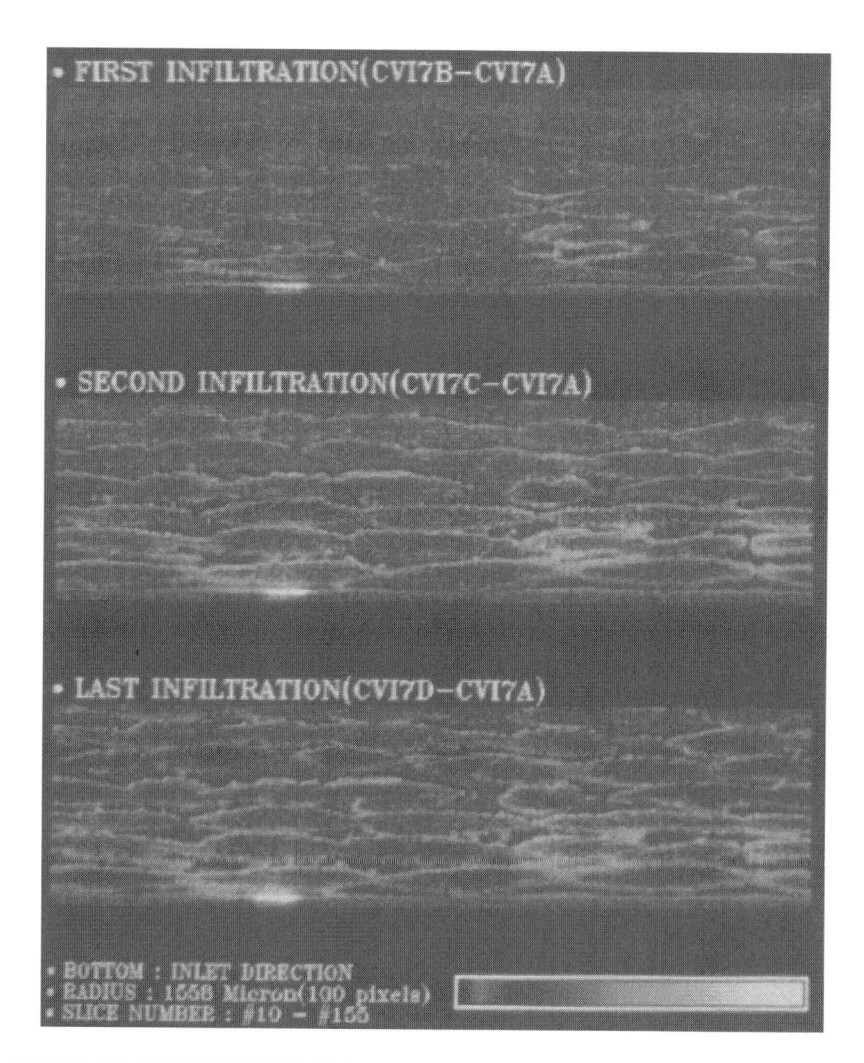

244 *MicroComputed Tomography: Methodology and Applications*

FIGURE 10.3 (SEE COLOR INSERT FOLLOWING PAGE 144.)
Circle cuts through a SiC/SiC cloth preform composite showing where SiC was deposited. (Top) Difference image showing SiC deposited in the first three hours of CVI. (Middle) Difference image showing the total SiC deposited after six hours. (Bottom) Difference image showing the total SiC deposited after nine hours. Different shades show increasing change from the preform. The open areas between tows, where there are no fibers onto which SiC can be deposited, remain red throughout. The gas enters the reaction chamber from the bottom, and the white shells seen on many tows show that the interior microporosity between individual ~15-µm-diameter fibers remains incompletely densified (Lee, 1993).

Warm drawing of a 6-mm-diameter PE rod down to ~5-mm diameter was studied with SAXS microCT (Schroer et al., 2006; Stribeck et al., 2006). Circular zones of differing lamellar sizes (longitudinal and lateral) were clear in the reconstructed slice presented (Stribeck et al., 2006), and undoubtedly much more can be done with this approach, especially in specimens that would appear featureless in absorption tomographs. It would be interesting to compare phase tomography to the SAXS-derived reconstruction.

Synchrotron microCT of an Al–Mg industrial alloy (AA5182) tracked the size distribution and spatial dispersion of intermetallic particles (iron rich, Mg_2Si) and of voids from as-cast + homogenized to hot rolled to cold rolled to tensile tested state (Maire et al., 2006). In addition to 3D views of a small portion of the structure (four thresholds with Al rendered transparent, voids as black, and the intermetallics in two different grays), numerical values for mean, minimum, and maximum equivalent radii of the intermetallic particles showed the progression of fragmentation expected for the different processing steps. Similar data for volume fraction, number density, and mean equivalent radii were measured for pores. One limitation in the study with 0.7-µm voxels was that very small particles (those with smallest dimension <1 µm) could not be included in the analysis; the authors expected that this would be improved with imaging with 0.3-µm voxels (Maire et al., 2006).

Movement of marker particles within metals or in unconsolidated powders can be used very effectively to map displacement fields in response to deformation (Nielsen et al., 2003, 2004; McDonald et al., 2006; Zettler et al., 2006). Specimens of Al–1 vol.% Ti (particle diameters between 1 and 10 µm) were imaged with synchrotron microCT at the 2-µm level, and displacement gradients for deformations up to 9.5 percent compression were quantified (Nielsen et al., 2003, 2004). Bulk material flow has also been studied in friction stirred welds in Al using the particle technique and synchrotron microCT (Zettler et al., 2006). Closed die compaction and sintering of powders have long been used to fabricate metal and ceramic components, but constitutive models for loose powder behavior under intense shear deformation need to be developed. McDonald et al. (2006) used lab microCT and image correlation to follow particle displacements when a cylindrical punch was pushed into a somewhat wider-diameter cylindrical die; they reported uncertainties in strains of ~0.05 percent and correlated particle displacement vectors with dilational strains calculated from the particle displacements (Figure 10.4). Generally speaking, these studies found that high-contrast particles (e.g., Ti, Sn, or W in Al) need to have diameters of several voxels for reliable detection and automated tracking of displacement.

Pressure gradients driving resin flow during liquid composite molding can be very low when large composite parts are manufactured, and capillary pressure can become the dominant force for tow infiltration, producing mesoscale voids. Mesostructure in glass fiber composites was

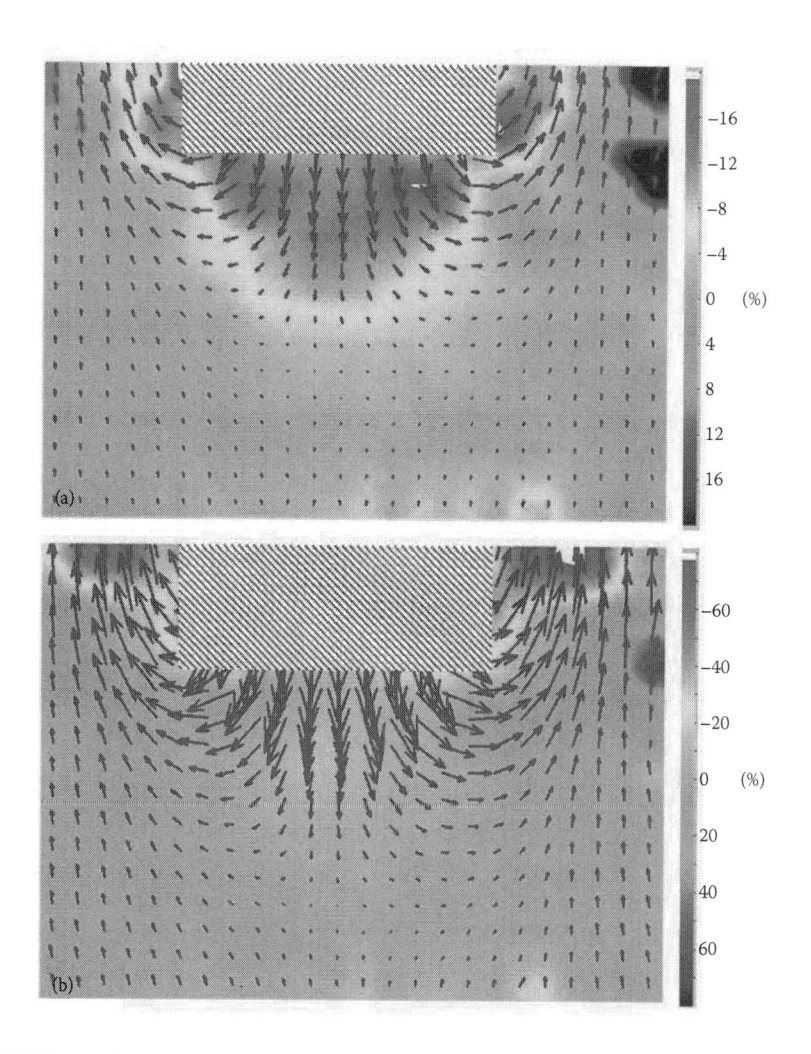

FIGURE 10.4 (SEE COLOR INSERT FOLLOWING PAGE 144.)
Powder displacement during compression. The arrows show particle displacement vectors around the downward moving punch calculated from image correlation over: (a) four 0.5 mm displacement steps (2.0 mm total) and (b) eight steps (4 mm total). The background gray scale represents the dilational (volumetric) strain calculated from the particle displacements, effectively showing the change in density across the diametral section. The powder immediately under the punch is compacted (negative strain), but dilation (positive strain) is observed as the loosely packed powder is sheared, particularly as it flows around the corners of the punch and upward against the sides of the container. (Reprinted from McDonald et al. (2006); © 2006, with permission from Elsevier.)

characterized (Schell et al., 2006), and experimental validation of a numerical method for predicting mesoscale voids in the glass-fiber, epoxy resin composite was provided by microCT, and, due to the voxel size of 6 μm, microvoids were ignored (Schell et al., 2007). Aspects of these two papers are reminiscent of the SiC/SiC CVI processing study described in Section 10.1.2.

Laser micromachining of hardened Portland cement paste was also studied with very high resolution synchrotron microCT (Trtik and Hauri, 2007). The 0.35-μm voxel data revealed that the hydrated cement was ablated but large unhydrated particles were relatively unaffected.

10.1.4 Particle Packing and Sintering

Particle shapes, size distributions, and packing are important in processing and also in fluid transport; discussion of those studies related to transport in the open phases appeared in Chapter 9. The 3D size and shape characteristics of collections of particles were studied with microCT (Lin and Miller, 2005; Thompson et al., 2006); note that microCT can eliminate the need to disperse particles originally in a dense assembly or packing or even in a somewhat agglomerated state and avoid artifacts inevitable in any physical separation process.

Thompson and coworkers developed their analysis algorithm in a dimensionless fashion, that is, in terms of voxels per unit particle diameter, so that it could be applied to different-sized particles' distributions studied with different voxel sizes (Thompson et al., 2006). The approach was based on simultaneous solid-phase and void-phase burn (the algorithm moves away from the interface and assigns a value to each voxel that equals the number of voxels it is away from the interface). The local maxima of burn number was used as a particle center, and all voxels previously identified as solid were assigned to one or another particle, with additional steps required to accurately partition contacting grains. Particle volume, surface area, orientation, aspect ratio, and contact statistics flowed directly from the assigned particles; computation performance metrics were supplied, and the results for computer-generated structures and for microCT of a standard sand showed the approach works quite well (Thompson et al., 2006). Lin and Miller (2005) characterized three collections of well-defined particles with lab microCT: nearly spherical beads, an isotropic quartz sand, and quite jagged rock fragments. The data showed the expected surface area versus volume behavior, and the plot for the irregularly shaped particles was offset from that of the beads and sand, as one would expect.

MicroCT datasets have been used as the basis for simulating random packing of polydisperse particles (Al-Raoush and Alsaleh, 2007). Fu and coworkers examined in situ compaction of powder with lab microCT (Fu et al. 2005, 2006), Richard et al. (2003) observed granular packing resulting from vibration with synchrotron microCT, and Seidler et al. (2000) characterized the distribution of granules packed under the influence of gravity. MicroCT-derived particle characteristics from a well-characterized ore were also compared with 2D measures (Gay and Morrison, 2006). Agglomeration of particles and the breakage of agglomerates during compression or extrusion have begun to be studied with microCT combined with numerical modeling (Golchert et al., 2004a,b; McGuire et al., 2007). Agglomerates, disruptions, and flashes were observed in micropowder injection molding of a silica model system (Heldele et al., 2006).

Sintering/cementation of powders has also been studied by microCT (Bernard et al., 2000; Lame et al., 2003, 2004; Bernard et al., 2005; Fu et al., 2005; Pfister et al., 2005; Vagnon et al., 2006; Vaucher et al., 2007); research areas include analysis of rapid prototyped material and evolution of materials with nanoparticulate precursors. Bernard and coworkers employed local tomography to quantify porosity elimination and neck evolution in a glass powder and in a lithium borate powder over 1.6×10^4 s at 700 and 720°C, respectively; growth of necks was particularly well illustrated by matched pairs of renderings, one showing the solid phase and the second the complementary void space (Bernard, 2005; Bernard et al., 2005).

Sintering of copper and steel (Distaloy AE) powders was also studied with synchrotron microCT, and changes in pore geometries in copper and elimination of very thin interparticle voids but not large pores in Distaloy were observed with increasing sintering time (Lame et al., 2003, 2004). Final dimensional changes are strongly anisotropic in Distaloy AE Densmix™ (axial swelling during delubrication and axial shrinkage during sintering), and repeated in situ synchrotron microCT observations during the different sintering stages (initial structure, lubricant burn-off, sintering, cooled structure) were used to investigate these changes (Vagnon et al., 2006). Analysis of the orientation distribution of three populations of pores (highly elongated, near circular, and intermediate geometry) was one probe; the second was an image correlation-based, local strain mapping comparison for the different directions in the compact.

10.2 Environmental Interactions

Whether the materials are natural or engineered or whether hard or soft tissue is affected, the interaction of these substances with their environment is an area of research where microCT offers important insights that are difficult or impossible to obtain by other methods. This is true in particular because these interactions often are very local or are in inaccessible locations, or at least begin that way. After discussing geological applications, construction materials are covered. The third subsection concerns degradation of biological structures and the fourth the corrosion of metals.

10.2.1 Geological Applications

Consideration of the transition of snow to firn to ice (porosities of >95 percent, ~40 percent, and <10 percent, respectively) follows naturally from the discussion of sintering of engineering materials. The particles (snow) are at a relatively high homologous temperature (i.e., a large fraction of melting temperature) and are frequently under pressure from the weight of subsequent snowfalls; thus, the process and the structures are very similar to those covered above. Properties of snow/firn/ice are important in understanding avalanches. If porosity is closed in specimens cored from ice, an air archive exists (atmospheric information) that can be compared to the ice archive (climatic information). Synchrotron microCT of a snow sample collected at the failure site of a slab avalanche, for example, showed a cohesive layer above a weak layer (Coléou and Barnola, 2001), a situation encountered in skier-triggered avalanches.

The metamorphosis of snow to ice depends on local 3D curvature and solid–vapor surface area; under isothermal conditions, minimization of local curvature (minimization of surface energy) is thought to govern the densification of the structure; that is, a structure with many sharp, flat grains transforms to a much more rounded structure. Results of synchrotron microCT of snow at different points during transformation have been compared with numerical models evolving from real (microCT-derived) 3D microstructures, and the agreement between actual and modeled structures was quite encouraging (Flin et al., 2003, 2005). One should note that considerable technical challenges were overcome in the preparation of unaltered snow specimens for microCT examination, but these are not discussed here (see Coléou and Barnola (2001)).

Understanding reactive percolation (CO_2) through porous limestone is important if, as some have proposed, rises in atmospheric CO_2 are to be combated by CO_2 sequestration in subsurface reservoirs. Synchrotron microCT followed changes in limestone porosity during flow of CO_2-charged water over 3×10^4 s and related changing microstructure to increasing permeability (Bernard, 2005). Other studies directly compared successive structures in the same solid–fluid system over 8×10^4 s (Noiriel et al., 2004, 2005). As limestone is an important construction material in historic structures, its attack by CO_2 and by industrial organic pollutants and repair of such damage is an important area of research (Brunetti et al., 2004). Vapor transport through two types of bricks was examined with synchrotron microCT, and the relationship among permeability, diffusivity, and pore size agreed with analytical expressions (Bentz et al., 2000). Water transport in the proton exchange membrane fuel cell was also tracked with microCT (Sinha et al., 2006).

Several studies of soils have appeared (Macedo et al., 1999; Hansel et al., 2002; Altman et al., 2005; Feeney et al., 2006; Hettiarachchi et al., 2006; Nunan et al., 2006). Metal contaminant absorption characterization is a current area of activity (Altman et al., 2005). In Altman et al. (2005), iron content was mapped, relative amounts of iron oxyhydroxides and iron-bearing clays were determined semiquantitatively, and the relationship of Cs adsorption with iron-bearing materials was imaged using absorption edge difference techniques. MicroCT showed that application of granule and liquid phosphorus-based fertilizer produced different effects on local density: over time, the former produced increased density within ~1 mm, but the latter did not (Hettiarachchi et al., 2006). Properties of harbor sediment were also studied (Jones et al., 2006).

Soil microbes and plant roots microengineer their habitats by changing porosity and pore cluster characteristics. Synchrotron microCT revealed soil modified by root action over 30 days had significantly greater pore volume and pore connectivity than soil without roots (Feeney et al., 2006; Nunan et al., 2006). Changing pore structure due to organic material decomposition was studied in several soils (De Gryze et al., 2006). Direct microCT observation over hours of the movement of the root-feeding clover weevil showed constant average speed toward root nodules, guided by a chemical signal released by the nodules, and indicated that the larvae respond to the gradient direction of the chemical signal but not its magnitude; this data was used to adjust numerical models of larva survival, a major negative effect on sustaining white clover in a mixed sward (Zhang et al., 2006). These and other results demonstrated that the soil ecosystem exhibits self-organization in relatively short periods of time. Some aquatic plant species sequester metal(loid) species on root surfaces, and synchrotron microCT and other techniques showed As associated with regions of enhanced Fe content (Hansel et al., 2002).

The location of the metal species of interest within or at the surface of particles governs the extent to which that metal may be recovered by leaching operations. In order to determine how well lab microCT might function as an assaying tool (i.e., predicting leaching performance for comparison with actual efficiency), Miller et al. (2003) quantified the volume fraction of copper-bearing minerals in contact with the surfaces of ore particles. Standards of natural particles of copper-bearing mineral phases (known to exist in the ore of interest) were used to set thresholds for segmentation and allowed 3D assessment of mineral exposure versus particle size to be determined for the ore of interest. MicroCT-predicted recovery generally underestimated actual recoveries from column leaching tests, an unsurprising result given the finite resolution and sensitivity of the microCT system and the additional exposure that partial leaching might produce (Miller et al., 2003). Silver is associated with pores in certain ores, and Chen et al. (2006) used lab microCT to determine that porosity was insufficient to allow silver to be leached from the interior of a coarsely ground ore sample.

10.2.2 Construction Materials

Construction materials frequently are exposed to aggressive environments over a period of years or even centuries, and, as much of the resulting damage is subsurface, a number of reports of microCT applied to degradation of building materials have appeared (e.g., Cnudde and Jacobs (2004)). One of the most important construction materials is concrete, and Portland cement used in this composite has been studied from several perspectives. A collection of useful baseline cement paste (synchrotron microCT) images was published (Bentz et al., 2002), and features such as unhydrated cement particles, regions of hydrated cement, and ettringite needles were identified. Based on this data, cement particle shapes were analyzed and were expected to improve computational models of cement hydration (Garboczi and Bullard, 2004) and to serve as a resource for those needing microstructures for modeling.

The curing of cement is perhaps more of a processing topic than an environmental interaction subject, but it seems more appropriate to deviate in this one particular from the overall plan and to mention a microCT study of the time course of the anhydrous cement particle: water interaction in conjunction with other studies of Portland cement. Gallucci and coworkers found there was a real conflict between resolving the narrow diameter of the capillary pores controlling the process (i.e., the required small specimen diameter) and obtaining representative volumes (Gallucci et al., 2007). This tension occurs again and again in microCT and nanoCT studies, and, for specimens with appreciable densities (i.e., those other than foams), progress in this area remains a challenge.

The pore structure of cement-based materials seriously affects resistance to environmental attack as well as mechanical properties. Often pozzolans (fine natural volcanic ash, fly ash from power generation, diatomaceous earth, etc.) are added to cement to produce a fine-structured composite. A study of the pore structure of Portland cement composites with pozzolan (neat cement vs. 25 wt.% fly ash vs. 10 wt.% metakaolin) found mean pore size and maximum pore diameters decreased for the composites compared to the simple cement (Rattanasak and Kendall, 2005). Chloride permeability is one cement durability issue, and microCT-measured pore structures of a reference concrete and of fly ash, silica fume, and slag modified concretes were compared with results of a rapid chloride permeability test (Lu et al., 2006). With 4-μm voxels, little pore connectivity over distances of 100–200 μm was documented for any of the four conditions; with 1-μm voxels the reference concrete showed deep pore penetration, whereas the modified concretes showed clear gaps in connected pore spaces. The somewhat limited data showed chloride permeability correlated clearly with disconnected pore distance (Lu et al., 2006).

Leaching of cement in mortar (sand plus Portland cement) was studied by repeated synchrotron microCT of the same specimen over 2.2×10^5 s (the four-day scheduling window at ESRF dictated the length of the experiment; Burlion et al. (2006)). Because of the time constraint, calcium efflux was increased by a factor of 300× compared to deionized water by using an ammonium nitrate solution (which produces the same mineral end products as water alone). Variation of linear attenuation coefficient μ in the cement phase was followed as a function of depth from the specimen surface for four exposure times. Within a zone near the surface, μ decreased rapidly during the first 24 hr (8.6×10^4 s), and the authors inferred that this was due to decalcification of Portlandite crystals (calcium hydroxide, CH) with little C–S–H (calcium silicate hydrate) involvement. Between 24 and 48 h of leaching (1.5×10^5 s), μ decreased more slowly, indicating that CH was completely removed from this portion of the specimen and that the less soluble (than CH) C–S–H was being removed, a process continuing to the end of the test. The variation of the thickness of the leached region agreed with diffusion-controlled kinetics (i.e., square root of time dependence).

Naik and coworkers studied sulfate attack of Portland cement paste using lab microCT and repeated observations on each specimen (Stock et al., 2002b; Naik, 2003; Jupe et al., 2004; Naik et al., 2004, 2006). Two cement types, different water-to-cement ratios, two different cation sulfates, and the presence/absence of aggregates were examined. Figure 10.5 shows how cracks develop in the same slice over 52 weeks sulfate exposure (pores within the cement paste are used as fiducials). Figure 10.6 shows 3D renderings of sulfate-induced cracking (that develops into spalling) in one specimen over 32 weeks of exposure. As the interpretation of the results depended to a significant extent on use of an additional x-ray modality (position-resolved energy-dispersive x-ray diffraction), further discussion is postponed until the section on multimode studies.

Synchrotron microCT has proven very valuable in studying wood degradation (Illman and Dowd, 1999). Fungi enzymes and metabolites degrade the structural integrity of tracheids (thick-walled, tubular structures with hollow centers containing air or water), and significant strength can be lost early in the decay process. Over the 96 hr following fungal inoculation, tracheid pore volume increased somewhat, as did pore interconnectivity (Illman and Dowd, 1999). Siloxanes/silanes mixtures are often applied as wood preservatives, and microCT was used to study penetration of the preservative into two types of wood using brominated silane as a contrast agent (Vetter et al., 2006). Boundaries between treated and untreated wood were clear, and one expects that with further work improved wood preservation protocols will result.

Evolution of Structures 253

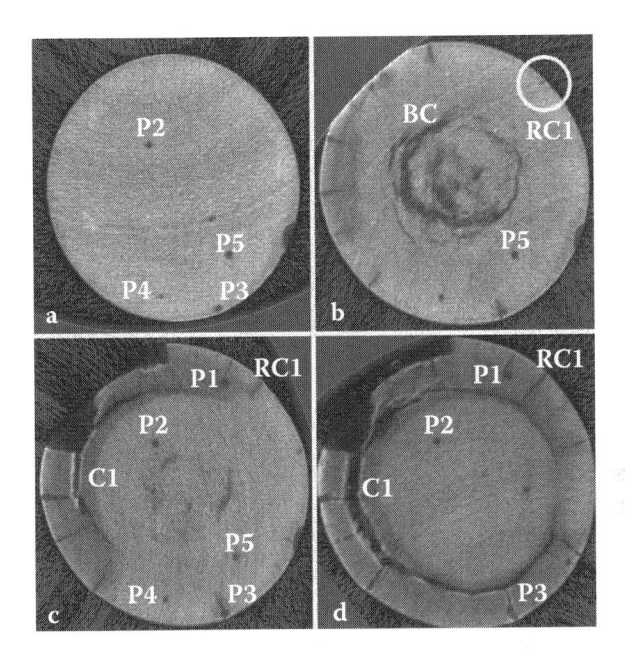

FIGURE 10.5
Matching lab microCT slices of a cement paste sample produced with a water-to-cement ratio of 0.485 and exposed to 10,000 ppm of sulfate ions in a sodium sulfate solution for (a) 21 weeks, (b) 36 weeks, (c) 42 weeks, and (d) 52 weeks. Crack C1, radial crack RC1, cracks within the body of the specimen BC, and pores P1–P5 are labeled. The horizontal field of view is (a) 15.3 mm, (b) 15.1 mm, (c) 16.1 mm, and (d) 15.2 mm; and reconstruction was with 37-µm isotropic voxels. The lighter the pixel is, the more absorbing the voxel. (Reprinted from Naik et al. (2006), © 2006, with permission from Elsevier.)

10.2.3 Degradation of Biological Structures

This fairly long subsection covers several topics, from dissolution of drug tablets to clinical and basic science aspects of tooth decay to degradation of implant materials. Coverage of topics such as formation of mineral at undesirable locations in soft tissue, growth of soft or hard tissue tumors, destruction of bone (development of high levels of porosity in cortical bone or erosion of bone surfaces), healing of damaged bone, and bone remodeling are covered in Section 10.3 below. Some aspects of bone loss and tumor formation were covered previously in Chapters 8 and 9, respectively.

Dissolution of drug tablets fits into this subsection in a very loose sense, but the microCT studies of this process deserve brief mention before treating the main subject. MicroCT and modeling were applied to granules and tablets (Ansari and Stepanek, 2006; Jia and Williams, 2007); as discussed above, this can be a very powerful combination. MicroCT was employed to study the surface membrane responsible for slow release rate characteristics in a tablet (Chauve et al., 2007).

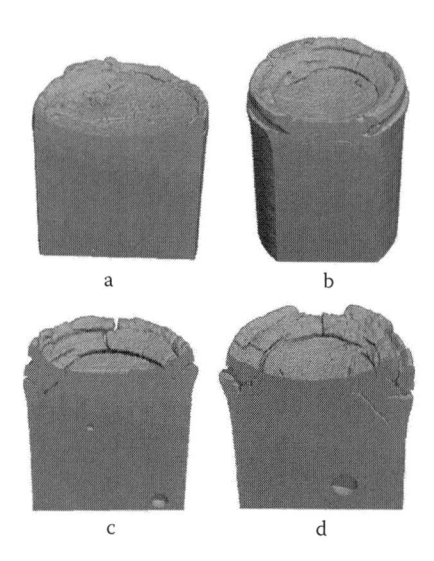

FIGURE 10.6
3D renderings of a single cement paste sample with water-to-cement ratio of 0.435 and exposed to the same solution as the specimens in Figure 10.5. (a) 9-weeks exposure showing onset of cracking, (b) 16-weeks exposure showing widening of pre-existing cracks and formation of new cracks, (c) 22-weeks exposure showing continued cracking, and (d) 32-weeks exposure showing spalling has begun. (Reprinted from Naik et al. (2006); © 2006, with permission from Elsevier.)

Teeth consist of an outer covering of enamel and an internal hard tissue dentin, and several microstructural features have been studied with microCT. Before reviewing microCT studies of changing tooth structure, it is sensible to consider microCT data on baseline microstructure and mineral levels.

The dentinoenamel junction (DEJ) divides enamel and dentin, has a graded interface structure, and, quite remarkably, couples these elastically very dissimilar materials under the application of complex loading patterns. Tafforeau and coworkers used synchrotron microCT and a multiresolution approach to study details of rhinoceros enamel structure (rods, incremental lines, mineral levels, etc.) and the structure of the DEJ (Tafforeau et al., 2007). Similar features were examined in hominid enamel with phase contrast microCT (Tafforeau and Smith, 2008).

Stock et al. (2008) studied the structure of bovine DEJ and surrounding tissue with the goal of understanding how the microstructure of the graded interface might contribute to its functionality. In dentin, tubules form during growth and run from the DEJ to the pulp chamber; a collar of hyper-mineralized dentin surrounds the fluid- or soft tissue–filled tubule and is the subject of considerable interest. Stock and coworkers were able to image the tubules, but higher-resolution microCT is needed to observe their structure. Phase microCT revealed tubules in dentin, features whose diameters are only slightly larger than the voxel size in the reconstructions (Zabler et al., 2006); submicrometer-sized tubule features were subsequently investigated by careful consideration of phase contributions to image formation (Zabler et al., 2007a).

By measuring the linear attenuation coefficients of enamel and dentin, Elliott et al. (1989) appear to have been the first to quantify differences in mineral content. Wong et al. (1991) demonstrated that a molar and developing incisor could be distinguished quantitatively within a mouse mandible. Mineral content of enamel and dentin have been compared in human molars and enamel pearls (Anderson et al., 1996). Gradients in mineral concentration in enamel but not in dentin were found from the apex to incisal end of a lower rat incisor (Wong et al., 1995a,b). Elliott et al. (1992) and Kinney et al. (1994b) studied demineralization at a carious lesion and subsequent remineralization at the same site. Deciduous enamel defects in low-birthweight children were easily distinguished from normal regions in the same teeth (Fearne et al., 1994).

Transparent dentin forms gradually with aging and can have different restoration-related fracture properties; measurement of mineral concentration from microCT, crystallite size from small angle scattering, and fracture toughness were combined in one study of this human tissue (Kinney et al., 2005). Mercer and Anderson (1995) reported effects of applying CO_2 laser pulses to human enamel. MicroCT's ability to quantify microstructure repeatedly can be key to following processes in biological structures whose microstructure varies greatly between individuals.

Wong and coworkers used microCT to address an interesting conundrum in clinical dentistry (Wong et al., 2006). Carious lesions in dentin have long been regarded as conical in shape with the base at the dentinoenamel junction. On the other hand, experience leads United Kingdom clinicians to teach students to remove softened carious dentin with a round bur or a curved excavator instead of a conical bur which would more closely resemble the purported shape of the lesion. MicroCT of ten carious primary molars revealed the lesions were bowl shaped, explaining the clinical tool preference; quite effective use was made of color 3D renderings showing sound dentin and enamel as nearly transparent grays and mineral levels of the lesions as different colors in the 3D interior of the tooth.

Environmental attack of apatite in tooth enamel has been studied with lab microCT and scanning microradiography for a number of years by Elliott and coworkers (Elliott et al., 2005). Over 70 days, these investigators periodically examined (with microCT) packed powders of carbonated apatite in an acidic buffer; these data were supplemented by infrared spectroscopy and Rietveld analysis of x-ray diffraction patterns of the dissected internal surface. This same group also carefully considered microCT-derived mineral levels in sound and carious enamel, and these papers will repay careful study (Elliott et al., 1998; Dowker et al., 2004). A de-/remineralization model using small coupons of bovine tooth (enamel plus dentin) was studied with synchrotron and lab microCT; analysis concentrated on quantifying the gradients of mineral content (Delbem et al., 2006; Vieira et al., 2006).

The papers cited at the end of the preceding paragraph illustrate two approaches to use of values of the linear attenuation coefficients in quantitative analyses. Elliott and coworkers took the fundamental approach of relating the linear attenuation coefficients back to mineral standards and expressing measured quantities in terms of absolute amounts of mineral present per unit volume of tissue (Elliott et al., 1998; Dowker et al., 2004). Similar approaches were utilized in studies of bone (Nuzzo et al., 2002; Borah et al., 2005). Although this approach allows for direct comparison between studies conducted under different conditions or with different techniques, extreme care must be taken to avoid systematic errors that might bias comparisons. The quantification of de-/remineralization of enamel employed an operational approach, assuming that linear attenuation coefficient values away from the surface were identical from specimen to specimen and scaling all values to this presumed reference (Delbem et al., 2006; Vieira et al., 2006). This might be thought to be a poor assumption because enamel mineral levels can differ by several percent or more between the tooth surface and volumes near the dentinoenamel junction, but the profiles were normalized to values of the linear attenuation coefficient at essentially the same depth from the tooth surface, thereby rendering this consideration moot. Very little of either sort of analysis has appeared to date, although, to be fair, some investigators explicitly verified that experimental values of linear attenuation corresponded to values expected for the material being studied (Kinney et al., 1990; Stock et al., 2002a).

The dentin–resin interface in tooth restorations is a zone of potential structural or chemical weakness; microhardness and microCT were combined to study this zone (Marshall et al., 1998). Another study of a dentin adhesive interface paired static loading (via synchrotron microCT) with FEM (Mollica et al., 2004); fatiguing in a solution of silver ions (and imaging at an energy just above the silver absorption edge) allowed interface leakage to be studied. Efficacy of endodontic seals was studied with synchrotron microCT (Contardo et al., 2005). Fissures in enamel and changes in mineralization were the subject of another report (Dowker et al., 2006).

Balloon dilation followed by stent implantation is a frequently applied percutaneous coronary intervention, but closure within the stented region occurs in a significant number of cases. Stent-related vessel closure may be triggered by introduction of foreign material or the presence of mechanical stresses from the rigid stent within a structure, the artery wall, designed to flex during pulsatile flow. Self-destructing stents that degrade before they are covered by the vascular cell walls are a possible solution, and microCT was used to follow Mg stent degradation over 56 days in minipigs (Loos et al., 2007). Note that the biologically necessary ion Mg^{2+} is well tolerated and features antithrombic activity which is helpful in vessel applications. The in vivo degradation of Mg implants in bone was also studied (Witte et al., 2006a). In vivo Mg corrosion rates were about four orders of magnitude lower than those in ASTM in vitro corrosion tests, and relative rates in two Mg implant alloys were opposite in in vivo versus in vitro tests (Witte et al., 2006b).

10.2.4 Corrosion of Metals

Materials transport is also important in solids that do not have well-defined porosity, and microCT was very effective for nondestructively quantifying the spatial distribution of corrosion products within printed wiring boards (Dollar, 1992; Stock et al., 1994). In the presence of high-voltage gradients, copper can diffuse and form conductive anodic filaments (CAF) and short out circuit elements, and microCT quantified the amount of copper within CAF. Results of microCT agreed with those of serial sectioning and with the amount of copper removed from the copper anode (Stock et al., 1994). It is important to emphasize that the copper-halide deposits constituting the CAF are typically 20 to 40 μm below the surface of the printed wiring board, that the translucent boards allow only indistinct visualization of the CAF with optical microscopy, and that an x-ray method is required for nondestructive characterization. Because the contrast from the printed wiring board is so variable and the contrast of the CAF along any particular viewing direction is so weak, radiography cannot provide images better than those from optical microscopy. This work demonstrated how subsurface CAF formation kinetics could be deduced nondestructively with microCT.

Repeated observations of the same specimen were also performed on stainless steel specimens undergoing localized corrosion (Connolly et al., 2006) and intergranular stress corrosion cracking (Babout et al., 2006; Marrow et al., 2006). Synchrotron microCT observed localized corrosion morphology within Al specimens exposed in situ to a chloride environment (Figure 10.7), and lab microCT investigated the morphology and quantified the transition from localized corrosion to stress corrosion cracking in steel specimens exposed ex situ to a simulated corrosive condensate environment (Connolly et al., 2006). A 302 stainless steel wire was heat treated to produce a stress-free, fully sensitized microstructure (i.e., one with grain boundary chromium carbides) and examined with synchrotron microCT after three increments of stress in an acidic environment (Babout et al., 2006). Analysis centered on identifying bridging ligaments formed during the first increment of crack propagation and on following the progressive failure of the ligaments using a combination of 2D sections and 3D renderings (Figure 10.8), and these authors noted the presence of unresolved cracks that could still be detected through their phase contrast (Babout et al., 2006). Three-dimensional finite element models were

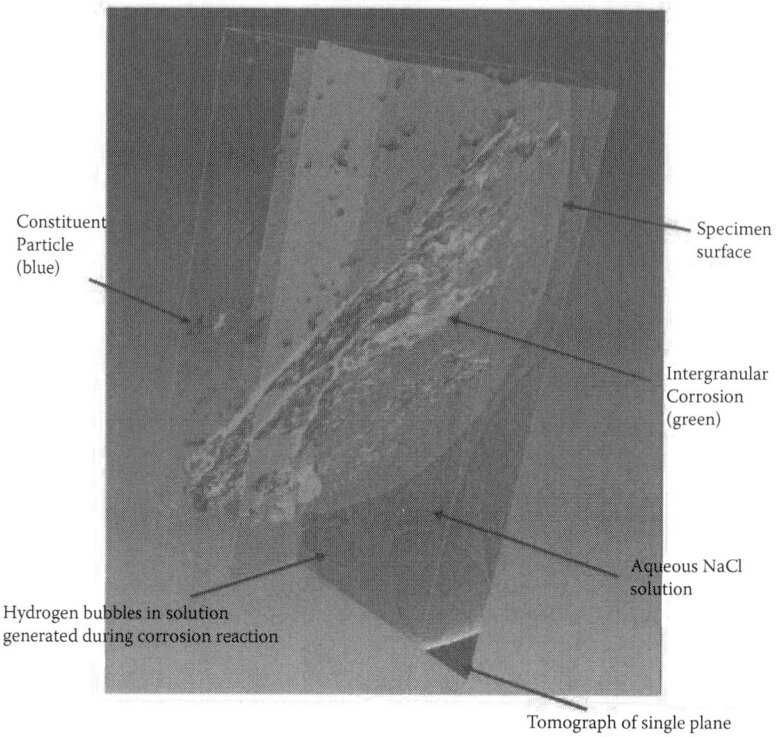

FIGURE 10.7 (SEE COLOR INSERT FOLLOWING PAGE 144.)
3D rendering of the heat-affected zone near a weld in AA2024 after 40 h exposure to 0.6 M NaCl. About 200 slices are used, showing the sample transparent except for intergranular corrosion and intermetallic particles (gray). The 0.5-mm-diameter specimen was imaged in situ with 20.5 keV synchrotron x-radiation and reconstructed with 0.7-µm isotropic voxels. (Reprinted from Connolly et al. (2006); © 2006, with permission of Maney Publishing (http://www.ingentaconnect.com/content/maney/mst).)

devised to investigate the development of crack bridging and its effects on crack propagation and crack coalescence in intergranular stress corrosion cracking (Marrow et al., 2006).

Synchrotron microCT was used to study the relationship between the spatial distribution of Y in Mg alloy WE43 and the morphology of corrosion attack (Davenport et al., 2007). Imaging above and below the Y K-edge (17 keV) increased sensitivity to this element. Repeated observations showed that Y-rich regions slowed local propagation of corrosion. Homogenization of the distribution of Y decreased the corrosion rate, but both general attack and pitting were still observed.

Liquid metal embrittlement was studied through use of the model system of liquid Ga applied to polycrystalline Al (Ludwig et al., 2000).

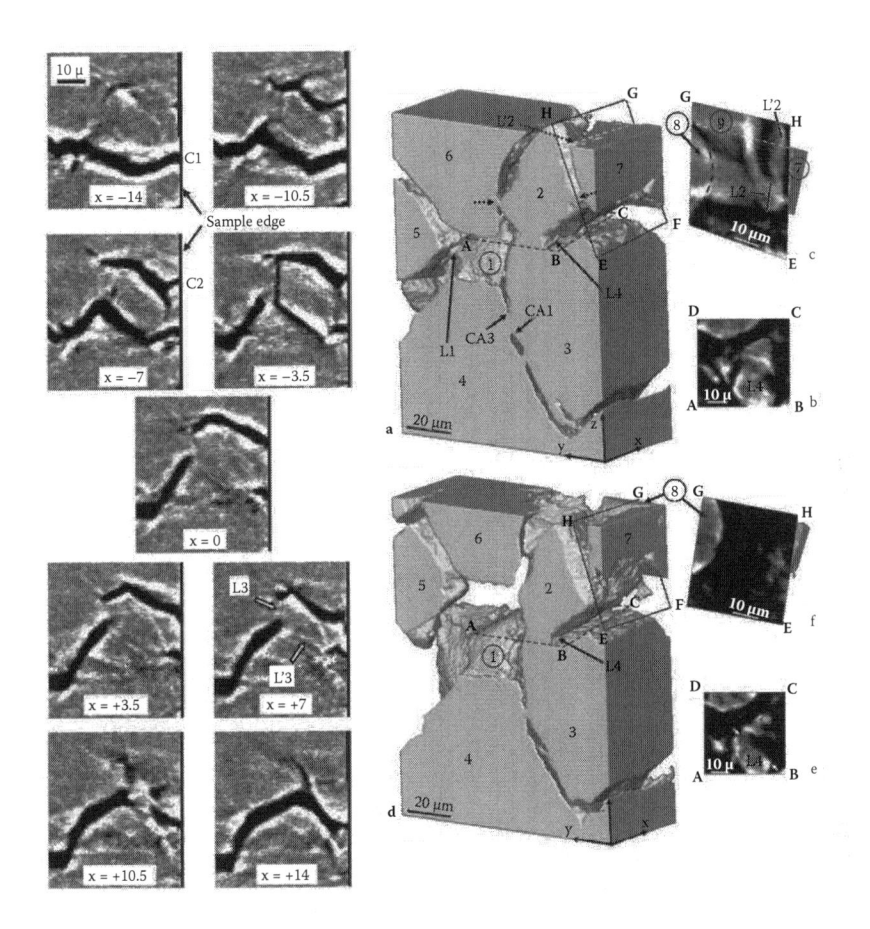

Evolution of Structures 259

FIGURE 10.8
Intergranular stress corrosion cracks in sensitized 302 stainless steel immersed in 0.15 mol $K_2S_4O_6$ acidifed with dilute H_2SO_4 to pH 2. Imaging was with 30 keV synchrotron x-radiation and with the detector 40 mm from the specimen, and the reconstruction was with 0.7-μm isotropic voxels. Labels have the following meanings: CA1, CA3 indicate positions of crack arrest; L1–L4 uncracked ligaments; C1, C2 crack segments. The nine rectangular images on the left side of the figure are parallel numerical sections (cuts) through the stack of slices; the vertical direction is the direction along which the stress was applied. These data were recorded after the initial increment of cracking under applied stress of 100 MPa and with an open electrical circuit. The relative separation between the cuts (in μm) is given by the value of x. The white blurred edging along the cracks is from phase contrast. The renderings in the right-hand portion of the figure show the same volume of the specimen after the second increment of crack growth, which was done with applied stress of 60 MPa ((a), top), and after the third increment under 40 MPa applied stress ((d), bottom). The circled numbers identify different grains, and the rectangular images at the far right are oblique cuts through the volume indicated on the renderings (i.e., oblique cut ABCD in (b) shows ligament L4 in (a)). (Reprinted from Babout et al. (2006); © 2006, with permission of Maney Publishing (http://www.ingentaconnect.com/content/maney/mst).)

As discussed elsewhere, Ga penetrated many but not all of the grain boundaries (Ludwig et al., 2003). Most studies of Ga on Al grain boundaries, however, focused on its use as a decoration so that grain boundary geometry could be correlated with fatigue crack path; see the deformation subsection above.

10.3 Bone and Soft Tissue Adaptation

This section has two parts. The first is on mineralized tissue and the second on soft tissue.

10.3.1 Mineralized Tissue: Implants, Healing, Mineral Levels, and Remodeling

Bone formation around implants and the resulting structural integrity (or lack thereof) is a biomedical engineering topic of increasing importance as the number of hip (and knee) replacements increases. Examples of microCT studies include the 3D analysis of bone formation around Ti implants (Bernhardt et al., 2006), study of bone around dental implants (Cattaneo et al., 2004), and investigation of human tooth–alveolar bone complex (Dalstra et al., 2006). One complication in many such imaging studies is that the high absorptivity of many implants such as stainless steel or Ti washes out contrast in bone. In the author's experience, this can cripple analysis if contrast is confined to linear, 256-level grayscale; such an 8-bit approach seems to be favored in many different analyses, so this is not an academic concern. Use of nonlinear contrast scales or high-end clipping can be effective; contrast enhancements for bone structures adjacent to implants are described elsewhere (Tesei et al., 2005; DaPonte et al., 2006). In synchrotron microCT, phase contrast at edges of different low-absorption tissue types may also aid segmentation in the presence of a very absorbing material.

Wear particles can produce inflammation at bone-implant surfaces and bone resorption or pain. Ren et al. (2006) found inflammation but not osteoclastic bone resorption in a study incorporating microCT. Cracks and the amount of wear in polyethylene from total disc replacements excised from patients after 1.8 to 16 years were characterized with microCT (Kurtz et al., 2007).

Weiss et al. (2007) performed human trials of an injectable bone substance (biphasic calcium phosphate in a water-soluble cellulose polymer) used to fill tooth sockets after extractions; after three years microCT revealed bone growth between the artificial mineral particles. Seashell material converted to hydroxyapatite was placed in femoral defects in rats, and in vivo microCT scanning (27-μm voxels) after implantation and after six weeks revealed the implants did not move, and bone grew toward the implant but untreated control defects remained empty (Vecchio et al., 2007). MicroCT showed thread integrity was maintained and bone contact established for titanium-coated PMMA screws placed into goat tibiae (Shalabi et al., 2007).

Healing processes in bone were also the subject of interesting microCT investigations. A rat ulnar loading model was studied with pQCT, with FEM, and with attached strain gauges (to assess load-sharing between ulna and radius), and greatest bone formation in response to fatigue loading was found in regions of high compressive strain (Kotha et al., 2004). Callus formation following tibial osteotomy and repair nailing was studied in mice; application of small amplitude cyclic loads after a short delay following surgery improved fracture healing as assessed by callus strength and by callus organization observed qualitatively by microCT (Gardner et al., 2006). Fracture healing in a rat model of bone union/nonunion was quantified using microCT (Schmidhammer et al., 2006).

Although bone healing stimulation was reported via administration of adenoviral vectors encoding bone morphogeneic protein-2 (BMP-2) in small animal models (Betz et al., 2007), Egermann et al. (2006) used microCT of standardized iliac crest defects and tibial osteotomies to find the opposite to be true in sheep. The role of osteopontin (OPN) on bone healing in a mouse femoral fracture model, specifically angiogenesis and fracture callus formation, was examined via microCT in wild-type and OPN$^{-/-}$ mice (Duvall et al., 2007). It is sometimes medically necessary to fuse the bones of the spine, and microCT was used to evaluate the effectiveness of a new bone growth factor in a rat spinal fusion model (Lu et al., 2007). MicroCT data suggested that controlled release of a thrombin peptide may enhance healing of critical and noncritical (sized) defects in rabbits (Sheller et al., 2004) and that BMP-silk composite matrices may be useful in healing critical-sized femoral defects in nude rats (Kirker-Head et al., 2006). MicroCT was also used to study an extreme case of healing, the regeneration of limbs in adult newts (Stock et al., 2003; Stock et al., 2004).

The material bone is a discontinuously reinforced composite of apatite (mineral) nanocrystallites dispersed in a matrix primarily of collagen, and it is important to realize that mineral content varies considerably from older mature areas (highest mineral content) to newly remodeled osteons (lower mineral content). The sensitivity limits of microCT make it ill-suited for quantifying small differences in composition, but a number of studies have shown that the mineralization levels in bone can be mapped. The cortices of rabbit tibiae have been imaged, and little, if any, change in mineral level was observed across the bone (Engelke et al., 1993). Spaceflight has a considerable effect on bone, and microCT was used to map the resulting "striking regional differences" in the distribution of mineral content as a function of distance from the end of femora of growing rats (Mechanic et al., 1990).

Synchrotron microCT of low-mineralized versus high-mineralized volumes in trabeculae as a function of bisphosphonate treatment was the subject of one study (Borah et al., 2005). Microradiography of thin sections of bone has long been used to show remodeling, and determination of the degree of bone mineralization via this method was compared to that with synchrotron microCT (Nuzzo et al., 2002). Synchrotron microCT was also used to study the degree of mineralization in human normal vertebra, osteoblastic metastases, and sites with degenerative osteosclerosis (Sone et al., 2004) and in different mouse genetic strains (Martín-Badosa et al., 2003; Bayat et al., 2005).

Cortical bone from human femora was the subject of a number of early microCT studies, and Haversian and Volkmann canals (for blood supply and innervation axial and transverse to long bone axes, respectively) were resolved. The Haversian and Volkman canals serve as sites for bone remodeling, that is, the replacement of old damaged bone with new bone. The process involves the concerted action of osteoclasts gouging out old bone, followed by osteoblasts depositing bone matrix and mineral. The cylinders of remodeled bone are termed *osteons* and, for some time after they are formed, contain lower mineral density than mature bone. The decreased mineral levels in newly formed osteons have long been imaged with microradiographs of thin sections of bone, and synchrotron microCT was used, more recently, to good effect to reveal the qualitative spatial distribution of mineral density.

There are a number of studies of mineral density differences in addition to those cited in the previous paragraph (Ritman et al., 1998; Nuzzo et al., 2003; Bossy et al., 2004; Bousson et al., 2004; Scherf et al., 2004). Dyck et al. (2002) discussed calibration of mineral levels in lab microCT for various calcified tissues, but it is normally difficult to see remodeled osteons in lab microCT. It should be possible to quantify differences in mineral density between osteons of varying ages within a bone specimen, but this appears still to be done. Standards measurements can be used to remove systematic biases in values of linear attenuation coefficients resulting from beam hardening (Prevrhal, 2004, 2006); although this is important for correctly interpreting mean mineral level, it does not help with detection of differences between recently formed and mature osteons.

Reports were published on determination of Ca/P ratios for bone from synchrotron microCT scans: linear attenuation coefficients from bone were compared to those of two phantoms of calcium phosphate salts with different Ca/P ratios (Speller et al., 2005; Tzaphlidou et al., 2006). Although differences in bone mineral density certainly appear to be supported by this group's data, interpretation of differences in terms of Ca/P ratios appears problematic at best. In bone, the relative amounts of collagen and mineral are expected to vary, the density of osteocyte lacunae can alter the apparent linear attenuation coefficients, and other variations will be present (see the discussion of Elliott et al. (1998), Kozul et al. (1999), and Dowker et al. (2004)).

Thin bands of high absorptivity were found in synchrotron microCT data of bone. In animals dosed with $SrCl_2$, these were interpreted as zones of bone with high Sr replacement of Ca in the mineral (Ritman et al., 1998). Similar bands of high absorptivity were observed in specimens of human femoral neck bone in extraosteonal areas near and parallel to the external surface of the cortex; these features in aged bone were interpreted as regions of high mineral content (Cooper et al., 2004). The author observed similar high-absorptivity bands within cortical bone of animals including newts and mice; examination of adjacent slices showed these could not be due to out-of-plane geometrical features producing unexpected phase contrast. The magnitude of the difference is quite striking: linear attenuation coefficients in the mouse cortical bone were 40 percent or more greater within the very small volume of hypermineralized tissue than in the "normal" bone, the difference being more than twice the standard deviation seen for areas of "normal" bone (Stock et al., 2006). It is difficult to imagine that these features, observed by several investigators working at different synchrotron radiation facilities, are an artifact of some sort, but their origin remains obscure.

The current model of bone remodeling holds that Haversian systems are most active where bone undergoes the most extreme loading, that is, where it accumulates the greatest amount of damage such as microcracking (see Schneider et al. (2006) for examples of synchrotron microCT imaging of microcracking associated with bone porosity including osteocyte lacunae). Several studies of channels in cortical bone focused on establishing the equivalence of microCT-based measures of porosity and osteon dimensions with long-standing methods such as microradiography or histology, for example, Wacheter et al. (2001) and Cooper et al. (2004). In a study of pore networks in human cortical bone, resolution was shown to have a significant biasing effect on various measurements when it was incommensurate with the structural scale of the pores (Cooper et al., 2007). Pore content and mineral levels determined from microCT correlated with axial ultrasound velocity; the data showed the structure in the outer 1 mm of cortex (about one-half of the cortex thickness) affected velocity and suggested clinical usefulness of this noninvasive monitoring method (Bossy et al., 2004).

Clustering of osteons might result from remodeling due to concentrated damage, particularly in bone of aged individuals, and the presence of such a significant stress concentrator could lead to unexpected fracture (Jordan et al., 2000). Intracortical porosity measured in compact tension samples of human bone was quantified and imported into a finite element model of aging bone; keeping nonporosity factors constant, the model showed that a 4 percent increase in porosity produced a 6 percent decrease in initiation toughness and a 62 percent decrease in propagation toughness, whereas incorporation of other age-related changes into the model affected primarily the initiation toughness (Ural and Vashishth, 2007).

The hypothesis that the density and range (length) of basic multicellular unit (BMU)-related resorption spaces in human cortical bone vary with age and sex was examined with lab microCT, and the investigators found that the range did not vary and that the age-dependent apparent decrease in resorption space density was due to cortical rarefaction leading to difficulty in detecting resorption spaces with microCT rather than to a decrease in remodeling (Cooper et al., 2006). Cooper et al. (2006) also showed a rendering of cutting cones. Nutrient canal network rarefaction (decreased canal cross-sectional area, sectional canal number density, etc.) in disuse-mediated versus control rat tibial diaphyses was confirmed using synchrotron microCT (Matsumoto et al., 2006); positional variation and changes with animal growth were documented in subsequent work (Matsumoto et al., 2007). Renders and coworkers determined the 3D distribution of the cortical canal network in the human mandibular condyle and related this to the directions of principal stresses and strains (Renders et al., 2007).

10.3.2 Soft Tissue and Soft Tissue Interfaces

Degradation of biological structures also occurs by the unwanted addition of mineral to soft tissue structures or erosion of bone at soft tissue interfaces. Cancers of various types affect bone and soft tissue sites; these microCT studies, especially those with a longitudinal component, have been very informative and are covered after discussion of errant calcification and bone erosion.

Lack of small-diameter synthetic vascular grafts limits the possibility of life- or limb-saving surgery for many individuals, and a mouse model for evaluation of these grafts was developed and assessed with microCT (Lopez-Soler et al., 2007). Bioprosthetic valves calcify, and assessing the tendency of new valve designs (including materials) to calcify in vitro is an important first step toward clinical use. MicroCT was used to assess porcine heart valve calcification in a more physically realistic accelerated testing model involving pH control of the mineralizing solution and millions of cycles of valve flexure (Krings et al., 2006). Calcification in excised human valves has been studied by microCT (Rajamannan et al., 2005a,b); see Figure 6.14. In vivo microCT of vascular calcification in a rat model of chronic renal failure produced interesting results (Persy et al., 2006). Microcalcifications in the cap of coronary plaques from autopsy specimens were observed with microCT, and numerical modeling suggested that they provide enough stress concentration to rupture the plaque in an unexpected location (Vengrenyuk et al., 2006).

Rheumatoid arthritis involves bone destruction, whereas bone is formed in inappropriate locations in osteoarthritis. Figure 4.6 compared matching tube-based and synchrotron microCT slices of the dysfunctional mineralized tissue formed in a mouse model of osteoarthritis. Appleton and coworkers studied a rat model of preclinical osteoarthritis using in vivo microCT at several time points up to 20 weeks and found forced mobilization after surgery (knee joint destabilization by anterior cruciate ligament transection and partial medial meniscectomy) accelerated pathogenesis (Appleton et al., 2007). The dose dependence of a therapeutic agent for rheumatoid arthritis was established in a rat model using microCT to measure changes in total bone volume (Schopf et al., 2006). Bone erosion in a preclinical model of rheumatoid arthritis was determined via local roughness quantification for the rat talus bone (Silva et al., 2006). The approach of Silva and coworkers, outlined in Chapter 6, appeared to be much more sensitive to the early stages of damage than measurements of quantities such as the total bone volume and seems to be quite robust; the method also allowed mapping on the surface of 3D renderings of the locations of greatest roughness.

There have been quite a number of microCT studies of tumor formation in animal models. Roudier and coworkers used microCT to examine a rat model of bone cancer pain; osteolytic bone loss in tibiae from the rat mammary tumor cells was examined (Roudier et al., 2006). In a murine model of cancer-induced bone pain, behavioral manifestations of pain were found to emerge in parallel with progression of bone destruction revealed by microCT, and the effectiveness of pain reduction drugs was assessed (El Mouedden and Meert, 2005). The effect of a noncalcemic vitamin D analogue on prostate cancer metastasis was examined in mice with microCT (Peleg et al., 2005). Squamous cell carcinoma and its spread to murine mandibles was the subject of another study (Henson et al., 2007). In a melanoma metastasis mouse model, PET and longitudinal microCT imaging were used to study development of lung, mandibular, long bone, and subcutaneous metastases (Winkelmann et al., 2006). Development of metastases in the murine liver was studied longitudinally with microCT using a hepatobiliary-specific contrast agent (Ohta et al., 2006).

In vivo microCT imaging is especially challenging for lung and heart studies, and many of these papers were reviewed in Chapter 9. Two studies that relate to damaged structures are mentioned here. Cody et al. (2004, 2005) used in vivo microCT to observe lung damage in a mouse model of pulmonary fibrosis; breath hold imaging and respiratory gating were two methods used to eliminate motion artifacts. Model murine myocardial infarct size was determined in another study using a gating approach and observations 5 or 35 days after the damage was introduced (Nahrendorf et al., 2007). The cardiac studies of mice were particularly challenging because of the small heart size and its rapid motion.

References

Al-Raoush, R. and M. Alsaleh (2007). Simulation of random packing of polydisperse particles. *Powder Technol* **176**: 47–55.

Altman, S.J., M.L. Rivers, M.D. Reno, R.T. Cygan, and A.A. Mclain (2005). Characterization of adsorption sites on aggregate soil samples using synchrotron x-ray computerized microtomography. *Env Sci Technol* **39**: 2679–2685.

Anderson, P., J.C. Elliott, U. Bose, and S.J. Jones (1996). A comparison of the mineral content of enamel and dentine in human premolars and enamel pearls measured by x-ray microtomography. *Archs Oral Biol* **41**: 281–290.

Ansari, M.A. and F. Stepanek (2006). Design of granule structure: Computational methods and experimental realization. *AIChE J* **52**: 3762–3774.

Appleton, C.T.G., D.D. McErlain, V. Pitelka, N. Schwartz, S.M. Bernier, J.L. Henry, D.W. Holdsworth, and F. Beier (2007). Forced mobilization accelerates pathogenesis: characterization of a preclinical surgical model of osteoarthritis. *Arth Res Therap* **9**: R13.

Atwood, R.C., P.D. Lee, R.V. Curtis, and D.M. Maijer (2007). Modeling the investment casting of a titanium crown. *Dent Mater* **23**: 60–70.

Babout, L., T.J. Marrow, D. Engelberg, and P.J. Withers (2006). X-ray microtomographic observation of intergranular stress corrosion cracking in sensitised austenitic stainless steel. *Mater Sci Technol* **22**: 1068–1075.

Baruchel, J., J.Y. Buffiere, P. Cloetens, M.D. Michiel, E. Ferrie, W. Ludwig, E. Maire, and L. Salvo (2006). Advances in synchrotron radiation microtomography. *Scripta Mater* **55**: 41–46.

Bayat, S., L. Apostol, E. Boller, T. Borchard, and F. Peyrin (2005). In vivo imaging of bone microarchitecture in mice with 3D synchrotron radiation microtomography. *Nucl Instrum Meth A* **548**: 247–252.

Bentz, D.P., S. Mizell, S. Satterfield, J. Devaney, W. George, P. Ketcham, J. Graham, J. Porterfield, D. Quenard, F. Vallee, H. Sallee, E. Boller, and J. Baruchel (2002). The visible cement data set. *J Res NIST* **107**: 137–148. See also visiblecement.nist.gov.

Bentz, D.P., D.A. Quenard, H.M. Kunzel, J. Baruchel, F. Peyrin, N.S. Martys, and E.J. Garboczi (2000). Microstructure and transport properties of porous building materials. II: Three-dimensional x-ray tomographic studies. *Mater Struct* **33**: 147–153.

Bernard, D. (2005). 3D quantification of pore scale geometrical changes using synchrotron computed microtomography. *Oil Gas Sci Technol* **60**: 747–762.

Bernard, D., D. Gendron, J.M. Heintz, S. Bordere, and J. Etourneau (2005). First direct 3D visualisation of microstructural evolutions during sintering through x-ray computed microtomography. *Acta Mater* **53**: 121–128.

Bernard, D., G.L. Vignoles, and J.M. Heintz (2000). Modelling porous materials evolution. *X-Ray Tomography in Materials Science*. J. Baruchel, J.Y. Buffière, E. Maire, P. Merle, and G. Peix (Eds.). Paris, Hermes Science: 177–192.

Bernhardt, R., D. Scharnweber, B. Müller, F. Beckmann, J. Goebbels, J. Jansen, H. Schliephake, and H. Worch (2006). 3D analysis of bone formation around Ti implants using microcomputed tomography (μCT). *Developments in X-Ray Tomography V*. U. Bonse (Ed.). Bellingham, WA, SPIE. *SPIE Proc Vol* **6318**: 631807–1 to –10.

Betz, V.M., O.B. Betz, V. Glatt, L.C. Gerstenfeld, T.A. Einhorn, M.L. Bouxsein, M.S. Vrahas, and C.H. Evans (2007). Healing of segmental bone defects by direct percutaneous gene delivery: Effect of vector dose. *Hun Gene Therap* **18**: 907–915.

Borah, B., E.L. Ritman, T.E. Dufresne, S.M. Jorgensen, S. Liu, J. Sacha, R.J. Phipps, and R.T. Turner (2005). The effect of risedronate on bone mineralization as measured by microcomputed tomography with synchrotron radiation: Correlation to histomorphometric indices of turnover. *Bone* **37**: 1–9.

Bossy, E., M. Talmant, F. Peyrin, L. Akrout, P. Cloetens, and P. Laugier (2004). An in vitro study of the ultrasonic axial transmission technique at the radius: 1 MHz velocity measurement are sensitive to both mineralization and intracortical porosity. *J Bone Miner Res* **19**: 1548–1556.

Bousson, V., F. Peyrin, C. Bergot, M. Hausard, A. Sautet, and J.D. Laredo (2004). Cortical bone in the human femoral neck: Three-dimensional appearance and porosity using synchrotron radiation. *J Bone Miner Res* **19**: 794–801.

Brunetti, A., S. Bidali, A. Mariani, and R. Cesareo (2004). X-ray tomography for the visualization of monomer and polymer filling inside wood and stone. *Developments in X-ray Tomography IV*. U. Bonse (Ed.). Bellingham, WA, SPIE. *SPIE Proc Vol* **5535**: 191–200.

Buffière, J.Y., S. Savelli, and E. Maire (2000). Characterisation of MMC$_p$ and cast aluminum alloys. *X-Ray Tomography in Materials Science*. J. Baruchel, J.Y. Buffière, E. Maire, P. Merle, and G. Peix (Eds.). Paris, Hermes Science: 103–114.

Burlion, N., D. Bernard, and D. Chen (2006). X-ray microtomography: Application to microstructure analysis of a cementitious material during leaching process. *Cement Concr Res* **36**: 346–357.

Butts, M.D. (1993). Nondestructive examination of nicalon fiber composite preforms using x-ray tomographic microscopy, M.S. Thesis. Atlanta, Georgia Institute of Technology.

Cattaneo, P.M., M. Dalstra, F. Beckmann, T. Donath, and B. Melsen (2004). Comparison of conventional and synchrotron-radiation-based microtomography of bone around dental implants. *Developments in X-ray Tomography IV*. U. Bonse (Ed.). Bellingham, WA, SPIE. *SPIE Proc Vol* **5535**: 757–764.

Chaijaruwanich, A., P.D. Lee, R.J. Dashwood, Y.M. Youssef, and H. Nagaumi (2007). Evolution of pore morphology and distribution during the homogenization of direct chill cast Al–Mg alloys. *Acta Mater* **55**: 285–293.

Chauve, G., F. Ravenelle, and R.H. Marchessault (2007). Comparative imaging of a slow-release starch excipient tablet: Evidence of membrane formation. *Carbohyd Poly* **70**: 61–67.

Chen, T.T., A. Sasov, J.F. Dutrizac, P. Kondos, and G. Poirier (2006). The x-ray tomography of a siliceous silver ore. *J Metals* **58**: 41–44.

Cnudde, V. and P.J.S. Jacobs (2004). Monitoring of weathering and conservation of building materials through non-destructive x-ray computed microtomography. *Env Geol* **46**: 477–485.

Cody, D., D. Cavanaugh, R.E. Price, B. Rivera, G. Gladish, and E. Travis (2004). Lung imaging of laboratory rodents in vivo. *Developments in X-ray Tomography IV*. U. Bonse (Ed.). Bellingham, WA, SPIE. *SPIE Proc Vol* **5535**: 43–52.

Cody, D.D., C.L. Nelson, W.M. Bradley, M. Wislez, D. Juroske, R.E. Price, X. Zhou, B.N. Bekele, and J.M. Kurie (2005). Murine lung tumor measurement using respiratory-gated microcomputed tomography. *Invest Radiol* **40**: 263–269.

Coléou, C. and J.M. Barnola (2001). 3-D snow and ice images by x-ray microtomography. *ESRF Newslett*: 24–26.

Connolly, B.J., D.A. Homer, S.J. Fox, A.J. Davenport, C. Padovani, S. Zhou, A. Turnbull, M. Preuss, N.P. Stevens, T.J. Marrow, J.Y. Buffiere, E. Boller, A. Groso, and M. Stampanoni (2006). X-ray microtomography studies of localised corrosion and transitions to stress corrosion cracking. *Mater Sci Technol* **22**: 1076–1085.

Contardo, L., M.D. Luca, M. Biasotto, R. Longo, A. Olivo, S. Pani, and R. Di Lenarda (2005). Evaluation of the endodontic apical seal after post insertion by synchrotron radiation tomography. *Nucl Instrum Meth A* **548**: 253–256.

Cooper, D., A. Turinsky, C. Sensen, and B. Hallgrimsson (2007). Effect of voxel size on 3D microCT analysis of cortical bone porosity. *Calcif Tiss Int* **80**: 211–219.

Cooper, D.M.L., J.R. Matyas, M.A. Katzenberg, and B. Hallgrimsson (2004). Comparison of microcomputed tomographic and microradiographic measurements of cortical bone porosity. *Calcif Tiss Int* **74**: 437–447.

Cooper, D.M.L., C.D.L. Thomas, J.G. Clement, and B. Hallgrimsson (2006). Three-dimensional micro computed tomography imaging of basic multicellular unit-related resoprtion spaces in human cortical bone. *Anat Rec* **288A**: 806–816.

Dalstra, M., P.M. Cattaneo, F. Beckmann, M.T. Sakima, G. Lemor, M.G. Laursen and B. Melsen (2006). Microtomography of the human tooth–alveolar bone complex. *Developments in X-Ray Tomography V*. U. Bonse (Ed.). Bellingham, WA, SPIE. *SPIE Proc Vol* **6318**: 631804–1 to –9.

DaPonte, J.S., M. Clark, P. Nelson, T. Sadowski, and E. Wood (2006). Quantitative confirmation of visual improvements to microCT bone density images. *Visual Information Processing XV*. Z. Rahman, S.E. Reichenbach, and M.A. Neifeld (Eds.). Bellingham, WA, SPIE. *SPIE Proc Vol* **6246**: 62460D–1 to –9.

Davenport, A.J., C. Padovani, B.J. Connolly, N.P.C. Stevens, T.A.W. Beale, A. Groso, and M. Stampanoni (2007). Synchrotron X-ray microtomography study of the role of Y in corrosion of magnesium alloy WE43. *Electrochem Sol State Lett* **10**: C5–C8.

De Gryze, S., L. Jassogne, J. Six, H. Bossuyt, M. Wevers, and R. Merckx (2006). Pore structure changes during decomposition of fresh residue: X-ray tomography analyses. *Geoderma* **134**: 82–96.

Delbem, A.C.B., A.E.M. Vieira, K.T. Sassaki, M.L. Cannon, S.R. Stock, X. Xiao, and F. De Carlo (2006). Quantitative analysis of mineral content in enamel using synchrotron microtomography and microhardness analysis. *Developments in X-Ray Tomography V*. U. Bonse (Ed.). Bellingham, WA, SPIE. *SPIE Proc Vol* **6318**: 631824–1 to –5.

Dobrich, K.M., C. Rau, and C. E. Krill III (2004). Quantitative characterization of the three-dimensional microstructure of polycrystalline Al-Sn using x-ray microtomography. *Metall Mater Trans A* **35**: 1953–1961.

Dollar, L.L. (1992). Evaluation of nondestructive x-ray techniques for electronic packaging materials, M.S. Thesis. Atlanta, Georgia Institute of Technology.

Dowker, S.E., J.C. Elliott, G.R. Davis, R.M. Wilson, and P. Cloetens (2006). Three-dimensional study of human dental fissure enamel by synchrotron x-ray microtomography. *Eur J Oral Sci* **114**: Suppl 1, 353–359; discussion 375–376, 382–383.

Dowker, S.E.P., J.C. Elliott, G.R. Davis, R.M. Wilson, and P. Cloetens (2004). Synchrotron x-ray microtomographic investigation of mineral concentrations at micrometer scale in sound and carious enamel. *Caries Res* **38**: 514–522.

Duvall, C.L., W.R. Taylor, D. Weiss, A.M. Wojtowicz, and R.E. Guldberg (2007). Impaired angiogenesis, early callus formation, and late stage remodeling in fracture healing of osteopontin-deficient mice. *J Bone Miner Res* **22**: 286–297.

Dyck, D., A. Postnov, S. Saveliev, A. Saso, and N.M.D. Clerck (2002). Definition of local density in biological calcified tissues using x-ray microtomography. *Medical Imaging 2002: Visualization, Image-Guided Procedures and Display.* S.K. Mun (Ed.). Bellingham, WA, SPIE. *SPIE Proc Vol* **4681**: 749–755.

Egermann, M., C.A. Lill, K. Griesbeck, C.H. Evans, P.D. Robbins, E. Schneider, and A.W. Baltzer (2006). Effect of BMP-2 gene transfer on bone healing in sheep. *Gene Therap* **13**: 1290–1299.

El Mouedden, M. and T.D. Meert (2005). Evaluation of pain-related behavior, bone destruction and effectiveness of fentanyl, sufentanil and morphine in a muring model of cancer pain. *Pharmacol Biochem Behavior* **82**: 109–119.

Elliott, J.C., P. Anderson, G.R. Davis, F.S.L. Wong, X.J. Gao, S.D. Dover, and A. Boyde (1992). X-ray microtomographic studies of bone and teeth. *X-ray Microscopy III.* A. Michette, G.R. Morrison, and C.J. Buckley (Eds.). New York, Springer: 461–464.

Elliott, J.C., F.R.G. Bollet-Quivogne, P. Anderson, S.E.P. Dowker, R.M. Wilson, and G.R. Davis (2005). Acidic demineralization of apatites studied by scanning x-ray microradiography and microtomography. *Mineralogical Mag* **69**: 643–652.

Elliott, J.C., T.G. Bromage, P. Anderson, G. Davis, and S.D. Dover (1989). Application of microtomography to the study of dental hard tissues. *Enamel V.* R.W. Feanhead (Ed.). Yokohama, Florence Publishers: 429–433.

Elliott, J.C., F.S.L. Wong, P. Anderson, G.R. Davis, and S.E.P. Dowker (1998). Determination of mineral concentration in dental enamel from x-ray attenuation measurements. *Conn Tiss Res* **38**: 61–72.

Engelke, K., W. Graeff, L. Meiss, M. Hahn, and G. Delling (1993). High spatial resolution imaging of bone mineral using computed microtomography: Comparison with microradiography and undecalcified histologic sections. *Invest Radiol* **28**: 341–349.

Fearne, J.M., J.C. Elliott, F.S. Wong, G.R. Davis, A. Boyde, and S.J. Jones (1994). Deciduous enamel defects in low-birth-weight children: correlated X-ray microtomographic and backscattered electron imaging study of hypoplasia and hypomineralization. *Anat Embryol* **189**: 375–381.

Feeney, D.S., J.W. Crawford, T. Daniell, P.D. Hallett, N. Nunan, K. Ritz, M. Rivers, and I.M. Young (2006). Three-dimensional microorganization of the soil-root-microbe system. *Microb Ecol* **52**: 151–158.

Flin, F., J.B. Brzoska, D. Coeurjolly, R.A. Pieritz, B. Lesaffre, C. Coléou, P. Lamboley, O. Teytaud, G.L. Vignoles, and J.F. Delesse (2005). Adaptive estimation of normals and surface area for discrete 3-D objects: Application to snow binary data from x-ray tomography. *IEEE Trans Image Process* **14**: 585–596.

Flin, F., J.B. Brzoska, B. Lesaffre, C. Coléou, and R.A. Pieritz (2003). Full three-dimensional modeling of curvature-dependent snow metamorphism: First results and comparison with experimental tomographic data. *J Phys D* **36**: A49–A54.

Fu, X., J.A. Elliott, A.C. Bentham, B.C. Hancock, and R.E. Cameron (2006). Application of x-ray microtomography and image processing to the investigation of a compacted granular system. *Part Part Syst Character* **23**: 229–236.

Fu, X., G.E. Milroy, M. Dutt, A.C. Bentham, B.C. Hancock, and J.A. Elliott (2005). Quantitative analysis of packed and compacted granular system by x-ray microtomography. *Medical Imaging 2005 — Image Processing*. J.M. Fitzpatrick and J.M. Reinhardt (Eds.). Bellingham, WA, SPIE. *SPIE Proc Vol* **5747, n III**: 1955–1964.

Gallucci, E., K. Scrivener, A. Groso, M. Stampanoni, and G. Margaritondo (2007). 3D experimental investigation of the microstructure of cement pastes using synchrotron X-ray microtomography (µCT). *Cement Concr Res* **37**: 360–368.

Garboczi, E.J. and J.W. Bullard (2004). Shape analysis of a reference cement. *Cement Concr Res* **34**: 1933–1937.

Gardner, M.J., M.C.H. van der Meulen, D. Demetrakopoulos, T.M. Wright, E.R. Myers, and M.P. Bostrom (2006). In vivo cyclic axial compression affects bone healing in the mouse tibia. *J Orthop Res* **24**: 1679–1686.

Gay, S.L. and R.D. Morrison (2006). Using two dimensional sectional distributions to infer three dimensional volumetric distributions — Validation using tomography. *Part Part Syst Charact* **23**: 246–253.

Golchert, D., R. Moreno, M. Ghadiri, and J. Litster (2004a). Effect of granule morphology on breakage behaviour during compression. *Powder Technol* **143–144**: 84–96.

Golchert, D.J., R. Moreno, M. Ghadiri, J. Litster, and R. Williams (2004b). Application of x-ray microtomography to numerical simulations of agglomerate breakage by distinct element method. *Adv Powder Technol* **15**: 447–457.

Hansel, C.M., M.J.L. Force, S. Fendorf, and S. Sutton (2002). Spatial and temporal association of As and Fe species on aquatic plant roots. *Env Sci Technol* **36**: 1988–1994.

Heldele, R., S. Rath, L. Merz, R. Butzbach, M. Hagelstein, and J. Haußelt (2006). X-ray tomography of powder injection moulded micro parts using synchrotron radiation. *Nucl Instrum Meth B* **246**: 211–216.

Henson, B., F. Li, D.D. Coatney, T.E. Carey, R.S. Mitra, K.L. Kirkwood, and N.J. D'Silva (2007). An orthotopic floor-of-mouth model for locoregional growth and spread of human squamous cell carcinoma. *J Oral Pathol Med* **36**: 363–370.

Hettiarachchi, G.M., E. Lombi, M.J. McLaughlin, D. Chittleborough, and P. Self (2006). Density changes around phosphorus granules and fluid bands in a calcareous soil *Soil Sci Soc Am J* **70**: 960–966.

Illman, B. and B. Dowd (1999). High resolution microtomography for density and spatial information about wood structures. *Developments in X-ray Tomography II*. U. Bonse (Ed.). Bellingham, WA, SPIE. *SPIE Proc Vol* **3772**: 198–204.

Jia, X. and R.A. Williams (2007). A hybrid mesoscale modelling approach to dissolution of granules and tablets. *Chem Eng Res Design* **85**(7A): 1027–1038.

Jones, K.W., H. Feng, E.A. Stern, U. Neuhäusler, N. Marinkovic, and Z. Song (2006). Properties of New York/New Jersey harbor sediments. *Acta Phys Polonica A* **109**: 279–286.

Jordan, G.R., N. Loveridge, K.L. Bell, K. Power, N. Rushton, and J. Reeve (2000). Spatial clustering of remodeling osteons in the femoral neck cortex: A cause of weakness in hip fracture? *Bone* **26**: 305–313.

Jupe, A.C., S.R. Stock, P.L. Lee, N.N. Naik, K.E. Kurtis, and A.P. Wilkinson (2004). Phase composition depth profiles using spatially resolved energy dispersive x-ray diffraction. *J Appl Cryst* **37**: 967–976.

Kakaboura, A., C. Rahiotis, D. Watts, N. Silikas, and G. Eliades (2007). 3D-marginal adaptation versus setting shrinkage in light-cured microhybrid resin composites. *Dent Mater* **23**: 272–278.

Kang, Z., R. Johnson, J. Mi, S. Bondi, M. Jiang, W.J. Lackey, S. Stock, and K. More (2004). Microstructure of carbon fibers prepared laser CVD. *Carbon* **42**: 2721–2727.

Kinney, J.H. and D.L. Haupt (1997). Evidence of critical scaling behavior during vapor phase synthesis of continuous filament composites. *J Mater Res* **12**: 610–612.

Kinney, J.H. and M.C. Nichols (1992). X-ray tomographic microscopy (XTM) using synchrotron radiation. *Annu Rev Mater Sci* **22**: 121–152.

Kinney, J.H., T.M. Breunig, T.L. Starr, D. Haupt, M.C. Nichols, S.R. Stock, M.D. Butts, and R.A. Saroyan (1993). X-ray tomographic study of chemical vapor infiltration processing of ceramic composites. *Science* **260**: 789–792.

Kinney, J.H., C. Henry, D.L. Haupt, and T.L. Starr (1994a). The topology of percolating porosity in woven fiber ceramic matrix composites. *Appl Compos Mater* **1**: 325–331.

Kinney, J.H., G.W. Marshall Jr., and S.J. Marshall (1994b). Three-dimensional mapping of mineral densities in carious dentin: Theory and method. *Scanning Microsc* **8**: 197–205.

Kinney, J.H., R.K. Nalla, J.A. Pople, T.M. Breunig, and R.O. Ritchie (2005). Age-related transparent root dentin: Mineral conchetnation, crystallite size and mechanical properties. *Biomater* **26**: 3363–3376.

Kinney, J.H., S.R. Stock, M.C. Nichols, U. Bonse, T.M. Breunig, R.A. Saroyan, R. Nusshardt, Q.C. Johnson, F. Busch, and S.D. Antolovich (1990). Nondestructive investigation of damage in composites using x-ray tomographic microscopy. *J Mater Res* **5**: 1123–1129.

Kirker-Head, C., V. Karageorgiou, S. Hofmann, R. Fajardo, O. Betz, H.P. Merkle, M. Hilbe, B. von Rechenberg, J. McCool, L. Abrahamsen, A. Nazarian, E. Cory, M. Curtis, D. Kaplan, and L. Meinel (2006). BMP-silk composite matrices heal critically sized femoral defects. *Bone* **41**: 247–255.

Kotha, S.P., Y.F. Hsieh, R.M. Strigel, R. Müller, and M.J. Silva (2004). Experimental and finite element analysis of the rat ulnar loading model — Correlations between strain and bone formation following fatigue loading. *J Biomech* **37**: 541–548.

Kozul, N., G.R. Davis, P. Anderson, and J.C. Elliott (1999). Elemental quantification using multiple-energy x-ray absorptiometry. *Meas Sci Technol* **10**: 252–259.

Krill III, C.E., K. Dobrich, D. Michels, A. Michels, C. Rau, T. Weitkamp, A. Snigirev, and R. Birringer (2001). Tomographic characterization of grain-size correlations in polycrystalline Al–Sn. *Developments in X-ray Tomography III.* U. Bonse (Ed.). Bellingham, WA, SPIE. *SPIE Proc Vol* **4503**: 205–212.

Krings, M., D. Kanellopoulou, D. Mavrilas, and B. Glasmacher (2006). In vitro pH-controlled calcification of biological heart valve prostheses. *Mat-wiss. Werkstofftech* **37**: 432–435.

Kulkarni, A., J. Gutleber, S. Sampath, A. Goland, W. Lindquist, H. Herman, A. Allen, and B. Dowd (2004a). Studies of the microstructure and properties of dense ceramic coatings produced by high-velocity oxygen-fuel combustion spraying. *Mater Sci Eng A* **369**: 124–137.

Kulkarni, A., H. Herman, F. De Carlo, and R. Subramanian (2004b). Microstructural characterization of electron beam physical vapor deposition thermal barrier coatings through high-resolution computed microtomography. *Metall Mater Trans* **35A**: 1945–1952.

Kulkarni, A., S. Sampath, A. Goland, H. Herman, and B Dowd (2000). Computed microtomography studies to characterize microstructure-property correlations in thermal sprayed alumina deposits. *Scripta Mater* **43**: 471–476.

Kulkarni, A.A., A. Goland, H. Herman, A.J. Allen, J. Ilavsky, G.G. Long, and F. De Carlo (2005). Advanced microstructural characterization of plasma-sprayed zirconia coatings over extended length scales. *J Therm Spray Technol* **14**: 239–250.

Kurtz, S.M., A. van Ooij, R. Ross, J. de Waal Malefijt, J. Peloza, L. Ciccarelli, and M.L. Villarraga (2007). Polyethylene wear and rim fracture in total disc arthroplasty. *Spine J* **7**: 12–21.

Lame, O., D. Bellet, M.D. Michiel, and D. Bouvard (2003). In situ microtomography investigation of metal powder compacts during sintering. *Nucl Instrum Meth B* **200**: 287–294.

Lame, O., D. Bellet, M.D. Michiel, and D. Bouvard (2004). Bulk observation of metal powder sintering by x-ray synchrotron microtomography. *Acta Mater* **52**: 977–984.

Lee, S.B. (1993). Nondestructive examination of chemical vapor infiltration of 0°/90° SiC/Nicalon composites, Ph.D. Thesis. Atlanta, Georgia Institute of Technology.

Lee, S.B., S.R. Stock, M.D. Butts, T.L. Starr, T.M. Breunig, and J.H. Kinney (1998). Pore geometry in woven fiber structures: 0°/90° plain-weave cloth lay-up preform. *J Mater Res* **13**: 1209–1217.

Lin, C.L. and J. Miller (2005). 3D characterization and analysis of particle shape using x-ray microtomography (XMT). *Powder Technol* **154**: 61–69.

Loos, A., R. Rohde, A. Haverich, and S. Barlach (2007). In vitro and in vivo biocompatibility testing of absorbable metal stents. *Macromol Symp* **253**: 103–108.

Lopez-Soler, R.I., M.P. Brennan, A. Goyal, Y. Wang, P. Fong, G. Tellides, A. Sinusas, A. Dardik, and C. Breuer (2007). Development of a mouse model for evaluation of small diameter vascular grafts. *J Surg Res* **139**: 1–6.

Lu, S., E.N. Landis, and D.T. Keane (2006). X-ray microtomographic studies of pore structure and permeability in Portland cement concrete. *Mater Struct* **39**: 611–620.

Lu, S.S., X. Zhang, C. Soo, T. Hsu, A. Napoli, T. Aghaloo, B.M. Wu, P. Tsou, K. Ting, and J.C. Wang (2007). The osteoinductive properties of Nell-1 in a rat spinal fusion model. *Spine J* **7**: 50–60.

Ludwig, O., M. Di Michiel, L. Salvo, M. Suery, and P. Falus (2005). In-situ three-dimensional microstructural investigation of solidification of an Al–Cu alloy by ultrafast x-ray microtomography. *Metall Mater Trans A* **36**: 1515–1523.

Ludwig, W., S. Bouchet, D. Bellet, and J.Y. Buffière (2000). 3D observation of grain boundary penetration in Al alloys. *X-Ray Tomography in Materials Science*. J. Baruchel, J.Y. Buffière, E. Marie, P. Merle, and G. Peix (Eds.). Paris, Hermes Science: 155–164.

Ludwig, W., J.Y. Buffière, S. Savelli, and P. Cloetens (2003). Study of the interaction of a short fatigue crack with grain boundaries in a cast Al alloy using x-ray microtomography. *Acta Mater* **51**: 585–598.

Macedo, A., C.M.P. Vaz, J.M. Naime, P.E. Cruvinel, and S. Crestana (1999). X-ray microtomography to characterize the physical properties of soil and particulate systems. *Powder Technol* **101**: 178–182.

Maire, E., J.C. Grenier, D. Daniel, A. Baldacci, H. Klöcker, and A. Bigot (2006). Quantitative 3D characterization of intermetallic phases in an Al–Mg industrial alloy by x-ray microtomography. *Scripta Mater* **55**: 123–126.

Marrow, T.J., L. Babout, A.P. Jivkov, P. Wood, D. Engelberg, N. Stevens, P.J. Withers, and R.C. Newman (2006). Three dimensional observations and modeling of intergranular stress corrosion cracking in austenitic stainless steel. *J Nucl Mater* **352**: 62–74.

Marshall, S.J., M. Balooch, T. Breunig, J.H. Kinney, A.P. Tomsia, N. Inai, L.G. Watanabe, I.C. Wu-Magidi, and G.W.J. Mashall (1998). Human dentin and the dentin-resin adhesive interface. *Acta Mater* **46**: 2529–2539.

Martin, C.F., J.L. Blandin, L. Salvo, C. Josserond, and P. Cloetens (2000a). Study of damage during superplastic deformation. *X-Ray Tomography in Materials Science*. J. Baruchel, J.Y. Buffière, E. Marie, P. Merle, G. Peix (Eds.). Paris, Hermes Science: 193–204.

Martin, C.F., C. Josserond, L. Salvo, J.J. Blandin, P. Cloetens, and E. Boller (2000b). Characterization by x-ray micro-tomography of cavity coalescence during superplastic deformation. *Scripta Mater* **42**: 375–381.

Martín-Badosa, E., D. Amblard, S. Nuzzo, A. Elmoutaouakkilo, L. Vico, and F. Peyrin (2003). Excised bone structures in mice: Imaging at three-dimensional synchrotron microCT. *Radiol* **229**: 921–928.

Matsumoto, T., M. Yoshino, T. Asano, K. Uesugi, M. Todoh, and M. Tanaka (2006). Monochromatic synchrotron radiation μCT reveals disuse-mediated canal network rarefaction in cortical bone of growing rat tibiae. *J Appl Physiol* **100**: 274–280.

Matsumoto, T., M. Yoshino, K. Uesugi, and M. Tanaka (2007). Biphasic change and disuse-mediated regression of canal network structure in cortical bone of growing rats. *Bone* **41**: 239–246.

McDonald, S.A., L.C.R. Schneider, A.C.F. Cocks, and P.J. Withers (2006). Particle movement during the deep penetration of a granular material studied by x-ray microtomography. *Scripta Mater* **54**: 191–196.

McGuire, P.A., S. Blackburn, and E.M. Holt (2007). An X-ray micro-computed tomography study of agglomerate breakdown during the extrusion of ceramic pastes. *Chem Eng Sci* **62**: 6451–6456.

Mechanic, G.L., S.B. Arnaud, A. Boyde, T.G. Bromage, P. Buckendahl, J.C. Elliott, E.P. Katz, and G.N. Durnova (1990). Regional distribution of mineral and matrix in the femurs of rats flown on Cosmos 1887 biosatellite. *FASEB J* **4**: 34–40.

Mercer, C.E. and P. Anderson (1995). X-ray microtomographic quantification of the effects on enamel following CO_2 laser application. *J Dent Res* **74**: 849.

Miller, J.D., C.L. Lin, C. Garcia, and H. Arias (2003). Ultimate recovery in heap leaching operations as established from mineral exposure analysis by x-ray microtomography. *Int J Miner Process* **72**: 331–340.

Mollica, F., R.D. Santis, L. Ambrosio, L. Nicolais, D. Prisco, and S. Rengo (2004). Mechanical and leakage behaviour of the dentin-adhesive interface. *J Mater Sci Mater Med* **15**: 485–492.

Nahrendorf, M., C. Badea, L.W. Hedlund, J.L. Figueiredo, D.E. Sosnovik, G.A. Johnson, and R. Weissleder (2007). High-resolution imaging of murine myocardial infarction with delayed-enhancement cine micro-CT. *Am J Physiol Heart Circ Physiol* **292**: H3172–H3178.

Naik, N. (2003). Sulfate attack on Portland cement-based materials: Mechanisms of damage and long term performance, Ph.D. Thesis. Atlanta, Georgia Institute of Technology.

Naik, N.N., A.C. Jupe, S.R. Stock, A.P. Wilkinson, P.L. Lee, and K.E. Kurtis (2006). Sulfate attack monitored by microCT and EDXRD: Influence of cement type, water-to-cement ratio, and aggregate. *Cement Concr Res* **36**: 144–159.

Naik, N.N., K.E. Kurtis, A.P. Wilkinson, A.C. Jupe, and S.R. Stock (2004). Sulfate deterioration of cement-based materials examined by x-ray microtomography. *Developments in X-ray Tomography IV*. U. Bonse (Ed.). Bellingham, WA, SPIE. *SPIE Proc Vol* **5535**: 442–452.

Nielsen, S.F., F. Beckmann, R.B. Godiksen, K. Haldrup, H.F. Poulsen, and J.A. Wert (2004). Measurement of the components of plastic displacement gradients in three dimensions. *Developments in X-Ray Tomography IV*. U. Bonse (Ed.). Bellingham, WA, SPIE. *SPIE Proc Vol* **5535**: 485–492.

Nielsen, S.F., H.F. Poulsen, F. Beckmann, C. Thorning, and J. Wert (2003). Measurements of plastic displacement gradient components in three dimensions using marker particles and synchrotron x-ray absorption microtomography. *Acta Mater* **51**: 2407–2415.

Noiriel, C., D. Bernard, P. Gouze, and X. Thibault (2005). Hydraulic properties and microgeometry evolution accompanying limestone dissolution by acidic water. *Oil Gas Sci Technol* **60**: 177–192.

Noiriel, C., P. Gouze, and D. Bernard (2004). Investigation of porosity and permeability effects from microstructure changes during limestone dissolution. *Geophys Res Lett* **31**: L24603/1–/4.

Nunan, N., K. Ritz, M. Rivers, D.S. Feeney, and I.M. Young (2006). Investigating microbial micro-habitat structure using x-ray computed tomography. *Geoderma* **133**: 398–407.

Nuzzo, S., M.H. Lafage-Proust, E. Martin-Badosa, G. Boivin, T. Thomas, C. Alexandre, and F. Peyrin (2002). Synchrotron radiation microtomography allows the analysis of three-dimensional microarchtiecture and degree of mineralization of human iliac crest biopsy specimens: Effects of etidronate treatment. *J Bone Miner Res* **17**: 1372–1382.

Nuzzo, S., C. Meneghini, P. Braillon, R. Bouvier, S. Mobilio, and F. Peyrin (2003). Microarchitectural and physical changes during fetal growth in human vertebral bone. *J Bone Miner Res* **18**: 760–768.

Ohta, S., E.W. Lai, J.C. Morris, D.A. Bakan, B. Klaunberg, S. Cleary, J.F. Powers, A.S. Tischler, M. Abu-Asab, D. Schimel, and K. Pacak (2006). MicroCT for high-resolution imaging of ectopic pheochromocytoma tumors in the liver of nude mice. *Int J Cancer* **119**: 2236–2241.

Peleg, S., F. Khan, N.M. Navone, D.D. Cody, E.M. Johnson, C.S. Van Pelt, and G.H. Posner (2005). Inhibition of prostate cancer-mediated osteoblastic bone lesions by the low-calcemic analog 1alpha-hydroxymethyl-16-ene-26,27-bishomo-25-hydroxy vitamin D_3. *J Steroid Biochem Mole Biol* **97**: 203–211.

Persy, V., A. Postnov, E. Neven, G. Dams, M. De Broe, P. D'Haese, and N. De Clerck (2006). High-resolution X-ray microtomography is a sensitive method to detect vascular calcification in living rats with chronic renal failure. *Arterio Thomb Vasc Biol* **26**: 2110–2116.

Pfister, A., U. Walz, A. Laib, and R. Mulhaupt (2005). Polymer ionomers for rapid prototyping and rapid manufacturing by means of 3D printing. *Macromol Mater Eng* **290**: 99–113.

Phillion, A.B., S.L. Cockcroft, and P.D. Lee (2006). X-ray micro-tomographic observations of hot tear damage in an Al-Mg commercial alloy. *Scripta Mater* **55**: 489–492.

Prevrhal, S. (2004). Beam hardening correction and quantitative microCT. *Developments in X-ray Tomography IV*. U. Bonse (Ed.). Bellingham, WA, SPIE. *SPIE Proc Vol* **5535**: 152–161.

Prevrhal, S. (2006). Simulation of trabecular mineralization measurements in microCT. *Developments in X-Ray Tomography V*. U. Bonse (Ed.). Bellingham, WA, SPIE. *SPIE Proc Vol* **6318**: 631808–1 to –10.

Rajamannan, N.M., T.B. Nealis, M. Subramaniam, S.R. Stock, K.I. Ignatiev, T.J. Sebo, J.W. Fredericksen, S.W. Carmichael, T.K. Rosengart, T.C. Orszulak, W.D. Edwards, R.O. Bonow, and T.C. Spelsberg (2005a). Calcified rheumatic valve neoangiogenesis is associated with VEGF expression and osteoblast-like bone formation. *Circulation* **111**: 3296–3301.

Rajamannan, N.M., M. Subramanium, S. Stock, F. Caira, and T.C. Spelsberg (2005b). Atorvastatin inhibits hypercholesterolemia-induced calcification in the aortic valves via the Lrp5 receptor pathway. *Circulation* **112** (suppl I): I–229–I–234.

Rattanasak, U. and K. Kendall (2005). Pore structure of cement/pozzolan composites by x-ray microtomography. *Cement Concr Res* **35**: 637–640.

Ren, W., B. Wu, X. Peng, J. Hua, H.N. Hao, and P.H. Wooley (2006). Implant wear induces inflammation, but not osteoclastic bone resorption, in RANK-/- mice. *J Orthop Res* **24**: 1575–1586.

Renders, G.A.P., L. Mulder, L.J. van Ruijven, and T.M.G.J. van Eijden (2007). Porosity of human mandibular condylar bone. *J Anat* **210**: 239–248.

Richard, P., P. Philippe, F. Barbe, S. Bourles, X. Thibault, and D. Bideau (2003). Analysis by x-ray microtomography of a granular packing undergoing compaction. *Phys Rev E* **68**: 020301/1–020301/4.

Ritman, E.L., M.E. Bolander, L.A. Fitzpatrick, and R.T. Turner (1998). MicroCT imaging of structure-to-function relationship of bone microstructure and associated vascular involvement. *Technol Health Care* **6**: 403–412.

Roudier, M.P., S.D. Bain, and W.C. Dougall (2006). Effects of the RANKL inhibitor, osteoprotegerin, on the pain and histopathology of bone cancer in rats. *Clin Exp Metast* **23**: 167–175.

Schell, J.S.U., M. Deleglise, C. Binetruy, P. Krawczak, and P. Ermanni (2007). Numerical prediction and experimental characterisation of meso-scale-voids in liquid composite moulding. *Compos Pt. A* **38**: 2460–2470.

Schell, J.S.U., M. Renggli, G.H. van Lenthe, R. Müller, and P. Ermanni (2006). Micro-computed tomography determination of glass fibre reinforced polymer meso-structure. *Compos Sci Technol* **66**: 2016–2022.

Scherf, H., F. Beckmann, J. Fischer, and F. Witte (2004). Internal channel structures in trabecular bone. *Developments in X-ray Tomography IV*. U. Bonse (Ed.). Bellingham, WA, SPIE. *SPIE Proc Vol* **5535**: 792–798.

Schmidhammer, R., S. Zandieh, R. Mittermayr, L.E. Pelinka, M. Leixnering, R. Hopf, A. Kroepfl, and H. Redl (2006). Assessment of bone union/nonunion in an experimental model using microcomputed technology. *J Trauma Inj Infec Crit Care* **61**: 199–205.

Schneider, P., R. Voide, M. Stauber, M. Stampanoni, L.R. Donahue, P. Wyss, U. Sennhauser, and R. Müller (2006). Assessment of murine bone ultrastructure using synchrotron light: Towards nanocomputed tomography. *Developments in X-Ray Tomography V*. U. Bonse (Ed.). Bellingham, WA, SPIE. *SPIE Proc Vol* **6318**: 63180C–1 to –9.

Schopf, L., A. Savinainen, K. Anderson, J. Kujawa, M. DuPont, M. Silva, E. Siebert, S. Chandra, J. Morgan, P. Gangurde, D. Wen, J. Lane, Y. Xu, M. Hepperle, G. Harriman, T. Ocain, and B. Jaffee (2006). IKKbeta inhibition protects against bone and cartilage destruction in a rat model of rheumatoid arthritis. *Arth Rheum* **54**: 3163–3173.

Schroer, C.G., M. Kuhlmann, T.F. Gunzler, B. Benner, O. Kurapova, J. Patormmel, B. Lengeler, S.V. Roth, R. Gehrke, A. Snigirev, I. Snigireva, N. Stribeck, A. Almendarez-Camarillo, and F. Beckmann (2006). Full-field and scanning microtomography based on parabolic refractive x-ray lenses. *Developments in X-ray Tomography V*. U. Bonse (Ed.). Bellingham, WA, SPIE. *SPIE Proc Vol* **6318**: 63181H–1 to –9.

Seidler, G.T., G. Martinez, L.H. Seeley, K.H. Kim, E.A. Behne, S. Zaranek, B.D. Chapman, S.M. Heald, and D.L. Brewe (2000). Granule-by-granule reconstruction of a sandpile from x-ray microtomography data. *Phys Rev E* **62**: 8175–8181.

Shalabi, M.M., J.G.C. Wolke, V.M.J.I. Cuijpers, and J.A. Jansen (2007). Evaluation of bone response to titanium-coated polymethylmethacrylate resin (PMMA) implants by x-ray tomography. *J Mater Sci Mater Med* **18**: 2033–2039.

Sheller, M.R., R.S. Crowther, J.H. Kinney, S. Yang, S. Di Jorio, T. Breunig, D.H. Carney, and J.T. Ryaby (2004). Repair of rabbit segmental defects with thrombin peptide, TP508. *J Orthop Res* **22**: 1094–1099.

Silva, M.D., J. Ruan, E. Siebert, A. Savinainen, B. Jaffee, L. Schopf, and S. Chandra (2006). Application of surface roughness analysis on micro-computed tomographic images of bone erosion: Examples using a rodent model of rheumatoid arthritis. *Mole Imaging* **5**: 475–484.

Sinha, P.K., P. Halleck, and C.Y. Wang (2006). Quantification of liquid water saturation in a PEM fuel cell diffusion medium using x-ray microtomography. *Electrochem Sol State Lett* **9**: A344–A348.

Sone, T., T. Tamada, Y. Jo, H. Miyoshi, and M. Fukunaga (2004). Analysis of three-dimensional microarchitecture and degree of mineralization in bone metastases from prostate cancer using synchrotron microcomputed tomography. *Bone* **35**: 432–438.

Speller, R., S. Pani, M. Tzaphlidou, and J. Horrocks (2005). MicroCT analysis of calcium/phosphorus ratio maps at different bone sites. *Nucl Instrum Meth A* **548**: 269–27.

Starr, T.L. (1992). Advances in modeling of the chemical infiltration process. *Chemical Vapor Deposition of Refractory Metals and Ceramics II*. T.M. Besman, B.M. Gallois and J. Warren (Eds.). *Mater Res Soc.* **250**: 207–214.

Stock, S.R., J. Barss, T. Dahl, A. Veis, and J.D. Almer (2002a). X-ray absorption microtomography (microCT) and small beam diffraction mapping of sea urchin teeth. *J Struct Biol* **139**: 1–12.

Stock, S.R., D. Blackburn, M. Gradassi, and H.G. Simon (2003). Bone formation during forelimb regeneration: A microtomography (microCT) analysis. *Dev Dynamics* **226**: 410–417.

Stock, S.R., L.L. Dollar, G.B. Freeman, W.J. Ready, L.J. Turbini, J.C. Elliott, P. Anderson, and G.R. Davis (1994). Characterization of conductive anodic filament (CAF) by x-ray microtomography and by seial sectioning. *Electronic Packaging Materials Science VII*. P. Børgesen, K.F. Jansen, and R.A. Pollak (Eds). Pittsburgh, *Mater Res Soc* **323**: 65–69.

Stock, S. R., A. Guvenilir, T.L. Starr, J.C. Elliott, P. Anderson, S.D. Dover, and D.K. Bowen (1989). Microtomography of silicon nitride / silicon carbide composites. *Ceram Trans* **5**: 161–170.

Stock, S.R., K.I. Ignatiev, H.G. Simon, and F.D. Carlo (2004). Newt limb regeneration studies with synchrotron microCT. *Developments in X-ray Tomography IV*. U. Bonse (Ed.). Bellingham, WA, SPIE. *SPIE Proc Vol* **5535**: 748–756.

Stock, S.R., N.N. Naik, A.P. Wilkinson, and K.E. Kurtis (2002b). X-ray microtomography (microCT) of the progression of sulfate attack of cement paste. *Cement Concr Res* **32**: 1673–1675.

Stock, S.R., A.E.M. Vieira, A.C.B. Delbem, M.L. Cannon, X. Xiao, and F. De Carlo (2008). Synchrotron microcomputed tomography of the bovine dentinoenamel junction. *J Struct Biol* **161**: 162–171.

Stock, S.R., X. Xiao, and F.D. Carlo (2006). Unpublished data 2-BM, APS.

Stribeck, N., A.A. Camarilla, U. Nochel, C. Schroer, M. Kuhlmann, S.V. Roth, R. Gehrke, and R.K. Bayer (2006). Volume-resolved nanostructure survey of a polymer part by means of SAXS microtomography. *Macromol Chem Phys* **207**: 1139–1149.

Tafforeau, P. and T.M. Smith (2008). Nondestructive imaging of hominoid dental microstructure using phase contrast x-ray synchrotron microtomography. *J Hum Evol* **54**: 272–278.

Tafforeau, P., I. Bentaleb, J.J. Jaeger, and C. Martin (2007). Nature of laminations and mineralization in rhinoceros enamel using histology and x-ray synchrotron microtomography: Potential implications for palaeoenvironmental isotopic studies. *Palaeogeo Palaeoclim Palaeoecol* **246**: 206–227.

Tesei, L., F. Casseler, D. Dreossi, L. Mancini, G. Tromba, and F. Zanini (2005). Contrast-enhanced x-ray microtomography of the bone structure adjacent to oral implants. *Nucl Instrum Meth A* **548**: 257–263.

Thompson, K.E., C.S. Willson, K.T.W Zhang, A.H. Reed, and L. Beenken (2006). Quantitative computer reconstruction of particulate materials from microtomography images. *Powder Technol* **163**: 169–182.

Trtik, P. and C.P. Hauri (2007). Micromachining of hardened Portland cement pastes using femtosecond laser pulses. *Mater Struct* **40**: 641–650.

Tzaphlidou, M., R. Speller, G. Royle, and J. Griffiths (2006). Preliminary estimates of the calcium/phosphorus ratio at different cortical bone sites using synchrotron microCT. *Phys Med Biol* **51**: 1849–1855.

Ural, A. and D. Vashishth (2007). Effects of intracortical porosity on fracture toughness in aging human bone: A μCT-based cohesive finite element study. *J Biomech Eng* **129**: 625–631.

Vagnon, A., O. Lame, D. Bouvard, M.D. Michiel, D. Bellet, and G. Kapelski (2006). Deformation of steel powder compacts during sintering: Correlation between macroscopic measurement and in situ microtomography analysis. *Acta Mater* **54**: 513–522.

Vaucher, S., P. Unifantowicz, C. Ricard, L. Dubois, M. Kuball, J.M. Catala-Civera, D. Bernard, M. Stampanoni, and R. Nicula (2007). On-line tools for microscopic and macroscopic monitoring of microwave processing. *Phys B* **398**: 191–195.

Vecchio, K.S., X. Zhang, J.B. Massie, M. Wang, and C.W. Kim (2007). Conversion of bulk seashells to biocompatible hydroxyapatite for bone implants. *Acta Biomater* **3**: 910–918.

Velhinho, A., P.D. Sequeira, R. Martins, G. Vignoles, F.B. Fernandes, J.D. Botas, and L.A. Rocha (2003). X-ray tomographic imaging of Al/SiC$_p$ functionally graded composites fabricated by centrifugal casting. *Nucl Instrum Meth B* **200**: 295–302.

Vengrenyuk, Y., S. Carlier, S. Xanthos, L. Cardoso, P. Ganatos, R. Virmani, S. Einav, L. Gilchrist, and S. Weinbaum (2006). A hypothesis for vulnerable plaque rupture due to stress-induced debonding around cellular microcalcifications in thin fibrious caps. *PNAS* **103**: 14678–14683.

Verrier, S., M. Braccini, C. Josserond, L. Salvo, M. Suèry, W. Ludwig, P. Cloetens, and J. Baruchel (2000). Study of materials in the semi-solid state. *X-Ray Tomography in Materials Science*. J. Baruchel, J.Y. Buffière, E. Maire, P. Merle, and G. Peix (Eds.). Paris, Hermes Science: 77–88.

Vetter, L.D., V. Cnudde, B. Masschaele, P.J.S. Jacobs, and J.V. Acker (2006). Detection and distribution analysis of organosilicon compounds in wood by means of SEM-EDX and micro-CT. *Mater Char* **56**: 39–48.

Vieira, A.E.M., A.C.B. Delbem, K.T. Sassaki, M.L. Cannon, and S.R. Stock (2006). Quantitative analysis of mineral content in enamel using laboratory microtomography and microhardness analysis. *Developments in X-Ray Tomography V*. U. Bonse (Ed.). Bellingham, WA, SPIE. *SPIE Proc Vol* **6318**: 631823-1 to –5.

Wacheter, N.J., P. Augat, G.D. Krischak, M. Mentzel, L. Kinzl, and L. Claes (2001). Prediction of cortical bone porosity in vitro by microcomputed tomography. *Calcif Tiss Int* **68**: 38–42.

Watson, I.G., M.F. Forster, P.D. Lee, R.J. Dashwood, R.W. Hamilton, and A. Chirrazi (2005). Investigation of the clustering behaviour of titanium diboride particles in aluminium. *Compos Pt A* **36**: 1177–1187.

Weiss, P., P. Layrolle, L.P. Clergeau, B. Enckel, P. Pilet, Y. Amouriq, G. Daculsi, and B. Giumelli (2007). The safety and efficacy of an injectable bone substitute in dental sockets demonstrated in a human clinical trial. *Biomater* **28**: 3295–3305.

Winkelmann, C.T., S.D. Figueroa, T.L. Rold, W.A. Volkert, and T.J. Hoffman (2006). Microimaging characterization of a B16-F10 melanoma metastasis mouse model. *Mole Imaging* **5**: 105–114.

Witte, F., J. Fischer, J. Nellesen, and F. Beckmann (2006a). Microtomography of magnesium implants in bone and their degradation. *Developments in X-Ray Tomography V*. U. Bonse (Ed.). Bellingham, WA, SPIE. *SPIE Proc Vol* **6318**: 631806-1 to –9.

Witte, F., J. Fischer, J. Nellesen, H.A. Crostack, V. Kaese, A. Pisch, F. Beckmann, and H. Windhagen (2006b). In vitro and in vivo corrosion measurements of magnesium alloys. *Biomater* **27**: 1013–1018.

Wong, F.S.L., J.C. Elliott, A. Anderson, and G.R. Davis (1995a). Mineral concentration in rat femoral diaphyses measured by x-ray microtomography. *Calcif Tissue Int* **56**: 62–70.

Wong, F.S.L., J.C. Elliott, P. Anderson, and G.R. Davis (1991). X-ray microtomographic study of the mineral content and structure of a mouse mandible. *J Dent Res* **70**: 691.

Wong, F.S.L., J.C. Elliott, P. Anderson, and G.R. Davis (1995b). Three dimensional mineral distribution in the dentine of a rat incisor measured by x-ray microtomography. *J Dent Res* **74**: 849.

Wong, F.S.L., N.S. Willmott, and G.R. Davis (2006). Dentinal carious lesion in three dimensions. *Int J Paed Dent* **16**: 419–423.

Youssef, Y.M., A. Chaijaruwanich, R.W. Hamilton, H. Nagaumi, R.J. Dashwood, and P.D. Lee (2006). X-ray microtomographic characterisation of pore evolution during homogenisation and rolling of Al-6Mg. *Mater Sci Technol* **22**: 1087–1093.

Zabler, S., P. Cloetens, and P. Zaslansky (2007a). Fresnel-propagated submicrometer x-ray imaging of water-immersed tooth dentin. *Opt Lett* **32**: 2987–2989.

Zabler, S., H. Riesemeier, P. Fratzl, and P. Zaslansky (2006). Fresnel-propagated imaging for the study of human tooth dentin by partially coherent x-ray tomography. *Optics Express* **14**: 8584–8597.

Zabler, S., A. Rueda, A. Rack, H. Riesemeier, P. Zaslansky, I. Manke, F. Garcia-Moreno, and J. Banhart (2007b). Coarsening of grain-refined semi-solid Al–Ge32 alloy: X-ray microtomography and in situ radiography. *Acta Mater* **55**: 5045–5055.

Zettler, R., T. Donath, J.F. d. Santos, F. Beckman, and D. Lohwasser (2006). Validation of marker material flow in 4mm thick friction stir welded Al 2024-T351 through computer microtomography and dedicated metallographic techniques. *Adv Eng Mater* **8**: 487–490.

Zhang, X., S.N. Johnson, P.J. Gregory, J.W. Crawford, I.M. Young, P.J. Murray, and S.C. Jarvis (2006). Modelling the movement and survival of the root-feeding clover weevil, *Sitona lepidus*, in the root-zone of white clover. *Eco Model* **190**: 133–146.

11

Mechanically Induced Damage, Deformation, and Cracking

This chapter covers mechanically induced deformation and damage (pore formation, cracking) as well as quantification of crack characteristics. Deformation of cellular materials (metal and ceramic foams, trabecular bone, etc.) and environmentally assisted cracking are not subjects here, as they were covered earlier (Chapters 7 and 10, respectively), land deformation-based materials processing is covered elsewhere as well (Chapter 10). Some material on measurement of the 3D spatial distribution of crack openings was presented in Chapter 6 and on detection limits of crack openings in Chapter 5.

Increasingly, investigators using microCT observe a given sample four or more times during its evolution; this is a thread running through the present chapter and it will be a continuing theme in future microCT studies. Accounts illustrating the scope of microCT of deformation open this chapter. Section 11.2 examines crack characterization with microCT. Section 11.3 covers microCT studies of deformation of composite materials, including cortical bone and tooth, and natural composites.

11.1 Deformation Studies

Strain localization, the concentration of deformation into narrow bands of intense shear, occurs not only in cellular solids, as noted in Chapter 8, but also in most geomaterials. Viggiani et al. (2004) observed a stiff soil at several strains with synchrotron microCT; a single shear band was formed at an axial strain of 2.7 percent. Development of shear bands in argillaceous rock during compressive triaxial deformation was followed with microCT using a volumetric digital correlation method (Lenoir et al., 2007).

In situ loading plus microCT of concrete has shown that the fracture processes must be treated not as a 2D crack but as a system of smaller 3D cracks (Landis and Keane, 1999). The nonrecoverable work of loading in compression of mortar, calculated using a linear elastic fracture mechanics approach, was determined by measuring the incremental changes in the crack surface area revealed by microCT as load was increased (Landis and Keane, 1999; Landis et al., 1999; Landis and Nagy, 2000); the incremental fracture energy rose with increasing damage indicating secondary toughening mechanisms such as friction made up a greater fraction of the measured energy (Landis et al., 2003, 2007).

Several microCT studies focused on the role of porosity in failure of metallic samples. Model copper specimens were manufactured with a regular array of interior pores, and their coalescence in the final stages preceding fracture was studied with synchrotron microCT (Weck et al., 2006). MicroCT revealed considerable void growth but not nucleation of new voids (i.e., no strain localization between the pre-existing manufactured voids). Evolution of pore size distribution during creep was studied by microCT and x-ray diffraction in a three-phase copper alloy (Pyzalla et al., 2005); details of this study are discussed later in the section on multimode studies. High-velocity impacts generate refraction waves within the target, waves that superimpose and generate zones of high density of pores; synchrotron microCT was applied to small specimens cut from impact-tested Ta disks (Bontaz-Carion and Pellegrini, 2006). These data showed that pore volume distribution obeyed a power law, at least for the larger pores, and that results from 2D methods, even with correction, did not extrapolate to the actual 3D distribution. Lab microCT of specimens from five positions in a high-pressure die-casting of Mg alloy AM60B revealed considerable variability in the amount and distribution of microporosity, location of the fracture plane, and the fracture strain; the fracture strain agreed with predictions of a critical strain model based on the initial pore distribution (Weiler et al., 2005). Void growth near a notch was followed at several loads with synchrotron microCT (~0.5-μm isotropic voxels) of a cast A356 Al specimen (Al grains surrounded by a Si-rich eutectic phase), and this study noted constraint effects on the specimen faces (compared to the middle of the specimen) as well as considerable void-Si particle spatial correlation (Qian et al., 2005).

Fatigued commercial bone cement (PMMA beads in a $BaSO_4$-filled PMMA matrix) was studied with synchrotron microCT; macroscopic failure was found to be linked to the presence of large voids, crack deflection was observed at matrix beads, and crack arrest was found within beads (Sinnett-Jones et al., 2005). In an Al engineering alloy, the distributions of pore sizes determined from individual lab microCT slices and from the 3D stack of slices were compared: the 2D measurement showed a significantly lower mean pore diameter, even after Saltykov-type correction for sampling biases (Underwood, 1968), than the 3D measurement (Li et al., 2006). The interrupted fatigue test specimens showed the crack tended to deviate toward pores that were near the transverse plane (normal to the load axis) that the crack was following across the specimen (Li et al., 2006).

Lab microCT with rather large voxels (0.08–0.12 mm) was used to investigate the mechanisms responsible for rising crack growth resistance with increased crack length (R curve behavior) in a fairly large specimen of polygranular graphite (Hodgkins et al., 2006). After crack extension, the crack was held open by inserting a wedge. Significant crack face contact was observed behind the crack tip, and the authors reported a zone of discrete, low attenuation features around the crack tip and in the crack wake, features that were interpreted as microcracks hypothesized to be responsible for the R-curve behavior. It would be interesting to confirm this interpretation by either performing phase-enhanced microradiography plus 3D stereometry (Ignatiev, 2004; Ignatiev et al., 2005) or performing synchrotron microCT of small sections cut from one of the samples.

11.2 Monolithic Materials: Crack Face Interactions and Crack Closure

Cracks in metals or ceramics are very high contrast features and are relatively easy to detect with microCT. Given the importance of cracking in engineered structures, it is not surprising that microCT imaging of small cracked specimens under load was performed and published very early on (Breunig et al., 1992b).

During fatigue crack propagation, application of a cyclically varying load drives crack extension, and many variables can affect crack growth rates. Processes such as fatigue crack closure greatly alter crack propagation rates, and understanding these has been the focus of considerable research. X-ray microCT with in situ loading of samples was used to study how, at what stress/stress intensities, and where, crack faces come into contact or separate during a fatigue cycle. After a brief description of the crack closure phenomenon, results obtained on an Al–Li alloy are discussed. Subsequent to this discussion, other results on cracks are reviewed.

Fatigue crack closure describes situations where the crack faces come into contact prematurely during unloading of a sample (i.e., before the minimum stress of a fatigue cycle is reached) or where the crack faces remain in contact much longer than expected during loading. Without crack closure, one would expect the crack faces either to contact when the cycle's minimum stress σ_{min} is reached or not to touch at all; without this effect, contact should occur over the entire crack face simultaneously. A variety of mechanisms of crack closure has been proposed, including: oxide- or (other) particle-induced closure (Christensen, 1963), plasticity-induced closure (Elber, 1971), and roughness-induced closure (Suresh and Ritchie, 1982). When portions of the crack faces touch at stress $\sigma > \sigma_{min}$, further displacements of the crack faces in the vicinity of the crack tip are resisted by this local contact, so that the driving "force" for crack extension decreases from the nominal value of the stress intensity range $\Delta K = K_{max} - K_{min}$ to $\Delta K_{eff} < \Delta K$. The fact that crack closure occurs and is important is widely accepted, even though details of what constitutes the closure load P_{Cl} used to calculate K_{eff} remain less than crystal clear. Normally, load-deflection curves of samples exhibiting crack closure show two stages, a lower slope at higher stresses and a higher slope at lower stresses; P_{Cl} is taken to be the point where the tangents to the upper and lower portions of the curves intersect.

A series of papers and theses employed microCT during in situ loading of specimens of AA2090 T8E41 to study the changing pattern of crack opening as a function of applied load, that is, at different points during a fatigue cycle. This alloy is particularly interesting to study because fatigue crack growth rates are much lower in the L–T orientation than in other Al alloys used in aerospace applications. Roughness-induced crack closure is very pronounced in this alloy and persists to much higher stress intensity ranges than in other alloys.

Notched tensile samples with 2.9-mm and ~2.0-mm gauge and notch tip diameters, respectively, were examined at four or five loads spanning the unloading portion of the fatigue cycle (Breunig et al., 1992b; Guvenilir, 1995; Stock et al., 1995; Guvenilir et al., 1997; Guvenilir and Stock, 1998; Guvenilir et al., 1999). One set of observations was done with 22-keV synchrotron x-radiation and the second with x-rays from an Ag-sealed source tube operated at 40 kV; both were reconstructed with ~6-µm isotropic voxels. The fatigue cracks were rough in some places and quite planar in other parts, as expected for this alloy. Crack openings were quantified as a function of position for each load (Figure 6.11 shows the openings measured for one sample projected onto a plane perpendicular to the load axis), and crack contact, even at the maximum stress, was observed behind the crack tip, particularly at some, but not all, positions where the crack was at its most nonplanar (compare the left and right sides of the valley labeled "d,s" in Figure 6.12).

The more planar sections of the cracks zipped shut well before the microscopic closure load was reached (e.g., at "c" in Figure 6.11), whereas portions of the cracks remained open in the nonplanar sections at stresses below the microscopic closure load. In other words, mixed mode contact was important, and this is best seen in three-dimensional meshes showing crack position on which the openings (indicated by different colors) are superimposed (Figure 6.12). Furthermore, the fraction of voxels (of the original crack) open were observed to remain nearly constant upon reducing the load from just above to just below the closure load; this was direct evidence that the mixed-mode surface began to carry significant load at the point where the load-displacement curve starts to deflect (Guvenilir et al., 1997), that is, where the samples started to stiffen during unloading. These observations could not have been made without microCT.

MicroCT quantification of fatigue crack opening as a function of position has also been performed in the more conventional compact tension sample geometry in AA2090 T8E41 (Guvenilir et al., 1994; Guvenilir, 1995; Morano, 1998; Morano et al., 2000; Ignatiev et al., 2006). Although these samples were much smaller than normal (2- to 2.5-mm thick and 25.4 mm from notch tip to back face), they were scaled according to ASTM specifications, and, because the entire width must be kept within the beam, this limited the voxel size to between 20 µm and 59 µm, depending on the data acquisition parameters. In these series of papers, fatigue cracks grown under different ratios of minimum to maximum applied stress were compared; patterns and amounts of crack opening were compared for other specimens before and after increments of extension.

The relationship of different spatial scales of crystallographic texture in AA2090 T8E41 to crack path has been examined in some detail. These studies, combining microCT and microbeam x-ray diffraction mapping, are described in Chapter 12.

A necessary component of the in situ microCT of crack closure is use of a loading apparatus. The standard load frame design of two posts on either side of the specimen blocks projections over a significant fraction of the 180° required for exact reconstruction. The simplest alternative is to use an x-ray transparent standoff to hold the two grips apart; thin-walled polycarbonate (Breunig et al., 1993a) or aluminum (Morano, 1998; Morano et al., 1999, 2000) tubes can provide enough rigidity and allow the sample to be viewed around 360°.

The force on the sample was, in the data presented above, applied pneumatically and can also be applied by a screw mechanism much in the fashion of commercial mechanical testing apparatus. Simple, small, and portable loading apparatus do not need to be designed for growing fatigue cracks, and the studies cited above, where the cracked specimen is dismounted from a conventional servohydraulic apparatus and remounted in the loading apparatus, demonstrated the efficacy and economy of this approach. Another approach employs support posts and a pair of synchronized rotators on either side of the sample; the rotators allow sample views over 360° to be obtained without moving the posts into the beam (Breunig et al., 1994; Hirano et al., 1995). The apparatus of Breunig et al. had the capability of testing samples to 15.6 kN.

One of the limitations of the AA2090 compact tension specimen studies was the limited crack opening sensitivity in absorption microCT dictated by the large voxel size. Ignatiev and coworkers indicated that phase contrast stereometry (tracking features' relative displacements vs. specimen rotation in multiple radiographs and computing 3D positions from this data; see Section 3.7) allowed one to measure (albeit laboriously) small changes in opening for special positions on the crack faces (Ignatiev, 2004; Ignatiev et al., 2005, 2006). Others addressed this sensitivity issue by cutting a small volume of material from around the crack tip and imaging this with the highest available spatial resolution (see the following paragraphs).

Several papers reported synchrotron microCT of a small section of Al specimens cut to contain the tip of the fatigue crack (Ludwig and Bellet, 2000; Ludwig et al., 2000, 2003; Toda et al., 2003, 2004; Ohgaki et al., 2005; Buffière et al., 2006; Khor et al., 2006; Zhang et al., 2007). The resulting small voxel size and the strong phase contrast in these reconstructions increased crack visibility enormously compared to reconstructions with pure absorption contrast and in intact specimens. One is never quite sure how much the change of constraint (from removed material) affects the observations, so some caution should be exercised in interpreting these results. Buffière et al. (2006) summarized their experience in visualizing cracks in these types of specimens, as well as the drawbacks and advantages of microCT in the presence of significant phase contrast, and a few comments on their studies and others' follow.

Toda et al. (2003, 2004) found that the large transients in contrast from the Fresnel fringes parallel to fatigue crack surfaces dictated that robust crack opening measurements required use of features somewhat displaced from the crack plane. Therefore, near-crack-tip opening displacements in in situ loaded AA2024-T351 were measured using small microvoids (a small distance away from the crack faces) as fiducials, in much the same way that Breunig and coworkers worked with C cores in SiC fibers on either side of a crack in an Al/SiC monofilament composite (Breunig et al., 1992a, 1993a). The high-resolution closure observations of Toda et al. were in agreement with earlier work (Guvenilir et al., 1997; Guvenilir and Stock, 1998; Guvenilir et al., 1999; Morano et al., 1999, 2000; Ignatiev et al., 2006); namely, loss of surface contact occurred gradually up to the maximum load of the fatigue cycle, mixed-mode surface contact was very important, and near-tip contact was suggested as producing crack growth resistance.

Decoration of grain boundaries in Al with Ga liquid (the melting point of gallium is 30°C) allowed grain boundary positions to be correlated with the 3D crack geometry and changes in crack path, without sectioning the specimen (Ludwig and Bellet, 2000; Ludwig et al., 2000; Babout et al., 2003; Ludwig et al., 2003; Ferrie et al., 2005; Ohgaki et al., 2005; Ferrie et al., 2006; Khor et al., 2006). In this approach, the authors first performed the in situ loading experiments on the material cut from larger specimens; subsequently, Ga was applied. Paths of short cracks are well known to be dominated by crystallography of the few grains cut by the crack, and electron backscattering diffraction (EBSD) was used to provide the crystallographic information needed to understand which changes in grain orientation produced large deflections in the crack path (Ludwig et al., 2003). Analysis of crystallographic character of several branches of a short crack in a cast AS7G03 Al–Si alloy specimen containing artificial pores illustrated the power of this approach (Ludwig et al., 2003).

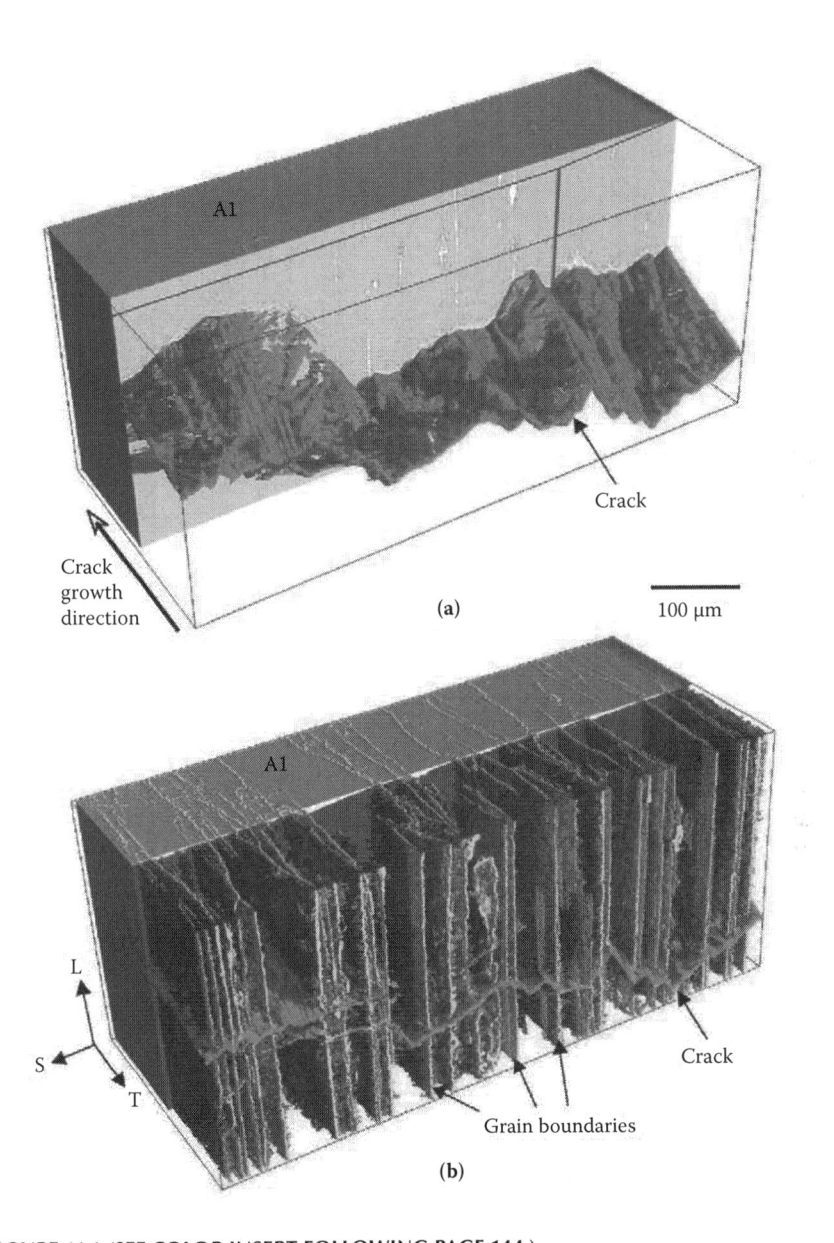

Mechanically Induced Damage, Deformation, and Cracking 287

FIGURE 11.1 (SEE COLOR INSERT FOLLOWING PAGE 144.)
Three-dimensional representations of a fatigue crack in AA2024-T351. (a) Crack volume (gray) and (b) grain boundaries decorated with Ga (dark gray). (Reprinted from Khor et al. (2006); © 2006, with permission from Elsevier.)

In the same material, 11 observations for different crack extensions were made of a short fatigue crack that had nucleated in a narrow ligament between a pore and the specimen surface (Ferrie et al., 2005), and the evolution of the crack front shape was interpreted with respect to the surrounding grain microstructure and pore positions. Figure 11.1 shows a 3D view of the tip of a fatigue crack in an AA2024-T351 specimen; the solid material is rendered transparent and only the Ga-labeled grain boundaries and crack are shown. Large portions of the crack (cut from the larger specimen to include only the near-tip region of the long fatigue crack) were within five degrees of {100} or {111}, with steeply inclined sections following {111} (Khor et al., 2006); these crack paths are those observed in AA2090 T8E41 (Yoder et al., 1989).

Ferrie et al. (2006) studied fatigue propagation in an ultra-fine-grained, powder metallurgy alloy (AA5091 material system with mechanical properties equivalent to the T1 condition). This alloy was selected because fatigue cracks follow highly planar paths, and an in situ fatiguing apparatus was mounted directly on the synchrotron microCT rotation stage. Radiography was used to monitor crack initiation, and, after each of nine increments of crack extension, data for reconstruction were collected with the specimen under maximum applied load. The crack grew more elliptical with increasing number of cycles with the major axis perpendicular to the specimen surface, and the authors attributed this to differences in closure stresses along the crack front. Local stress intensity range ΔK was calculated via FEM for the portion of the crack growing parallel to the surface and for the portion growing perpendicular to the surface. Plots of crack growth rate da/dN versus ΔK showed comparable power law exponents, but the surface curve was displaced to lower ΔK relative to the bulk curve and the latter followed the experimental long crack growth curve. The authors concluded that, for the specimen geometry studied, a single Paris equation can predict the observed crack growth anisotropy, provided variation in closure stress along the crack front was assumed. Experiments of the sort described in this paragraph are enormously difficult to conduct and are not to be undertaken lightly.

Zhang and coworkers characterized the interaction of a fatigue crack with pores resulting from flow forming of AA A356 T6 (Zhang et al., 2007). Marrow and coworkers performed synchrotron microCT on a very small ductile cast iron specimen and focused on characterizing changing crack front geometry as the short cracks interacted with pores and graphite nodules (Marrow et al., 2004). A gauge diameter of ~0.35 mm was required to give adequate transmission through the iron specimen at the highest photon energy that was practical for use with the high-resolution x-ray detector (30 keV). The difficulty of working with such fragile specimens is undoubtedly the reason that relatively few experiments are performed on steels, copper, or still more attenuating metals or composites. Even Ti poses challenges for microCT imaging.

11.3 Composite Systems

Metal matrix composites employ a ductile matrix (metal) and stiff fiber or particulate reinforcements; the resulting properties can be tailored by design of the composite system (e.g., by aligning fibers along the direction of desired high tensile stiffness). The reinforcing phase is generally a ceramic such as SiC or Al_2O_3, materials that have high elastic moduli but very low toughness. The metal matrix surrounds and protects the reinforcements and carries only a small fraction of the applied stress. In the case of fiber reinforcement, cracks generated by low-amplitude fatigue loading can propagate in the metal and bypass the fibers; the intact fibers bridging the crack will restrain the crack faces from opening as far as they otherwise would and thereby lower the stress intensity.

In ceramic matrix composites, the reinforcements provide increased toughness and damage tolerance and are normally in the form of fibers. One strategy is to use the fibers to absorb energy by causing a network of microcracks to form, energy which might otherwise contribute to catastrophic propagation of a single main crack. A second strategy is to use the fibers to bridge cracks. Ceramic matrix composites have reinforcements and matrices with negligible ductility; therefore, the faces of microcracks and large cracks tend to resume contact once applied loads decrease to zero. Even with load applied, crack openings typically remain below a few micrometers; being able to detect such small features has been the main impediment to applying microCT to quantification of damage in ceramic matrix composites.

This section follows the rather traditional scheme of separating the material into particle- and fiber-reinforced subsections. It may be possible to organize the material another way, but the kinds of questions asked (and the analysis approaches adopted) strongly favor the approach used here.

11.3.1 Particle-Reinforced Composites

In an early microCT study, the pore structure in degassed and nondegassed reaction-bonded silicon nitride/silicon carbide was studied with 10-µm voxels in a 1 mm × 1 mm cross-section, and it is not surprising that the 15-µm-diameter Nicalon fibers could not be resolved (Breunig et al., 1990). Other early studies include those of metal matrix composites reinforced with short, small-diameter fibers (Bonse et al., 1991) and with particulates (Mummery et al., 1995; Peix et al., 1997); the latter study followed damage accumulation.

A pencil beam microtomography system was used to investigate several aspects of ceramic particulate reinforced aluminum composites (Mummery et al., 1995). Within a given cross-section of the powder processed composite, the content of 12-μm TiB_2 particles was found to deviate substantially from the nominal 20 vol.% of reinforcement: contents as low as 10 vol.% were reported. The local void volume fraction within the necked regions of a set of composite samples with 5, 10, and 20 vol.% 30-μm SiC particles was also measured as a function of true strain; this type of measurement focused attention on the portion of the sample where damage was concentrated, offered a very sensitive probe of damage, and allowed the same volume to be interrogated in three dimensions multiple times during its evolution. Unfortunately, the sections were not completely contiguous, a limitation imposed by the quite low data acquisition rates with pencil beam systems.

The characteristics of a model discontinuously reinforced composite system (0, 5, and 10 vol.% Ni particles blended with AA2124 powder and hot extruded) were studied with lab microCT (Watson et al., 2006). Particle clustering was quantified by 2D (SEM) and 3D (microCT) methodologies, and good agreement was found. Subvolumes of the microCT-reconstructed volume were meshed and incorporated into FE simulations of the three materials, and the actual and simulated stress–strain curves showed quite good agreement (Watson et al., 2006).

Buffière et al. (1999) used synchrotron microCT to study a more challenging composite system than Ni–Al: 10 vol.% SiC particles in a matrix of AA6061-T4. As the mass attenuation coefficients of SiC and Al differ by less than 3 percent at 23 keV, phase-enhanced imaging (detector–specimen separation of 830 mm instead of a few millimeters) was used to provide contrast between particle and matrix and to enhance crack visibility. The same volume was compared at five strains (initial microstructure, at the yield point of the stress–strain curve, and at three strains up to 13 percent), the fraction of broken particles was greater in the bulk than near the surface, and FE calculations (normal stress, total stored elastic energy in particles) were supplied to explain the observations (Buffière et al., 1999). The spatiotemporal distribution of fractured SiC particles was mapped in a subsequent study (Buffière et al., 2000). Interest in microCT-based FEM analyses of particulate-reinforced composites continues (Crostock et al., 2006).

The association between deformation-induced porosity in several Al matrices and particulate reinforcements (ZrO_2) was investigated with lab microCT using a dual energy reconstruction technique (Justice et al., 2000, 2003). This was an example where an energy-sensitive x-ray detector (and a translate-rotate or pinhole data collection scheme) was required. Variance analysis was used to show that little, if any, clustering of the particles was present and to determine that voids were also not clustered.

A direct relationship between volume fraction of particles and void volume fraction was demonstrated (Justice et al., 2000, 2003). Synchrotron microCT and in situ loading were used to study damage in aluminum–zirconia composites (Babout et al., 2003, 2004).

Before turning the discussion to the most widespread discontinuously reinforced composite, the cAp–collagen system in tooth dentin and bone, it is interesting to examine a microCT study of a model for an energetic material (read explosive), a nonstructural material whose high strain rate deformation properties determine its functionality. McDonald and coworkers examined the shock response of a model energetic material consisting of a polymer matrix loaded with two sizes of glass spheres, 300-µm and 30-µm diameters (McDonald et al., 2007). They found very few of the spheres were damaged in processing prior to testing. The microCT-derived microstructure of a representative volume was incorporated into a numerical simulation, and the investigators found that the shock front was not planar (on the order of 100 µm) and that flow was inhomogeneous at this length scale.

Most microCT deformation studies of mineralized tissue, for example, bone and dentin, both nanocrystalline particle-reinforced composites, concern trabecular bone; these were covered in Chapter 8. There have been a number of microCT studies of the deformation, cracking, and fracture of cortical bone and dentin; these have been particularly important, as they have the potential to directly test the extent to which microcracking or crack bridging toughens these materials.

One bovine cortical bone study used lab microCT to examine short-rod chevron-notched tension specimens for fracture toughness determination (Santis et al., 2000). The V-shaped notch allowed steady-state crack propagation in a sample diameter rather smaller than a standard compact tension specimen, an important advantage given limited dimensions available, even in the long bones of large animals. In principle, fracture toughness for this specimen geometry does not require measurement of the crack length, but practically realizable geometry did not meet the assumption for the calculation and compliance tests and crack length measurements (via microCT) were used for more robust determination of the plain strain–stress intensity factor (Santis et al., 2000).

Nanoindentation is increasingly popular for applications such as quantification of the anisotropy of elastic moduli in bone, and such moduli were shown to agree well with moduli from load-deflection curves (Hengsberger et al., 2003). Some assumptions are required in the analysis, and Hengsberger et al. used synchrotron microCT to provide some of this information (specimen mineral levels, porosity, and cross-sectional dimensions).

Cortical bone exhibits good toughness, and two views of the source of toughness are: bridging by uncracked ligaments in the crack wake and microcracking ahead of the crack tip. Both absorb energy that would otherwise be used to extend the crack. Establishing the relative importance

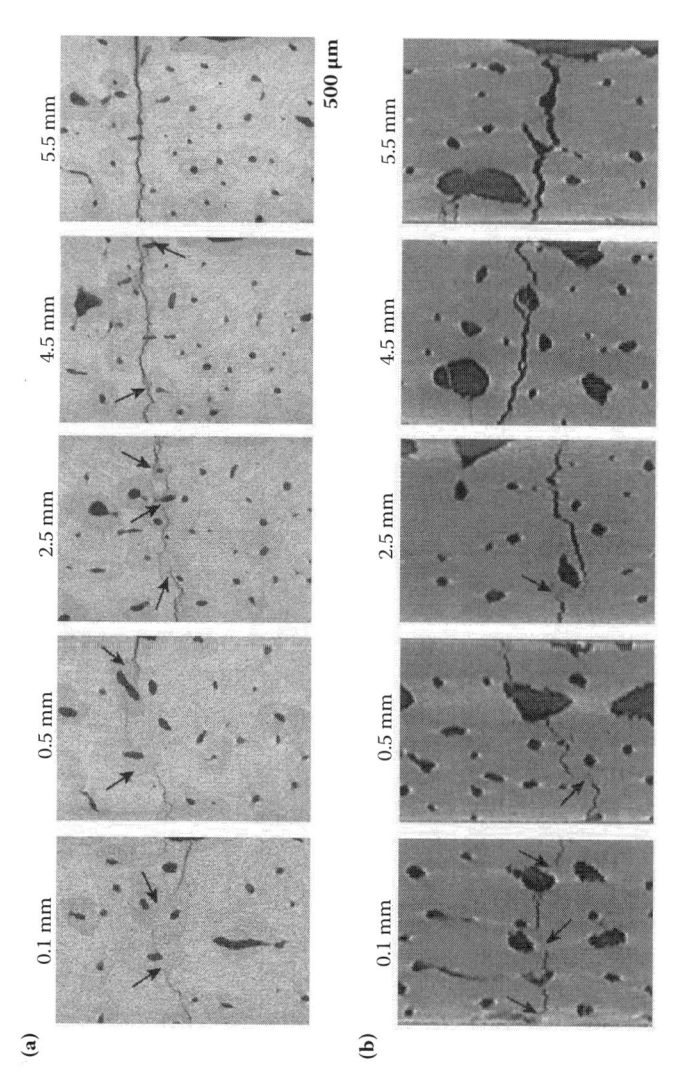

FIGURE 11.2 (SEE COLOR INSERT FOLLOWING PAGE 144.)
Synchrotron microCT sections through human cortical bone showing typical cracks in (a) young (34 years) and (b) aged (85 years) groups. The numbers above each column of images give the distance from the crack tip, and the black arrows indicate uncracked ligaments bridging the crack. The darker the shade, the lower is the x-ray absorption and mineral content (Nalla et al., 2004).

of these mechanisms would suggest treatment strategies for osteoporosis prevention. Nalla and coworkers examined mechanistic aspects of crack growth resistance in human cortical bone by determining crack growth resistance curves (R-curves) and using synchrotron microCT to image the 3D crack structure (Nalla et al., 2004, 2005).

Fracture toughness rose linearly with crack length, but there were clear differences in behavior between bone from young mature adults (age <41 yr, described as young bone below) and that from aged individuals (age >85 yr, described as aged bone in what follows). Figure 11.2 shows differences in crack paths revealed by synchrotron microCT for the young bone compared to the aged bone (Nalla et al., 2004). Ex vivo crack initiation toughness decreased 40 percent from young to aged bone, and crack growth toughness present in the young bone was essentially eliminated over this period. Quantification of the amount of crack bridging versus crack extension (practical only with microCT) revealed considerable initial bridging for both young and aged bone (Figure 11.3); after some crack extension, bridging for young bone remained comparable to the initial levels but for the aged bone was virtually absent (Nalla et al., 2004; Ritchie et al., 2006).

MicroCT showed that bridges were present throughout the specimen thickness (demonstrating that SEM data for bridging at specimen surfaces are valid) and that cracks tended to follow cement lines bordering

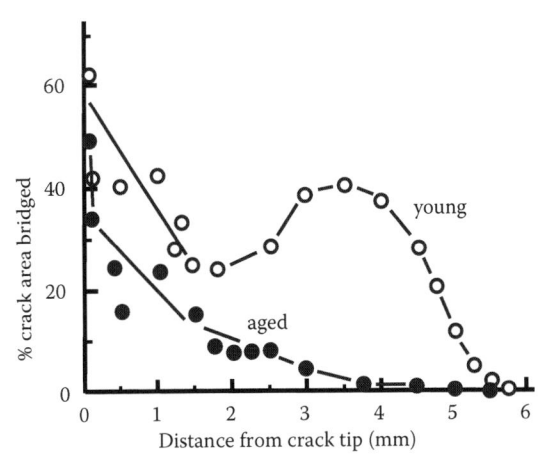

FIGURE 11.3
The fraction of crack-bridging zones in human cortical bone measured from microCT images as a function of distance from the crack tip. The plots for young and aged bone demonstrate the decline in size and percentage area of the cracked bridged zones in the older bone, and particularly the much lower bridging in the aged bone beyond ~2 mm from the crack tip. (Plotted from data in Nalla et al. (2004) and Ritchie et al. (2006), with the trend lines shown here differing from what is shown in the original presentation.)

osteons (Nalla et al., 2003). The bridging zone length was on the order of 5.5 mm long for this human cortical bone. Although toughness values for bone and dentin, related collagen–apatite composites, were comparable and were thought to reflect the nanoscale structure, differences in time-dependent crack blunting between the two mineralized tissues were thought to reflect the very different micrometer-level structures. Fracture, aging, and disease in bone were also discussed in another review by the same group (Ager et al., 2006).

Quantifying damage (cracking) produced by in vivo loading of bones is important in understanding the etiology of stress fractures and may be important in osteoporotic fractures as well. The in vivo ulnar loading model for rats allows controlled levels of damage to be applied, and this is an important advantage if the response of bone to damage is to be followed longitudinally. Lab microCT was used to quantify deformation-induced cortical bone cracking following ulnar fatigue with displacement amplitudes corresponding to 30, 45, 65 and 85 percent of fracture (Uthgenannt and Silva, 2007). Control ulnae and bones loaded to 30 percent of the fracture displacement did not show cracking; nearly one-half of the 45 percent displacement group exhibited detectable cracks; and all of the 65 and 85 percent displacement specimens were cracked, in particular, large branching cracks in the medial side of the 85 percent bones. Furthermore, statistically significant differences in crack length density were found for the four displacement groups. In another in vivo ulnar loading study employing microCT, woven bone repair of fatigue damage was found to restore whole bone strength and to enhance resistance to further fatigue damage (Silva and Touhey, 2007).

Diffraction-enhanced imaging (microradiography) was used along with bone's diffraction peak widths in an attempt to identify damage in cortical bone (Connor et al., 2005). Neither method revealed damage: even with the less than tenfold increase in sensitivity to small cracks compared to absorption-based imaging, this is not surprising because microcracks are very tiny features and the effect of overlapping depths will obscure even larger features. This conclusion should not be taken to demonstrate that x-ray diffraction cannot reveal useful information for damaged bone; as examples in Chapter 12 on multimode studies demonstrate, this is not the case.

11.3.2 Fiber-Reinforced Composites

Images of a SiC monofilament Si_3N_4 (91 wt.% Si_3N_4, 6 wt.% Y_2O_3, 3 wt.% Al_2O_3) composite were produced in an early microCT study (Hirano et al., 1989); in 111-μm thick slices, the 140-μm-diameter fibers and their 30-μm-diameter cores were quite visible, and the radial variation of linear attenuation coefficient at 24 keV for the SiC fibers agreed with others' 21-keV measurements of similar fibers (Kinney et al., 1990). Thermomechanical fatigue of ceramic as well as metal matrix composites was investigated by others (Baaklini et al., 1995).

Indentation damage was studied in carbon–fiber-reinforced plastic composites, but this study was limited to qualitative comparisons with results of ultrasonic characterization (Symons, 2000). A more complete focus on detection limits for different types of damage in fiber-reinforced polymer matrix composites was provided by a second study that examined the same cracks before and after (high x-ray absorption) dye penetrants were added (Schilling et al., 2005); in this lab microCT study, the authors reported crack detection limits without penetrant that were similar to those determined by Breunig et al. (1992a,1993a), but cracks open 0.5–1 μm in ~20-μm voxels (opening <5 percent of the voxel size) could be detected when penetrant was added. Other polymeric composite systems containing cracks and studied with microCT were the elastomeric material of auto tires (Dunsmuir et al., 1999) and aged dental composites (Drummond et al., 2006). An optical-fiber sensor was processed into a carbon–fiber thermoplastic composite, and microCT and the fiber sensor monitored strain during fatigue testing (De Baere et al., 2007).

Fracture of unidirectional fiber composites is generally thought to occur when a cluster of broken fibers reaches a critical number N*, and microCT is an ideal tool for assessing whether the critical cluster concept is valid and, if so, what N* might be for a given composite system. Synchrotron microCT of uniaxially aligned quartz fiber, epoxy matrix composites investigated this concept with in situ loading, and simple stochastic failure models were reported to under-predict N* by a factor of 3–5 (Aroush et al., 2006).

Early reports of microCT of metal matrix composites focused on Al and Ti matrices and on SiC monofilament reinforcements (Armistead, 1988; Hirano et al., 1989; Breunig et al., 1990; Elliott et al., 1990; Kinney et al., 1990; London et al., 1990; Stock, 1990; Breunig et al., 1991; Kinney et al., 1991; Breunig, 1992; Breunig et al., 1992a; Stock et al., 1992; Breunig et al., 1993a; Baaklini et al., 1995; Hirano et al., 1995; Peix et al., 1997), not only because these composites offer important performance gains over other composites but also because the size of the fibers allows unambiguous identification of damage modes at individual fibers and because these results can be linked to work on damage mode characterization in optically transparent glass composites employing the same monofilaments.

MicroCT of SiC monofilament Ti matrix composites presents a somewhat different challenge from imaging small-diameter particulate or chopped fiber Al matrix composites or, for that matter, imaging the same monofilament in an Al matrix. Although there is relatively little contrast between Al and SiC (their linear attenuation coefficients are quite similar; Kinney et al. (1990)), the large absorption of Ti makes it difficult to detect changes within the SiC or C portions of the monofilaments (Stock et al., 1992). Matrices of Ti$_3$Al and Ti-6-4 (6% Al, 4% V) with Textron SCS-6 SiC monofilament reinforcements were imaged using tube-generated (London et al., 1990) or synchrotron x-radiation (Stock et al., 1992). In these three- and eight-ply, aligned-fiber composites, matrix cracks and broken fibers were clearly seen.

The bulk of the microCT work on monofilament-reinforced composites, however, was on Al matrix materials and included at least one PhD thesis (Breunig, 1992). With the exception of one damaged* sample of $[0_2/\pm 45]_s$ eight-ply SCS-2 SiC/Al (Stock et al. 1992), the samples consisted of aligned monofilaments. In an eight-ply as-processed SCS-8 SiC/6061-0 Al matrix composite (see Figure 5.10 for a more recently recorded slice of this same material), considerable intraply, processing-related porosity was noted in reconstructions (Breunig et al., 1990; Kinney et al., 1990; Breunig et al., 1991; Breunig, 1992). Monochromatic synchrotron x-radiation with energy between 20 and 22 keV was used to collect the data for these reconstructions with 6-µm isotropic voxels. The contrast between SiC and Al was quite small at these energies, so that porosity and the monofilaments' carbon cores were the most visible features, and differentiating between fiber cores and porosity required careful scrutiny and a close eye for the weak contrast between SiC and Al.

Based on examination of many slices, loading to 828 MPa eliminated much of the porosity. Likewise, measurement of the separation between fiber cores of first- and higher-order neighbors revealed that the fiber centers became more regularly spaced as the load increased. Thus, as monotonic load increased, the porosity disappeared and fiber spacing became more regular (Breunig et al., 1991; Breunig, 1992). It is doubtful whether fiber rearrangement and porosity elimination would have been uncovered in studies of polished sections: too much labor would be required to interrogate enough serial sections.

Reconstructions with isotropic 6- to 7-µm voxels of ~1.5 mm × ~1.5 mm cross-sections cut from coupons of the same composite panel (as described above) revealed increased mechanical stiffness after the first few fatigue cycles. This corresponded to elimination of the processing-related matrix porosity (initially, 2 to 7 vol.%) and to displacement of the fibers from somewhat irregular arrangement into a more nearly hexagonal array (Breunig, 1992; Breunig et al., 2006). This study showed the fibers rearranged and the porosity disappeared by the time the load reached 828 MPa. Fracture of the C cores of the SiC fibers appeared to occur before 828 MPa and to nucleate the subsequent SiC fracture, but SiC cracking could not be observed except after fracture (σ = 1,448 MPa), no doubt because the cracks are pulled closed by the relaxation of adjacent fibers at lower stresses.

---

* The sample (the outer two plies on each side of the sample were parallel to and the four central plies were oriented at $\pm 45°$ to the load axis) experienced 5.0×10^6 cycles with R = $\sigma_{min}/\sigma_{max}$ = 0.3 and load range of 413 MPa.

Mechanically Induced Damage, Deformation, and Cracking

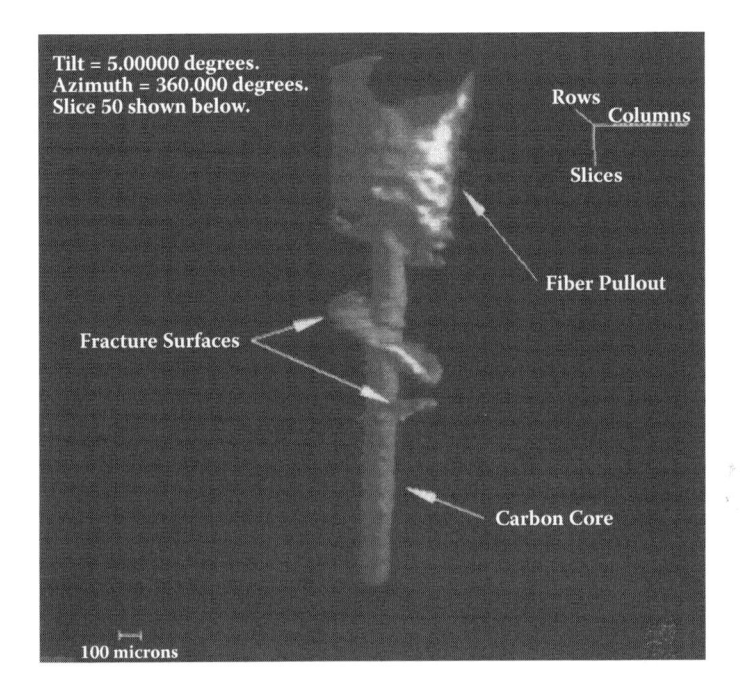

FIGURE 11.4 (SEE COLOR INSERT FOLLOWING PAGE 144.)
Three-dimensional rendering showing the surface of low-absorption voxels within a SiC monofilament in a fractured Al–SiC composite specimen. The fracture surface is at the top, and the C fiber core extends down from the fracture surface. The larger-diameter cylinder at the top and concentric with the C core shows where the SiC fiber pulled from the Al surface in this half of the specimen. A spiral crack appears midway down the fiber (Breunig, 1992; Kinney and Nichols, 1992).

Spiral and planar cracks within SiC fibers were visible with microCT in fractured samples of the Al/SiC metal matrix composite (Breunig 1992; Kinney and Nichols, 1992); Figure 11.4 includes only the low-absorption voxels from within the volume containing a single SiC fiber. The carbon core extends vertically to the sample surface (top) where the fiber pulled out of the matrix. The observation of C core fracture in the SiC fibers illustrates one of the advantages of tomography: this observation could not be made with serial sectioning and optical or scanning electron microscopy because any such fractures observed would undoubtedly be attributed to polishing damage.

Three-point bending of a notched eight-ply SiC/Al monofilament composite produced broken fibers, matrix cracking, and fiber–matrix disbonds that were observed with microCT (Breunig et al., 1990,1991; Breunig, 1992). The notch, was 0.6 mm deep and ~0.8 mm wide, and it was on the tensile side of the 1.5 mm × 1.5 mm cross-section of the sample. Outside the notch, only a small amount of fiber disbonding was present, adjacent to the end of the notch. Most of the damage was confined to within the material ahead of the notch. Large lengths of fibers pulled from the matrix, and the authors noted that the matrix–fiber shear zone extended 200–300 μm beyond both ends of fractured fibers. The fiber fracture surfaces were not planar, and significant plastic deformation was inferred to have occurred after fiber fracture and pullout.

In situ observation of monotonic deformation in an Al/SiC monofilament composite also was reported (Hirano et al., 1995). In this study, the 140-μm-diameter fibers were cut into 1-mm lengths and included in a 1-mm-diameter sample at a volume fraction of 0.10. These dimensions are rather unrealistic when compared to typical monofilament-reinforced composites, but it appears that they were convenient for microtomography and, in the case of the monofilament length, avoided the need to apply large loads and avoided unstable fracture once the ultimate tensile stress was exceeded. Observations were reported at zero stress, at a stress just before failure, and after the sample failed, but only 40 slices were recorded at each stress. The slice width and interval between slices were 31 μm and 79 μm, respectively.

Despite the difficulty of working with a set of noncontiguous slices and the decreased crack sensitivity due to the relatively large slice thickness, the authors reported observing fiber–matrix disbonding, fiber pullout, and matrix cracking only at the maximum stress (and, of course, in the failed condition). From observation of the failed composite, the authors inferred that the matrix crack, imaged at the maximum stress, propagated close to several monofilament segments and led to the sample's failure. Because of the small monofilament lengths, the relatively low volume fraction of reinforcement, and the relatively small sample diameter (compared to the monofilament diameter), it is unclear whether the conclusions represent what occurs during fracture of composites containing the much longer monofilaments typically employed.

The rapid change of attenuation coefficients across most interfaces poses a particular challenge for accurate reconstruction. This is a greater problem for SiC reinforcements in a Ti matrix than in an Al matrix, for example, and an experimenter's choice of x-ray energy needs to consider the desired difference in contrast between reinforcement and matrix as well as optimum sample transmissivity. There is a large area of internal interfaces, and detecting cracks at these interfaces, an important failure mode, presents a very real challenge. Little can be done if the interface cracks are tightly closed, except if in situ loading can be applied to open the cracks or if phase contrast is signficant. Even when these cracks are open, it is difficult to detect the change of attenuation from the crack superimposed on the transition in absorption across the interface. Considerable caution must be exercised, therefore, in quantification of interface cracking using x-ray microCT reconstructions.

In a uniaxial monofilament Al/SiC composite, microCT of mechanically induced damage was compared with unloading modulus (Breunig, 1992;

Breunig et al., 2006). Macroscopic measures of damage (changes in unloading compliance and in unrecovered strain) correlated with microCT quantification of microstructural changes (fiber separation, fiber misorientation relative to the load axis, fiber carbon core fracture). More recent generations of SiC monofilaments have improved properties, so these results are not indicative of current performance. Few complete fractures of SiC fibers were observed except after specimen failure; the authors concluded SiC fiber fractures were responsible for decreased compliance but, upon unloading, residual stresses from intact fibers presumably pulled fracture surfaces back together in the damaged fibers. Because these synchrotron microCT data were obtained under imaging conditions where phase contrast was negligible (i.e., during the early 1990s at CHESS and SSRL), crack visibility in the SiC fibers was substantially lower than that in the studies reported in the following paragraph, and it is not surprising that tightly closed cracks might be invisible.

Uniaxially aligned monofilament Ti/SiC specimens were imaged with synchrotron microCT under in situ loads, and fiber fracture geometry and spatial distribution were characterized (Maire and Buffière, 2000; Maire et al., 2001; McDonald et al., 2002, 2003; Preuss et al., 2003; Sinclair et al., 2004, 2005; Withers and Lewandowski, 2006). Single-fiber, single-ply, and multiple-ply specimens were studied; artificially fractured fibers within the one-ply specimen and fiber bridging across a fatigue crack in the multiple-ply specimen were studied. The SiC fiber fractures were similar to what had been reported for Al/SiC monofilament composites, namely wedge cracks and spiral cracks.

Careful consideration of synchrotron phase-enhanced microCT renderings of the fractured fibers identified with wedge cracks (e.g., Figures 5b and 6 of McDonald et al. (2003)) revealed complex contrast between the wedge edges (where contrast was strongest), and the discussion in this paper clearly identified small fragments that gave rise to the complex contrast. The fainter contrast regions suggested that the SiC material between the wedge edges contained additional (albeit more tightly closed) crack segments (small fragments of the locally shattered fiber?). Although this may seem to be a minor point, the fine details of fiber fracture could provide important insight into interface bonding or into stress wave interactions during fiber fracture. The longitudinal sections through the fiber centers also revealed that the fibers curve along the length of the specimen (smaller apparent width of the fiber's core at top and bottom of the sections). As an integral part of the study was microbeam diffraction mapping of strains in the Ti matrix and of the longitudinal fiber strains, more detailed discussion is postponed until Chapter 12.

References

Ager, J.W., G. Balooch, and R.O. Ritchie (2006). Fracture, aging, and disease in bone *J Mater Res* **21**: 1878–1892.

Armistead, R.A. (1988). CT: Quantitative 3-D inspection. *Adv Mater Process inc Met Prog* (Mar): 41–49.

Aroush, D.R.B., E. Maire, C. Gauthier, S. Youssef, P. Cloetens, and H.D. Wagner (2006). A study of fracture of unidirectional composites using in situ high-resolution synchrotron x-ray microtomography. *Compos Sci Technol* **66**: 1348–1353.

Baaklini, G.Y., R.T. Bhatt, A.J. Eckel, P. Engler, M.G. Castelli, and R.W. Rauser (1995). X-ray microtomography of ceramic and metal matrix composites. *Mater Eval* **53**: 1040–1044.

Babout, L., W. Ludwig, E. Maire, and J.Y. Buffière (2003). Damage assessment in metallic structural materials using high resolution synchrotron x-ray tomography. *Nucl Instrum Meth B* **200**: 303–307.

Babout, L., E. Maire, and R. Fougères (2004). Damage initiation in model metallic materials: X-ray tomography and modelling. *Acta Mater* **52**: 2475–2487.

Bonse, U., R. Nusshardt, F. Busch, R. Pahl, J.H. Kinney, Q.C. Johnson, R.A. Saroyan and M.C. Nichols (1991). X-ray tomographic microscopy of fibre-reinforced materials *J Mater Sci* **26**: 4076–4085.

Bontaz-Carion, J. and Y.P. Pellegrini (2006). X-ray microtomography analysis of dynamic damage in tantalum. *Adv Eng Mater* **8**: 480–486.

Breunig, T.M. (1992). Nondestructive evaluation of damage in SiC/Al metal/matrix composite using x-ray tomographic microscopy, Ph.D. Thesis. Atlanta, Georgia Institute of Technology.

Breunig, T.M., J.C. Elliott, P. Anderson, G. Davis, S.R. Stock, A. Guvenilir, and S.D. Dover (1990). Application of x-ray microtomography to the study of SiC/Al metal matrix composite material. *New Materials and Their Applications*. D. Holland (Ed.). London, Institute of Physics: **111**: 53–60.

Breunig, T.M., J.C. Elliott, S.R. Stock, P. Anderson, G.R. Davis, and A. Guvenilir (1992a). Quantitative characterization of damage in a composite material using x-ray tomographic microscopy. *X-ray Microscopy III* A.G. Michette, G.R. Morrison, and C.J. Buckley (Eds.). New York, Springer: 465–468.

Breunig, T.M., J.H. Kinney, and S.R. Stock (2006). MicroCT (microtomography) quantification of microstructure related to macroscopic behavior Part 2 — Damage in SiC-Al monofilament composites tested in monotonic tension and fatigue. *Mater Sci Technol* **22**: 1059–1067.

Breunig, T.M., M.C. Nichols, J.S. Gruver, J.H. Kinney, and D.L. Haupt (1994). Servo-mechanical load frame for in situ, non-invasive, imaging of damage development. *Ceram Eng Sci Proc* **15**: 410–417.

Breunig, T.M., S.R. Stock, and R.C. Brown (1993a). Simple load frame for in situ computed tomography and x-ray tomographic microscopy. *Mater Eval* **51**: 596–600.

Breunig, T.M., S.R. Stock, S.D. Antolovich, J.H. Kinney, W.N. Massey, and M.C. Nichols (1992b). A framework relating macroscopic measures and physical processes of crack closure of Al–Li Alloy 2090. *Fracture Mechanics: Twenty-Second Symposium* (Vol. 1). H.A. Ernst, A. Saxena, and D.L. McDowell (Eds.). Philadelphia, ASTM. *ASTM STP* **1131**: 749–761.

Breunig, T.M., S.R. Stock, A. Guvenilir, J.C. Elliott, P. Anderson, and G.R. Davis (1993b). Damage in aligned fibre SiC/Al quantified using a laboratory x-ray tomographic microscope. *Composites* **24**: 209–213.

Breunig, T.M., S.R. Stock, J.H. Kinney, A. Guvenilir, and M.C. Nichols (1991). Impact of X-ray tomographic microscopy on deformation studies of a SiC/Al MMC. *Advanced Tomographic Imaging Methods for the Analysis of Materials*. J.L. Ackerman and W.A. Ellingson (Eds.). Pittsburgh, Mater Res Soc **217**: 135–141.

Buffière, J.Y., E. Ferrie, H. Proudhon, and W. Ludwig (2006). Three-dimensional visualization of fatigue cracks in metals using high resolution synchrotron x-ray micro-tomography. *Mater Sci Technol* **22**: 1019–1024.

Buffière, J.Y., E. Maire, P. Cloetens, G. Lormand, and R. Fougeres (1999). Characterization of internal damage in a MMCp using x-ray synchrotron phase contrast microtomography. *Acta Mater* **47**: 1613–1625.

Buffière, J.Y., S. Savelli, and E. Maire (2000). Characterisation of MMC$_P$ and cast aluminum alloys. *X-Ray Tomography in Materials Science*. J. Baruchel, J.Y. Buffière, E. Maire, P. Merle, and G. Peix (Eds.). Paris, Hermes Science: 103–114.

Christensen, R.H. (1963). Fatigue crack growth affected by metal fragments wedged between opening-closing crack surfaces. *Appl Mater Res* October: 207–210.

Connor, D.M.D. Sayers, D.R. Sumner, and Z. Zhong (2005). Identification of fatigue damage in cortical bone by diffraction enhanced imaging. *Nucl Instrum Meth A* **548**: 234–239.

Crostock, H.A., J. Nellesen, G. Fischer, S. Schmauder, U. Weber, and F. Beckmann (2006). Tomographic analysis and FE-simulations of MMC microstructures under load. *Developments in X-Ray Tomography V*. U. Bonse (Ed.). Bellingham, WA, SPIE. *SPIE Proc Vol* **6318**: 63181A–1 to –12.

De Baere, I., E. Voet, W. Van Paepegem, J. Vlekken, V. Cnudde, B. Masschaele, and J. Degrieck (2007). Strain monitoring in thermoplastic composites with optical fiber sensors: Embedding process, visualization with micro-tomography, and fatigue results. *J Thermoplast Comp Mater* **20**: 453–472.

Drummond, J.L., F. De Carlo, K.B. Sun, A. Bedran-Russo, P. Koin, M. Kotche, and B. Super (2006). Tomography of dental composites. *Developments in X-Ray Tomography V*. U. Bonse (Ed.). Bellingham, WA, SPIE. *SPIE Proc Vol* **6318**: 63182B–1 to –8.

Dunsmuir, J.H., A.J. Dias, D.G. Peiffer, R. Kolb, and G. Jones (1999). Microtomography of elastomers for tire manufacture. *Developments in X-ray Tomography II*. U. Bonse (Ed.). Bellingham, WA, SPIE. *SPIE Proc Vol* **3772**: 87–96.

Elber, W. (1971). The significance of fatigue crack closure. Damage tolerance in aircraft structures. M.S. Rosenfeld (Ed.). West Conshohocken, PA, ASTM. *STP* **486**: 230–242.

Elliott, J.C., P. Anderson, S.D. Dover, S.R. Stock, T.M. Breunig, A. Guvenilir, and S.D. Antolovich (1990). Application of X-ray microtomography in materials science illustrated by a study of a continuous fiber metal matrix composite. *J X-ray Sci Technol* **2**: 249–258.

Ferrie, E., J.Y. Buffière, and W. Ludwig (2005). 3D characterisation of the nucleation of a short fatigue crack at a pore in a cast Al alloy using high resolution synchrotron microtomography. *Int J Fatigue* **27**: 1215–1220.

Ferrie, E., J.Y. Buffière, W. Ludwig, A. Gravouil, and L. Edwards (2006). Fatigue crack propagation: In situ visualization using x-ray microtomography and 3D simulation using the extended finite element method. *Acta Mater* **54**: 1111–1122.

Guvenilir, A. (1995). Investigation into asperity induced closure in an Al–Li alloy using X-ray tomography, Ph.D. Thesis. Atlanta, Georgia Institute of Technology.

Guvenilir, A. and S.R. Stock (1998). High resolution computed tomography and implications for fatigue crack closure modeling. *Fatigue Fract Eng Mater Struct* **21**: 439–450.

Guvenilir, A., T.M. Breunig, J.H. Kinney, and S.R. Stock (1997). Direct observation of crack opening as a function of applied load in the interior of a notched tensile sample of Al–Li 2090. *Acta Mater* **45**: 1977–1987.

Guvenilir, A., T.M. Breunig, J.H. Kinney, and S.R. Stock (1999). New direct observations of crack closure processes in Al–Li 2090 T8E41. *Phil Trans Roy Soc (Lond)* **357**: 2755–2775.

Guvenilir, A., S.R. Stock, M.D. Barker, and R.A. Betz (1994). *Aluminum Alloys: Their Physical Properties and Mechancial Properties*. T.H. Sanders and E.A. Starke. Atlanta, Georgia Institute of Technology. II: 413–419.

Hengsberger, S., J. Enstroem, F. Peyrin, and P. Zysset (2003). How is the indentation modulus of bone tissue related to its macroscopic elastic response? A validation study. *J Biomech* **36**: 1503–1509.

Hirano, T., K. Usami, and K. Sakamoto (1989). High resolution monochromatic tomography with x-ray sensing pickup tube. *Rev Sci Instrum* **60**: 2482–2485.

Hirano, T., K. Usami, Y. Tanaka, and C. Masuda (1995). In situ x-ray CT under tensile loading using synchrotron radiation. *J Mater Res* **10**: 381–385.

Hodgkins, A., T.J. Marrow, P. Mummery, B. Marsden, and A. Fok (2006). X-ray tomography observation of crack propagation in nuclear graphite. *Mater Sci Technol* **22**: 1045–1051.

Ignatiev, K.I. (2004). Development of x-ray phase contrast and microtomography methods for the 3D study of fatigue cracks, Ph.D. Thesis. Atlanta, Georgia Institute of Technology.

Ignatiev, K.I., G.R. Davis, J.C. Elliott, and S.R. Stock (2006). MicroCT (microtomography) quantification of microstructure related to macroscopic behaviour Part 1 — Fatigue crack closure measured in situ in AA 2090 compact tension samples. *Mater Sci Technol* **22**: 1025–1037.

Ignatiev, K.I., W.K. Lee, K. Fezzaa, and S.R. Stock (2005). Phase contrast stereometry: Fatigue crack mapping in 3D. *Phil Mag* **83**: 3273–3300.

Justice, I., B. Derby, G. Davis, P. Anderson, and J. Elliott (2000). Characterisation of void and reinforcement distributions by edge contrast. *X-Ray Tomography in Materials Science*. J. Baruchel, J.Y. Buffière, E. Maire, P. Merle, and G. Peix (Ed.). Paris, Hermes Science: 89–102.

Justice, I., B. Derby, G. Davis, P. Anderson, and J. Elliott (2003). Characterisation of void and reinforcement distributions in a metal matrix composite by x-ray edge-contrast microtomography. *Scripta Mater* **48**: 1259–1264.

Khor, K.H., J.Y. Buffière, W. Ludwig, and I. Sinclair (2006). High resolution x-ray tomography of micromechanisms of fatigue crack closure. *Scripta Mater* **55**: 47–50.

Kinney, J.H. and M.C. Nichols (1992). X-ray tomographic microscopy (XTM) using synchrotron radiation. *Annu Rev Mater Sci* **22**: 121–152.

Kinney, J.H., R.A. Saroyan, W.N. Massey, M.C. Nichols, U. Bonse, and R. Nusshardt (1991). X-ray tomographic microscopy for nondestructive characterization of composites. *Rev Prog Quant NDE* **10A**: 427–433.

Kinney, J.H., S.R. Stock, M.C. Nichols, U. Bonse, T.M. Breunig, R.A. Saroyan, R. Nusshardt, Q.C. Johnson, F. Busch, and S.D. Antolovich (1990). Nondestructive investigation of damage in composites using x-ray tomographic microscopy. *J Mater Res* **5**: 1123–1129.

Landis, E. and D.T. Keane (1999). X-ray microtomography for fracture studies in cement-based materials. *Developments in X-ray Tomography II*. U. Bonse (Ed.). Bellingham, WA, SPIE. *SPIE Proc Vol* **3772**: 105–113.

Landis, E.N. and E.N. Nagy (2000). Three-dimensional work of fracture for mortar in compression. *Eng Fract Mech* **65**: 223–234.

Landis, E.N., E.N. Nagy, and D.T. Keane (2003). Microstructure and fracture in three dimensions. *Eng Fract Mech* **70**: 911–925.

Landis, E.N., E.N. Nagy, D.T. Keane, and G. Nagy (1999). Technique to measure 3D work-of-fracture of concrete in compression. *J Exp Mech* **125**: 599–605.

Landis, E.N., T. Zhang, E.N. Nagy, G. Nagy, and W.R. Franklin (2007). Cracking, damage and fracture in four dimensions. *Mater Struct* **40**: 357–364.

Lenoir, N., M. Bornert, J. Desrues, P. Bésuelle, and G. Viggiani (2007). Volumetric digital image correlation applied to X-ray microtomography images from triaxial compression tests on Argillaceous rock. *Strain* **43**: 193–205.

Li, P., P.D. Lee, T.C. Lindley, D.M. Maijer, G.R. Davis, and J.C. Elliott (2006). X-ray microtomographic characterisation of porosity and its influence on fatigue crack growth. *Adv Eng Mater* **8**: 476–479.

London, B., R.N. Yancey, and J.A. Smith (1990). High-resolution x-ray computed tomography of composite materials. *Mater Eval* **48**: 604–608.

Ludwig, W. and D. Bellet (2000). Penetration of liquid gallium into the grain boundaries of aluminum: A synchrotron radiation microtomographic investigation. *Mater Sci Eng A* **281**: 198–203.

Ludwig, W., S. Bouchet, D. Bellet, and J.Y. Buffière (2000). 3D observation of grain boundary penetration in Al alloys. *X-Ray Tomography in Materials Science*. J. Baruchel, J.Y. Buffière, E. Marie, P. Merle, and G. Peix (Ed.). Paris, Hermes Science: 155–164.

Ludwig, W., J.Y. Buffière, S. Savelli, and P. Cloetens (2003). Study of the interaction of a short fatigue crack with grain boundaries in a cast Al alloy using x-ray microtomography. *Acta Mater* **51**: 585–598.

Maire, E. and J. Buffière (2000). X-ray tomography of aluminium foams and Ti/SiC composites. *X-Ray Tomography in Materials Science*. J. Baruchel, J.Y. Buffière, E. Maire, P. Merle, and G. Peix (Ed.). Paris, Hermes Science: 115–126.

Maire, E., A. Owen, J.Y. Buffière, and P.J. Withers (2001). A synchrotron X-ray study of a Ti/SiC$_f$ composite during in situ straining. *Acta Mater* **49**: 153–163.

Marrow, T.J., J.Y. Buffière, P.J. Withers, G. Johnson, and D. Engelberg (2004). High resolution x-ray tomography of short fatigue crack nucleation in austempered ductile cast iron. *Int J Fatigue* **26**: 717–725.

McDonald, S.A., J.C.F. Millett, N.K. Bourne, K. Bennett, A.M. Milne, and P.J. Withers (2007). The shock response, simulation and microstructural determination of a model composite material. *J Mater Sci* **42**: 9671–9678.

McDonald, S.A., M. Preuss, E. Maire, J.Y. Buffière, P.M. Mummery, and P.J. Withers (2002). Synchrotron x-ray study of micromechanics of Ti/SiC$_f$ composites with fibres containing defects introduced by laser drilling. *Mater Sci Technol* **18**: 1497–1503.

McDonald, S.A., M. Preuss, E. Marie, J.Y. Buffière, P.M. Mummery, and P.J. Withers (2003). X-ray tomographic imaging of Ti/SiC composites. *J Microsc* **209**: 102–112.

Morano, R. (1998). Effect of R-ratio on crack closure in Al–Li 2090 T8E41, investigated nondestructively with x-ray microtomography, M.S. Thesis. Atlanta, Georgia Institute of Technology.

Morano, R., S.R. Stock, G.R. Davis, and J.C. Elliott (1999). Macrotexture-related fatigue crack closure in Al–Li 2090 studied by x-ray microtomography. *Proceedings of the Twelfth International Conference on Textures of Metals*, Vol. 2. J.A. Szpunar (Ed.). National Research Council of Canada, Ottawa: 1106–1111.

Morano, R., S.R. Stock, G.R. Davis, and J.C. Elliott (2000). X-ray microtomography of fatigue crack closure as a function of applied load in Al–Li 2090 T8E41 samples. Warrendale, PA, MRS. *MRS Symp Proc* Vol **591**: 31–35.

Mummery, P.M., B. Derby, P. Anderson, G.R. Davis, and J.C. Elliott (1995). X-ray microtomographic studies of metal matrix composites using laboratory x-ray sources. *J Microsc* **177**: 399–406.

Nalla, R.K., J.H. Kinney, and R.O. Ritchie (2003). Mechanistic fracture criteria for the failure of human cortical bone. *Nature Mater* **2**: 164–168.

Nalla, R.K., J.J. Kruzic, J.H. Kinney, and R.O. Ritchie (2004). Effect of aging on the toughness of human cortical bone: Evaluation by R-curves. *Bone* **35**: 1240–1246.

Nalla, R.K., J.J. Kruzic, J.H. Kinney, and R.O. Ritchie (2005). Mechanistic aspects of fracture and R-curve behavior in human cortical bone. *Biomater* **26**: 217–231.

Ohgaki, T., H. Toda, I. Sinclair, J.Y. Buffière, W. Ludwig, T. Kobayashi, M. Niinomi, and T. Akahori (2005). Quantitative assessment of liquid Ga penetration into an aluminium alloy by high-resolution x-ray tomography. *Mater Sci Eng A* **406**: 261–267.

Peix, G., P. Cloetens, M. Salome, J.Y. Buffière, J. Baruchel, F. Peyrin, and M. Schlenker (1997). Hard x-ray phase tomographic investigation of materials using Fresnel diffraction of synchrotron radiation. *Developments in X-ray Tomography*. U. Bonse (Ed.). Bellingham, WA, SPIE. *SPIE Proc Vol* **3149**: 149–159.

Preuss, M., G. Rauchs, T.J.A. Doel, A. Steuwer, P. Bowen, and P.J. Withers (2003). Measurements of fibre bridging during fatigue crack growth in Ti/SiC fibre metal matrix composites. *Acta Mater* **51**: 1045–1057.

Pyzalla, A., B. Camin, T. Buslaps, M.D. Michiel, H. Kaminiski, A. Kottar, A. Pernack, and W. Reimers (2005). Simultaneous tomography and diffraction analysis of creep damage. *Science* **308**: 92–95.

Qian, L., H. Toda, K. Uesugi, T. Kobayashi, T. Ohgaki, and M. Kobayashi (2005). Application of synchrotron x-ray microtomography to investigate ductile fracture in Al alloys. *Appl Phys Lett* **87**: 241907.

Ritchie, R.O., R.K. Nalla, J.J. Kruzic, J.W. Ager III, G. Balooch, and J.H. Kinney (2006). Fracture and ageing in bone: Toughness and structural characterization. *Strain* **42**: 225–232.

Santis, R.D., P. Anderson, K.E. Tanner, L. Ambrosio, L. Nicolais, W. Bonfield, and G.R. Davis (2000). Bone fracture analysis on the short rod chevron-notch specimens using the x-ray computer micro-tomography. *J Mater Sci Mater Med* **11**: 629–636.

Schilling, P.J., B.P.R. Karedla, A.K. Tatiparthi, M.A. Verges, and P.D. Herrington (2005). X-ray computed microtomography of internal damage in fiber reinforced polymer matrix composites. *Compos Sci Technol* **65**: 2071–2078.

Silva, M.J. and D.C. Touhey (2007). Bone formation after damaging in vivo fatigue loading results in recovery of whole-bone monotonic strength and increased fatigue life. *J Orthop Res* **25**: 252–261.

Sinclair, R., M. Preuss, and P.J. Withers (2005). Imaging and strain mapping fibre by fibre in the vicinity of a fatigue crack in a Ti/SiC fibre composite. *Mater Sci Technol* **21**: 27–34.

Sinclair, R., M. Preuss, E. Marie, J.Y. Buffière, P. Bowen, and P.J. Withers (2004). The effect of fibre fractures in the bridging zone of fatigue cracked Ti-6Al-4V/SiC fibre composites. *Acta Mater* **52**: 1423–1438.

Sinnett-Jones, P.E., M. Browne, W. Ludwig, J.Y. Buffière, and I. Sinclair (2005). Microtomography assessment of failure in acrylic bone cement. *Biomater* **26**: 6460–6466.

Stock, S.R. (1990). X-ray methods for mapping deformation and damage. *Micromechanics: Experimental Techniques*. W.N.J. Sharpe (Ed.). New York, ASME. *AMD* **102**: 147–162.

Stock, S.R., T.M. Breunig, A. Guvenilir, J.H. Kinney, and M.C. Nichols (1992). Nondestructive X-ray tomographic microscopy of damage in various continuous-fiber metal matrix composites. *Damage Detection in Composite Materials*. J.E. Masters (Ed.). West Conshohocken, PA, ASTM. *STP* **1128**: 25–34.

Stock, S.R., A. Guvenilir, T.M. Breunig, J.H. Kinney, and M.C. Nichols (1995). Computed tomography Part III: Volumetric, high-resolution x-ray analysis of fatigue crack closure. *J Metals* Jan: 19–14.

Suresh, S. and R.O. Ritchie (1982). A geometric model for fatigue crack closure induced by fracture surface roughness. *Met Trans* **13A**: 1627–1631.

Symons, D.D. (2000). Characterization of indentation damage in 0/90 lay-up T300/914 CFRP. *Compos Sci Technol* **60**: 391–401.

Toda, H., I. Sinclair, J.Y. Buffière, E. Maire, T. Connolley, M. Joyce, K.H. Khor, and P. Gregson (2003). Assessment of the fatigue crack closure phenomenon in damage-tolerant aluminium alloy by in-situ high-resolution synchrotron x-ray microtomography. *Phil Mag* **83**: 2429–2448.

Toda, H., I. Sinclair, J.Y. Buffière, E. Maire, K.H. Khor, P. Gregson, and T. Kobayashi (2004). A 3D measurement procedure for internal local crack driving forces via synchrotron x-ray microtomography. *Acta Mater* **52**: 1305–1317.

Underwood, E.E. (1968). Particle size distribution. *Quantitative Microscopy*. F.N.R.R.T. DeHoff (Ed.). New York, McGraw-Hill: 149–200.

Uthgenannt, B.A. and M.J. Silva (2007). Use of the rat forelimb compression model to create discrete levels of bone damage in vivo. *J Biomech* **40**: 317–324.

Viggiani, G., N. Lenoir, P. Bésuelle, M.D. Michiel, S. Marello, J. Desrues, and M. Kretzschmer (2004). X-ray microtomography for studying localized deformation in fine-grained geomaterials under triaxial compression. *C R Mech* **332**: 819–826.

Watson, I.G., P.D. Lee, R.J. Dashwood, and P. Young (2006). Simulation of the mechanical properties of an aluminum matrix composite using x-ray microtomography. *Metall Mater Trans A* **37**: 551–558.

Weck, A., D.S. Wilkinson, H. Toda, and E. Maire (2006). 2D and 3D visualization of ductile fracture. *Adv Eng Mater* **8**: 469–472.

Weiler, J.P., J.T. Wood, R.J. Klassen, E. Maire, R. Berkmortel, and G. Wang (2005). Relationship between internal porosity and fracture strength of die-cast magnesium AM60B alloy. *Mater Sci Eng A* **395**: 315–322.

Withers, P.J. and J.J. Lewandowski (2006). Three-dimensional imaging of materials by microtomography. *Mater Sci Technol* **22**: 1009–1010.

Yoder, G.R., P.S. Pao, M.A. Imam, and L.A. Cooley (1989). Micromechanisms of fatigue fracture in Al-Li Alloy 2090. *Aluminum-Lithium Alloys, Proceedings of the Fifth Aluminum-Lithium Conference*. J.T.H. Sanders, E.A. Starke, Jr. (Ed.). Birmingham, UK, Materials and Component Engineering: 1033–1041.

Zhang, H., H. Toda, H. Hara, M. Kobayashi, T. Kobayashi, D. Sugiyama, N. Kuroda, and K. Uesugi (2007). Three-dimensional visualization of the interaction between fatigue crack and micropores in an aluminum alloy using synchrotron X-ray microtomography. *Met Mater Trans A* **38**: 1774–1785.

12

Multimode Studies

A number of groups have employed x-ray microCT and another x-ray or non-x-ray modality to gain a more complete understanding than either method could have provided separately. Here, the first example is microstructural characterization of sea urchin teeth: x-ray microbeam mapping and precision lattice parameter determination were combined with microCT. The second example is sulfate attack of Portland cement paste, where energy dispersive x-ray diffraction mapping of reaction products was used along with microCT. Three combined diffraction and microCT studies of mechanical responses of specimens provide the next examples: x-ray mesotexture analysis of crack path and microCT quantification of changes in fatigue crack opening, microCT plus x-ray diffraction of creep damage, and diffraction-based strain mapping plus microCT of load redistribution in monofilament metal matrix composites. The composite material bone is covered in Section 12.6: internal stress analysis plus microCT of loaded bone and microCT plus diffraction or strain gauge analysis or other modalities. The final section looks at networks analyzed with microCT plus other methods.

12.1 Sea Urchin Teeth

Sea urchin teeth were discussed in some detail in Section 7.2.3, and the interested reader is referred to that section and the references therein for a start to reading about these fascinating structures. What is important here is that sea urchin teeth mineralize in two stages. The primary, secondary, and carinar process plates as well as the needles/prisms are termed the primary mineralized tissue and, in *L. variegatus*, are nonequilibrium calcite $Ca_{1-x}Mg_xCO_3$ with $x \approx 0.13$, as determined by synchrotron x-ray diffraction (Stock et al., 2002). At about the point where the keel develops, the primary skeletal elements begin to be linked by the secondary skeletal elements, which are termed columns or disks. The columns are much higher Mg calcite with $x \approx 0.33$ (Stock et al., 2002). These columns, linking adjacent primary structural elements into a rigid structure, were often described as being polycrystalline, but the transmission x-ray diffraction data clearly showed that the high Mg ($x \approx 0.13$) and very high Mg ($x \approx 0.33$)

phases had their crystallographic axes identically aligned. The view of Stock and coworkers, therefore, is that the tooth is a compositionally modulated crystal much like multiple quantum well structures from molecular beam epitaxy (Stock et al., 2002), although, of course, of much lower crystal quality than the artificial material. X-ray diffraction mapping (precision lattice parameter and crystallite size/microstrain broadening determinations) provided additional structural information supplementing 3D microCT-derived geometric information.

12.2 Sulfate Ion Attack of Portland Cement

Study of sulfate ion attack of Portland cement via microCT was introduced briefly in Chapter 10, where it was noted that microCT showed the results of the attack but provided little information about the reaction phases producing the damage (i.e., the softening, cracking, loss of adhesion, etc. of the cement). Position-resolved x-ray diffraction with high-energy synchrotron x-radiation is a good method of mapping phase content as a function of depth, and using this method and microCT provided much more information than either technique by itself (Naik, 2003; Jupe et al., 2004; Wilkinson et al., 2004; Naik et al., 2006). Energy dispersive x-ray diffraction was used instead of the more normal single wavelength methods because the former allowed more precise definition of the sampling volume combined with simultaneous collection of diffraction patterns from multiple phases within this same volume. The reader is directed elsewhere for more details specific to this application (Jupe et al., 2004).

Sulfate attack is (simplistically) described in the literature by one of two classes of reaction and associated damage. The first is gypsum formation, which is associated with loss of adhesion and strength. The second is ettringite formation associated with expansion and cracking. After considerable sulfate exposure, energy dispersive x-ray diffraction identified an ettringite-rich, gypsum-free layer outside cylindrical cracks paralleling the outer surface of the cylindrical specimens (i.e., C1 in Figure 10.5). Inside the crack, that is, closer to the cylinder center, a gypsum-containing volume was identified (Naik, 2003; Jupe et al., 2004; Wilkinson et al., 2004; Naik et al., 2006). Although the same identification might have been performed by destructive specimen preparation (with considerably more effort), the results could have been criticized as affected by exposure to the atmosphere and the like.

12.3 Fatigue Crack Path and Mesotexture

MicroCT of cracked AA2090 specimens revealed complex 3D patterns of crack face contact as a function of applied load (see the deformation subsection). Roughness of the crack faces produced the closure effects and is intrinsically related to the low fatigue crack propagation rate for this material. The underlying question is what drives the crack to assume this highly nonplanar path: fracture mechanics indicates that the energetically favorable path would be more or less directly across the specimen (i.e., a path perpendicular to the applied tensile load). Yoder and coworkers related the average texture or macrotexture to the faces of asperities (large peaks) on the fracture surface (Yoder et al., 1989); although this data explained the geometry of asperities and why they, on the average, formed, these observations did not identify the cause of the transition between an asperity and a relatively planar section of the crack. Microbeam Laue pattern mapping revealed the scale of crystallographic texture between that of individual grains (microtexture) and the average specimen texture (Haase et al., 1998, 1999). This particular type of mesotexture consisted of groups of 5–10 adjacent pancake-shaped grains with nearly identical orientations; that is, these adjacent grains constituted near-single crystal volumes. Asperities formed when the fatigue crack passed through the border between near-single crystal volumes with different orientations (Figure 12.1). Furthermore, a large fraction of the volume of the plate centers of AA2090 T8E41 consisted of near-single crystal domains, and this differentiated AA2090 from other Al–Li alloys with similar macrotextures (Ignatiev et al., 2000) and produced decreased fatigue crack growth rates compared to the other alloys.

12.4 Creep Damage

Pyzalla et al. (2005) studied creep of a three-phase copper alloy (Pb particles in a mixture of α- and β-brass) using synchrotron microCT and x-ray diffraction. Three detector systems were positioned so that microCT, energy-dispersive x-ray diffraction, and angle-dispersive x-ray diffraction could be performed sequentially without realignment or recalibration. The microCT-determined pore size distribution was reported for ten time intervals and agreed with an exponential growth dependence. The decrease in diffraction peak FWHM ended when the voids started to reach appreciable size, and changes in peak intensity after this point in the creep test revealed texture formation.

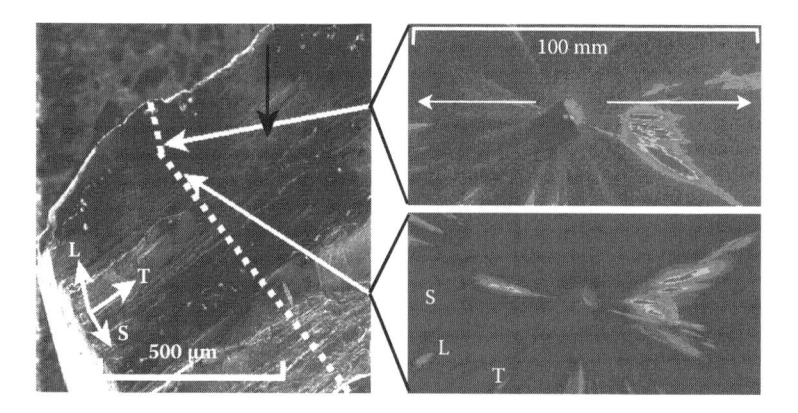

FIGURE 12.1 (SEE COLOR INSERT FOLLOWING PAGE 144.)
SEM fractograph (left) of a fatigue crack surface in a small compact tension specimen of AA2090 viewed at a large angle of tilt. A large asperity on the crack face appears at the top. The directions indicate directions in the plate from which the specimen was machined: longitudinal (L), transverse (T), and short transverse (S) directions. The dashed line indicates the positions at which a series of (synchrotron x-ray) microbeam Laue patterns were recorded. The central portions of two transmission Laue patterns, from the positions indicated by the arrows, are shown at the right. The streaks are 111 reflections, and increasing x-ray diffracted intensity is indicated by the different shades. The diamond shape near the center of the image is the beam stop. The orientation of the streaks changes abruptly from within the asperity to outside the asperity. See Haase et al. (1998).

12.5 Load Redistribution in Damaged Monofilament Composites

Failure of uniaxially aligned, monofilament-reinforced composites depends on many factors. MicroCT allows one to study where and at what applied stresses the reinforcements fail; repeated observations of the same specimen are particularly important because fibers such as the SCS series SiC monofilaments will often fracture several times within a 10-mm gauge section; the location of each successive break is an important input for modeling. The increasing strain within the fiber, longitudinal and transverse strains within the matrix, strain relaxation to either side of fiber fractures, and the fiber matrix interface strength are other important quantities that microCT alone cannot define. As demonstrated by a series of reports on Ti/SiC$_f$ composites, combining high-energy x-ray microbeam diffraction mapping with microCT was a powerful approach to measuring these quantities (McDonald et al., 2002; Preuss et al., 2002; McDonald et al., 2003; Sinclair et al., 2004, 2005; Aroush et al., 2006; Withers et al., 2006).

Preuss et al. (2002) studied deformation of a single SiC fiber in a Ti-6Al-4V matrix: synchrotron microCT revealed the position and morphology of SiC fiber fractures and, as a function of applied stress, microbeam diffraction quantified the matrix and fiber strains. At each of 19 loading steps, mapping with transmission x-ray diffraction along the length of the fiber (100 diffraction profiles spaced by 50-μm steps at each load) revealed SiC longitudinal strains of at least 1.5 percent before the fiber cracked at the first point (equivalent to a failure stress of at least 6 GPa for E = 400 GPa). Above the nominal yield stress for the matrix, strains in the matrix became only slightly nonlinear, but the fiber longitudinal strain rose very rapidly.

At 790 MPa, the load preceding first fracture, two local strain maxima were observed along the length of the monofilament; the next deformation increment produced a load drop and local strains approaching zero at the positions of the two maxima, positions that corresponded to fiber fracture revealed in microCT. Longitudinal strains in the matrix were relatively uniform at 790 MPa but, at the next deformation state, rose sharply at the positions where the fiber fractured. In other words, localized strain concentration occurred in the matrix in the vicinity of the SiC breaks. Fitting the data to a partial sliding model allowed the authors to estimate a constant interfacial friction shear stress of ~200 MPa that was significantly higher than results from fiber push-out tests. The authors note that blind application of conventional full-fragmentation post mortem analysis of fragment lengths would suggest a significantly higher interfacial strength (~700 MPa) and suggest that fiber strength decreases after the first fast fracture event.

Well-defined defects were introduced into a single-ply Ti/SiC$_f$ composite, and redistribution of loads from damaged fibers to neighboring ones was investigated with microbeam diffraction mapping (McDonald et al., 2002). Load redistribution around damage sites increased the load in the nearest neighbor fibers by ~25 percent and second nearest neighbors by ~10 percent. The interfacial fractional shear stress was found to be similar or slightly larger than that cited in the previous paragraph. Reverse sliding was observed during unloading and produced compressive residual stresses near the fiber ends. Wedge crack geometry was frequently observed (McDonald et al., 2002, 2003).

In the examples of a single-fiber and a single-ply composite described above, simple phase-enhanced microradiography would have sufficed to correlate fiber fracture and maxima/minima in the fiber and matrix strain profiles. In multiple-ply composites, the overlapping fiber images necessitated use of microCT, and microCT plus microbeam diffraction mapping was applied to multiple-ply Ti/SiC$_f$ to determine the stress partition between fiber bridging a fatigue crack and broken fibers (Sinclair et al., 2004, 2005; Withers et al., 2006). Initially, strain mapping averaged over the entire thickness of the specimen (Sinclair et al., 2004), but subsequent experiments used a narrow receiving slit and 2θ-scanning to limit the gauge volume to a single SiC fiber plus the surrounding matrix material (Sinclair et al., 2005).

Strain distribution in an intact fiber in the crack wake was compared for maximum and minimum applied loads; for example, strain distribution as a function of distance from the crack plane was analyzed using a partial debonding shear lag model (Sinclair et al., 2005). Measurements of crack opening displacements showed that the fatigue crack front bowed out between fibers when it emerged from a ply and advanced preferentially toward fibers when the front was between plies (Withers et al., 2006). Furthermore, the 3D distributions of crack opening were measured for three stress intensities characteristic of a fatigue cycle (K_{max}, K_{min}, and K_{mid}), and very little variation in crack opening was observed parallel to the crack front irrespective of the proximity to bridging fibers (Withers et al., 2006).

12.6 Bone

X-ray scattering measurement of internal strains (and conversion to internal stresses) in loaded bone (or tooth dentin and enamel) is a relatively uninvestigated research area and one where a combination with microCT will provide valuable insight. For bone, the collagen D-period (~67 nm) along the fibril axis produces SAXS peaks, and the Angstrom-level periodicities of carbonated apatite (cAp) crystallites produce diffraction peaks in the WAXS regime. Although the mineral nanoparticles in long bone have a pronounced crystallographic texture, there are still enough orientations present to produce more-or-less complete cones of diffracted intensity for monochromatic x-rays; Debye cones from different hkl exist simultaneously and produce rings of increased intensity on area detectors. Force applied to a specimen distorts the unit cells and alters the Debye cones. Hydrostatic applied stresses (those with equal magnitude in all directions) uniformly alter the diameter of cones, whereas deviatoric stresses (those with directionality) change the shape of diffraction rings. Similarly, SAXS peak positions from (the collagen D-period) alter in response to applied stress.

X-ray scattering measures quantities such as d_{hkl} in cAp or the D-period in collagen, and changes in these quantities define the internal strain imposed during loading; that is, strain in cAp is $\varepsilon_{cAp} = (d - d_{initial})/d_{initial}$ and in collagen is $\varepsilon_{collagen} = (D - D_{initial})/D_{initial}$. Internal stress is a quantity derived from internal strain, and stress σ_{ij} and strain ε_{kl} are second-rank tensors related through the fourth-rank elastic constants C_{ijkl}; that is, $\sigma_{ij} = C_{ijkl}\, \varepsilon_{kl}$. Describing the conversion of deviatoric strains to deviatoric stresses is beyond the scope of this chapter, and the reader is directed elsewhere for details as pertains to bone internal stress measurements (Almer and Stock, 2005, 2007).

The inverse of the slope of macroscopic strain ε_{macro} (measured by an attached strain gauge) as a function of $\sigma_{applied}$ (measured by the load cell of the mechanical testing apparatus) is Young's modulus for the specimen. Such slopes for the WAXS and SAXS data (ε_{cAp} and $\varepsilon_{collagen}$ vs. $\sigma_{applied}$) reflect Young's modulus for the individual constituent phase of bone. This last extension may not be strictly correct; however, it does provide a numerical operational probe of how the individual phases differ from pure (inorganic) cAp or collagen. For a section of canine fibula, the resulting moduli (90 percent confidence limit) were: E_{macro} = 24.7(0.2) GPa, E_{cAp} = 41(1.0) GPa, and $E_{collagen}$ = 18(1.2) GPa (Almer and Stock, 2007). The value for E_{macro} was in good agreement with moduli of similar bone types reported in the literature; for cAp it was about one-third of that of inorganic apatite; for collagen it was at least nine times higher than one would expect (see Almer and Stock (2007) for details). The data demonstrated the extent to which the local environment affects the different phases' responses to applied load.

In the studies described in the preceding pair of paragraphs, microCT was used to measure cross-sectional area and to account for internal porosity. This is a rather trivial use of microCT because of the simple geometry. In specimens containing bone in complex geometries (i.e., specimens containing trabecular bone or cortical plus trabecular bone), determination of the spatial distribution of bone segments and of their orientation relative to the load axis will be essential for proper interpretation of the scattering data. Obtaining such 3D maps is impractical except through microCT, especially when one considers that the different bone segments may suffer significant relative displacements during loading, displacements that may change from load to load and that may not be preserved during post-testing serial sectioning.

X-ray diffraction and synchrotron microCT were combined in a study of changing mineralization during human fetal vertebrae growth (Nuzzo et al., 2003). From microCT, trabecula were much thicker and more widely spaced in the interior of the vertebrae than in the peripheral volume, and bone volume fraction increased linearly over gestational ages of 16 through 24 weeks. X-ray diffraction revealed a linear increase in crystallite size with age and a linear increase in lattice parameter ratio c/a (both over the same gestational range as above). As complete understanding of bone mechanical properties depends not only on properly incorporating microarchitecture but also on inclusion of proper materials properties (in bone these include crystallite dimensions and distortions of the apatite unit cell), more studies of this sort need to be completed and the data imported into numerical models of elastic moduli and the like.

The role of insulinlike growth factor-1 (IGF-1) was studied in fetal mouse bone with microCT, with Fourier transform infrared spectroscopy (FTIR), and von Kossa staining, a histology technique designed to reveal bone (Burghardt et al., 2007). Tibiae and lumbar vertebrae were examined from knock-out (IGF-1 −/−) and wild-type animals at the 18th gestational day.

The degree of mineralization (in terms of mg hydroxyapatite per cubic centimeter) and the morphology of the bone were determined with synchrotron microCT. FTIR was used to infer the ratio of mineral to matrix using the ratio of the phosphate peak integral to that of the amide I band and to measure the carbonate-to-phosphate ratio using a carbonate band, all standard methods in bone research. FTIR data at both bone sites and both animal populations were characteristic of poorly crystalline mineral. The mineral-to-matrix ratio was systematically lower in IGF-1 –/– animals than wild types for both sites. The carbonate to phosphate ratio was significantly lower in the IGF-1 –/– spine but not the tibia. Interestingly, von Kossa staining failed to reveal mineral at the spinal ossification center of the IGF-1 –/– animals, even though the synchrotron microCT clearly revealed high attenuation components within this structure, indicating a subtle difference in the mineral's environment that affects the staining process.

The last result cited in the previous paragraph, von Kossa staining failing to reveal mineral shown to be present from two other techniques, suggests that the results of studies that employ only von Kossa staining and that fail to find mineral despite the presence of bone matrix need to be re-examined. Further investigation of the necessary environment for the staining to occur may reveal a new way to probe the mineral's local environment.

In another study, diffraction with small-diameter synchrotron x-ray beams was combined with synchrotron microCT to characterize coronary atherosclerosis in vitro (Jin et al., 2002). It may be that diffraction-based measures of crystallite size, microstrain, or a/c ratio will help define the natural history of such pathological mineralization processes or even the rate of development of these dangerous structures.

The role of sutures between bones was discussed earlier in the context of analysis of effect on jaw structure of the hardness or softness of food. Another study employed microCT, strain gauges, and high-speed photography to investigate in vivo the mechanical role of cranial sutures and their morphology during feeding in a fish; the goal was to determine what inferences, if any, could be gathered about skull function in living and fossil fish (Markey et al., 2006). These investigators studied *Polypterus*, a fish that feeds by suction or by biting, and determined that peak suture strains are higher during suction than biting. They also found that interfrontal and frontoparietal sutures, typically loaded in tension, were less interdigitated in cross-section than the interparietal suture which experiences compression.

MicroCT can often provide data to validate the results of other techniques. Ultrasonic analysis, a noninvasive, inexpensive, and nonharmful modality, is being developed as a clinical tool, for example, not only for fetal imaging and for heart morphology and fluid flow imaging, but also for cortical and cancellous bone analysis. Interpretation of in vivo ultrasonic velocity measurements in cortical bone depends on microCT calibration, an active area of research. MicroCT data also were used to validate conclusions developed from lower-resolution MRI studies of cancellous bone (Krug et al., 2007).

In animal models, increasing use is being made of bioluminescence imaging, that is, detection of photons emitted through the skin by luciferase-expressing cells in the living animal. Recently, this approach was demonstrated for tumor progression in deep tissues including bone. MicroCT plus bioluminescence were combined in a bone tumor study (Fritz et al., 2007), and one expects the number of such studies to grow substantially.

As was mentioned in Chapter 8, scaffold design and evaluation of bone growth into scaffolds is an active area of research. Cancedda et al. reviewed application of microCT and of microdiffraction to this problem; only a very few studies combined the two modalities. Combined SAXS and WAXS were used to map the orientations of the cAp crystallites' c-axes and the collagen fibril axes relative to the hydroxyapatite scaffold material seeded with bone stromal cells (Cedola et al., 2006; Cancedda et al., 2007). This approach extended to 3D diffraction mapping, for example, as described elsewhere (Stock et al., 2008) and, combined with microCT, should be very powerful.

12.7 Networks

Fibers form the basis of some network solids, and biodegradable fibers are used in applications as diverse as fishing line and sutures. Tanaka and coworkers applied synchrotron microCT and SAXS to study the internal structure of as-formed and isothermally crystallized fibers of poly[(R)-hydroxybutyrate] (Tanaka et al., 2007). Without crystallization, the drawn fibers produced strong SAXS peaks along the meridian (fiber axis), suggesting that there were lamellar crystals of systematic long period along the fiber axis. The meridian reflections could barely be detected in the crystallized fibers, but equatorial streak scattering was much stronger than for the uncrystallized fibers, suggesting voids were an important component of these fibers. MicroCT revealed the inner structure of the 100–120-μm-diameter fibers with 1-μm spatial resolution. Few micrometer-sized voids were seen in the uncrystallized material. In the crystallized fiber, voids constituted nearly 50 percent of the cross-sectional area, 3D renderings revealed the voids were highly elongated along the drawing direction, and the voids were principally 1–6 μm in cross-section, with a mean of 2.3 μm and a strong peak in the size distribution of ~1.5 μm.

Blood vessels in the brain constitute a network of great interest in a variety of disciplines. Dorr et al. (2007) combined MRI and microCT imaging of the brains of CBA mice to produce a comprehensive vasculature atlas of this organ, emphasizing the location of vessels with respect to neuroanatomical structures and watershed regions associated with specific arteries. Given the much higher spatial resolution of microCT, it was used to image the blood vessels that had previously been filled with an x-ray absorbing polymer (Microfil), and MRI mapped the different neuroanatomical structures using another contrast agent.

References

Almer, J.D. and S.R. Stock (2005). Internal strains and stresses measured in cortical bone via high-energy x-ray diffraction. *J Struct Biol* **152**: 14–27.

Almer, J.D. and S.R. Stock (2007). Micromechanical response of mineral and collagen phases in bone. *J Struct Biol* **157**: 365–370.

Aroush, D.R.B., E. Maire, C. Gauthier, S. Youssef, P. Cloetens, and H.D. Wagner (2006). A study of fracture of unidirectional composites using in situ high-resolution synchrotron x-ray microtomography. *Compos Sci Technol* **66**: 1348–1353.

Burghardt, A.J., Y. Wang, H. Elalieh, X. Thibault, D. Bikle, F. Peyrin, and S. Majumdar (2007). Evaluation of fetal bone structure and mineralization in IGF-I deficient mice using synchrotron radiation microtomography and Fourier transform infrared spectroscopy. *Bone* **40**: 160–168.

Cancedda, R., A. Cedola, A. Giuliani, V. Komlev, S. Lagomarsino, M. Mastrogiacomo, F. Peyrin, and F. Rustichelli (2007). Bulk and interface investigations of scaffolds and tissue-engineered bones by x-ray microtomography and x-ray microdiffraction. *Biomater* **28**: 2505–2524.

Cedola, A., M. Mastrogiacomo, M. Burghammer, V. Komlev, P. Giannoni, A. Favia, R. Cancedda, F. Rustichelli, and S. Lagomarsino (2006). Engineered bone from bone marrow stromal cells: A structural study by an advanced x-ray microdiffraction technique. *Phys Med Biol* **51**: N109–N116.

Dorr, A., J.G. Sled, and N. Kabani (2007). Three-dimensional cerebral vasculature of the CBA mouse brain: A magnetic resonance imaging and micro computed tomography study. *Neuroimage* **35**: 1409–1423.

Fritz, V., P. Louis-Plence, F. Apparailly, D. Noël, R. Voide, A. Pillon, J.C. Nicolas, R. Müller, and C. Jorgensen (2007). Micro-CT combined with bioluminescence imaging: A dynamic approach to detect early tumor–bone interaction in a tumor osteolysis murine model. *Bone* **40**: 1032–1040.

Haase, J.D., A. Guvenilir, J.R. Witt, and S.R. Stock (1998). X-ray microbeam mapping of microtexture related to fatigue crack asperities in Al–Li 2090. *Acta Mater* **46**: 4791–4799.

Haase, J.D., A. Guvenilir, J.R. Witt, M.A. Langøy, and S.R. Stock (1999). Microtexture, asperities and crack deflection in Al–Li 2090 T8E41. *Mixed-Mode Crack Behavior*. K. Miller and D.L. McDowell (Eds.). West Conshohocken, PA, ASTM. *ASTM STP* **1359**: 160–173.

Ignatiev, K., Z.U. Rek, and S.R. Stock (2000). X-ray microbeam diffraction comparison of mesostructures in plates of three aluminum alloys. *Adv X-ray Anal* **44**: 56–61.

Jin, H., K. Ham, J.Y. Chan, L.G. Butler, R.L. Kurtz, S. Thiam, J.W. Robinson, R.A. Agbaria, I.M. Warner, and R.E. Tracy (2002). High resolution three-dimensional visualization and characterization of coronary atherosclerosis in vitro by synchrotron radiation x-ray microtomography and highly localized x-ray diffraction. *Phys Med Biol* **47**: 4345–4356.

Jupe, A.C., S.R. Stock, P.L. Lee, N.N. Naik, K.E. Kurtis, and A.P. Wilkinson (2004). Phase composition depth profiles using spatially resolved energy dispersive x-ray diffraction. *J Appl Cryst* **37**: 967–976.

Krug, R., J. Carballido-Garmio, A.J. Burghardt, S. Haase, J.W. Sedat, W.C. Moss, and S. Majumdar (2007). Wavelet-based characterization of vertebral bone structure from magnetic resonance images at 3T compared with micro-computed tomographic measurements. *Mag Res Imaging* **25**: 392–398.

Markey, M.J., R.P. Main, and C.R. Marshall (2006). In vivo cranial suture function and suture morphology in the extant fish Polypterus: Implications for inferring skull function in living and fossil fish. *J Exp Biol* **209**: 2085–2102.

McDonald, S.A., M. Preuss, E. Maire, J.Y. Buffiere, P.M. Mummery, and P.J. Withers (2002). Synchrotron x-ray study of micromechanics of Ti/SiC$_f$ composites with fibres containing defects introduced by laser drilling. *Mater Sci Technol* **18**: 1497–1503.

McDonald, S.A., M. Preuss, E. Marie, J.Y. Buffiere, P.M. Mummery, and P.J. Withers (2003). X-ray tomographic imaging of Ti/SiC composites. *J Microsc* **209**: 102–112.

Naik, N. (2003). Sulfate attack on Portland cement-based materials: Mechanisms of damage and long term performance, Ph.D. Thesis. Atlanta, Georgia Institute of Technology.

Naik, N.N., A.C. Jupe, S.R. Stock, A.P. Wilkinson, P.L. Lee, and K.E. Kurtis (2006). Sulfate attack monitored by microCT and EDXRD: Influence of cement type, water-to-cement ratio, and aggregate. *Cement Concr Res* **36**: 144–159.

Nuzzo, S., C. Meneghini, P. Braillon, R. Bouvier, S. Mobilio, and F. Peyrin (2003). Microarchitectural and physical changes during fetal growth in human vertebral bone. *J Bone Miner Res* **18**: 760–768.

Preuss, M., P.J. Withers, E. Maire, and J.Y. Buffiere (2002). SiC single fibre full-fragmentation studied during straining in a Ti-6Al-4V matrix studied by synchrotron x-rays. *Acta Mater* **50**: 3175–3190.

Pyzalla, A., B. Camin, T. Buslaps, M. D. Michiel, H. Kaminiski, A. Kottar, A. Pernack, and W. Reimers (2005). Simultaneous tomography and diffraction analysis of creep damage. *Science* **308**: 92–95.

Sinclair, R., M. Preuss, and P.J. Withers (2005). Imaging and strain mapping fibre by fibre in the vicinity of a fatigue crack in a Ti/SiC fibre composite. *Mater Sci Technol* **21**: 27–34.

Sinclair, R., M. Preuss, E. Marie, J.Y. Buffiere, P. Bowen, and P.J. Withers (2004). The effect of fibre fractures in the bridging zone of fatigue cracked Ti-6Al-4V/SiC fibre composites. *Acta Mater* **52**: 1423–1438.

Stock, S.R., J. Barss, T. Dahl, A. Veis, and J.D. Almer (2002). X-ray absorption microtomography (microCT) and small beam diffraction mapping of sea urchin teeth. *J Struct Biol* **139**: 1–12.

Stock, S.R., F. De Carlo, and J.D. Almer (2008). High energy x-ray scattering tomography applied to bone. *J Struct Biol* **161**: 144–150.

Tanaka, T., K. Uesugi, A. Takeuchi, Y. Suzuki, and T. Iwata (2007). Analysis of inner structure in high-strength biodegradable fibers by X-ray microtomography using synchrotron radiation. *Polymer* **48**: 6145–6151.

Wilkinson, A.P., A.C. Jupe, K.E. Kurtis, N.N. Naik, S.R. Stock, and P.L. Lee (2004). Spatially resolved energy dispersive x-ray diffraction (EDXRD) as a tool for nondestructively providing phase composition depth profiles on cement and other materials. *Applications of X-Rays in Mechanical Engineering 2004*. New York, ASME: 49–52.

Withers, P.J., J. Bennett, Y.C. Hung, and M. Preuss (2006). Crack opening displacements during fatigue crack growth in Ti-SiC fibre metal matrix composites by x-ray tomography. *Mater Sci Technol* **22**: 1052–1058.

Yoder, G.R., P.S. Pao, M.A. Imam, and L.A. Cooley (1989). Micromechanisms of fatigue fracture in Al–Li Alloy 2090. *Aluminum–Lithium Alloys, Proceedings of the Fifth Aluminum–Lithium Conference*. J.T.H. Sanders and E.A. Starke, Jr. (Eds.). Birmingham, UK, Materials and Component Engineering: 1033–1041.

Index

('f' indicates a figure)

A

Absorbing power, scintillator property, 48
Absorption edge comparison, 13, 14f, 20, 45
Absorption microCT methods, 39
   cone beam, 40f, 41
   fan beam, 40f, 40–41
   parallel-beam, 40f, 41
   pencil beam, 39–40, 40f
Absorption tomography, 21
Absorptivity, specimen, 116
Accuracy assessment, microCT, 96–98
Advanced Photon Source (APS), 4, 60, 105, 153
   bending magnet, 11, 12f
Advanced Photon Source (APS), reviews, 58t
Algebraic reconstruction technique (ART), 23–24, 24f
Almer, J. D., 312, 313
Altman, S, J., 104, 218, 250
Aluminum (Al)
   corrosion studies, 257–258, 258f
   crack opening, 134, 135f, 136, 284, 285, 286
   matrix composites, 149, 296
Aluminum (Al) alloys
   AA-A356 cracking, 288
   AA2024-T351 fatigue cracking, 286, 287f, 288
   AA2090-T8E41 cracking, 284, 285, 309
   AA5091 cracking, 288
   Ag$_2$Al precipitates, 148, 149f
   crack opening, 134, 135f, 136
Aluminum (Al) foams, internal architecture, 174–175, 175f
Aluminum-magnesium (Al-Mg) alloy
   hot tear damage, 239–240
   intermetallic particle tracing, 245
Aluminum-silicon (Al-Si) composite
   propagation contrast, 66, 67f
   structure evolution, 239
Aluminum-silicon carbon (Al-SiC)
   cracking, 296–298, 297f
   high-contrast sensitivity, 99f, 99–100, 100f
Alveolar bone, remodeling, 187
Alyssum murale, metal storage, 156
Analyisis approaches, simplicity, 145
Angular undersampling, 98, 99f
Anisotrophy, 127, 130
   cellular solids, 173
   human bone, 183
   materials, 3
   measurement, 118
   metal foams, 176
   microCT studies, 153
Aposto, L., 184
Arabidopsis, iron in seeds, 156
Artifacts
   motion, 85–86, 86f
   ring, 86–87, 87f
   source to detector output, 50
Ashby, A. F., 171
Assignment by inspection, 121
Asthenosoma varium, demipyramid, 133f
Asymmetric Bragg magnifiers, 60
Asymmetrically cut crystal, 19, 42
Attenuation of x-ray beam equation, 13
Auditory apparatus, studies, 159
Automatic tracking, evolving structures, 128
Avalanche evolution, 249

B

Babout, L., 257, 291
Back-projection
   circular object demonstration, 24–29, 25f, 26f
   filtering, 26–27, 28f
   grid, 26, 27f
Basic multicellular unit (BMU)-related resorption, 264

Beam hardening, 18, 89–90, 90f, 191
Beam spreaders, 19, 19f
Beam time, synchrotron, 70–71
Bending magnets, synchrotron, 11, 11f
Bentley, M. D., 226
Berea sandstones, porosity, 221
Bernard, D., 242, 248
Binary segmentation, 119–120, 120f
Bioactive glass scaffolds, 196, 197f
Biodegradable fibers, 315
Biological structure degradation, 253–257
Biological tissues, as phases, 151–152
Bioluminescence imaging, 315
Biomedical fluorescence/phase microCT, 68
Biophosphonate treatment, bone study, 262
Bioprosthetic valves, 264
Blob visualization, 222
Blurred profile, back-projection, 26, 27
Boltzmann simulations, network microCT, 217, 221
Bone
  basic multicellular unit resorption, 264
  Ca/P ratios, 262
  components, 261
  cortical, 13, 14f, 224f, 226–227, 262, 263, 291
  intracortical porosity, 263–264
  multimode studies, 312–315
  spaceflight impact, 261–262
  sutures, 314
Bone cement fatigue, 282
Bone erosion, 265
Bone healing, 261
Bone loading, 187
Bone mineral density (BMD), and fractures, 182
Bone replacement
  implants, 260–261
  ingrowth, 196–197
  scaffolds, 195–196, 315
Bonse, U., 3, 149, 183, 184
Bonse-Hart interferometry, 63f, 64
Boundary-following algorithms, 124
Bouxsein, M. L., 186
Bovine cortical bone study, 291

Bragg plane, synchrotron micrcoCT system, 59–60
Bragg's law, scattering, 18
Brain model, vasculature corrosion casting, 226, 227f
Bremsstrahlung (continuous spectrum), 9, 10f, 44
Breunig, T. M., 283, 289, 295, 296
Bubble characteristics, basalts, 147
Bubble formation
  breads, 177
  liquid foams, 176–177
Buffière, J. Y., 290

C

Cadmium tungstate, 48–49, 59
Calcein labeling, bone structures, 96
Calcifications, 151
Calcified human heart valve, 3D rendering, 137, 138f
Calcium (Ca), mass attenuation coefficient, 13, 14f
Callovo-Oxfordian argillite, 148
Camarodonts, teeth, 158, 159f
Cancedda, R., 196, 315
Cancellous (trabecular) bone, 171, 172, 182–184
  growth and aging, 184–190
Cancers, studies, 265
Canine fibula, contrast sensitivity, 102f
Carbon composite, 150, 151f
Carbonated apatite (cAp), bone and tooth, 181
Case, Tom, 43f
CAT-scan, 1
CCD (charge coupled device) detectors, 46–48
  characteristics, 48t
Cellular solids
  cancellous bone studies, 182–184
  definition, 171
  microstructural characteristics, 172
  mineralized tissue, 181
  nonbiological, 176–181
  static structures, 173–176
Cellulosic fibrous networks, 215
Cement curing, 251
Cement leaching in mortar, 252

Index

Center of rotation errors, 87, 88f, 89f
Centrostephanus rodersii, spine slices, 131f
Ceramic studies, 3
 cracking, 289
 pore topology, 240
Channel structures, 217
Chemical tomographic mapping, 68
Chemical vapor infiltration (CVI), ceramic composites, 240, 241
Chen, T. T., 250
Chu, T. M. G., 197
Circular objects, back-projection, 24–26, 25f, 26f
Circulatory system, microCT studies, 223–227, 224f, 227f
Clarke, A. R., 216
Cloetens, P, 63, 66, 67
Closed-cell aluminum foam, 175
Cluster labeling, 221
Clyne, T. W., 217
CO2 sequestration, 249
Cody, D. D., 265
Collateral vessel development, 226
Colloid transport, porosity, 223
Color imaging, microCT, 137
Commercial microCT systems, 52, 54t–55t
Compendex, microCT citations, 4, 5f
Composites systems, damage and deformation, 3, 289–299
Compressive fatigue testing, trabecular bone, 194
Compton scattering, 13
Computed tomography (CT), 2
Conductive anodic filaments (CAF), copper, 257
Cone beam geometry
 reconstruction, 43f, 44
 tube-based microCT, 23, 52
 volumetric CT, 40f, 41
Constraints, specimen, 116
Construction materials, environmental impact, 251–252
Consumer electronics, 3–4
Contaminant classification, cotton fabric, 217
Contrast, 15f, 15–17
Contrast sensitivity, design, 116

Contrast-detail-dose curve, 17
Convolution, equation, 17
Convolution operation, back-projection, 27–28
Cool, Steven, 44t
Cooper, D. M. L., 263, 264
Copper, corrosion studies, 257
Cormack, A. M., 2
Coronary atherosclerosis, 314
Coronary circulation, 224–225
Cortical bone
 canal system, 224f, 226–227, 262, 263
 cracking, 291–293, 292f, 293f, 294
 mass attenuation coefficient, 13, 14f
Counting statistics, feature visibility, 101, 102f
Crack opening
 aluminum specimens, 134, 135f, 136, 284, 285, 286
 analysis, 134, 135f
 2D rendering,134, 135f, 136
 3D rendering,134, 136f
Crack position, mesh map,136, 136f
Cracking
 engineered structures, 283
 human cortical bone, 291–293, 292f, 293f
 stainless steel wire, 257, 259f
Cracks
 closure, 283–285
 detectability, 94–95
 tortuosity, 134
Creep damage, copper alloy, 309
Crystal monochromator, 18, 19, 42
Cullity, B. D., 9, 13, 17, 29, 126
Cylindrical specimen geometry, 117

D

Dalstra, M., 187
Damage propagation, 3
Dark field correction, 50, 51f
Data acquisition, challenge, 104–105
Data analysis, 117–118, 122–129
 binary segmentation, 119–120, 120f
 distance transformation, 121, 122, 123f
 evolving structure tracking, 128–129

image texture, 127
segmentation, 121–122
watershed segmentation, 122
Data collection reproducibility, microCT, 97
Datasets, microCT, 127
David, V., 187, 189
Davis, G. R., 42, 47, 50, 97
DeCarlo, F., 51, 60, 87, 102f, 106
Deer antler, porosity, 184
Deformation studies, 281–283
Dentineonamel junction (DEJ), microCT studies, 153, 154, 254–255
Detection limits, 94–96
enhancement, 20
Detectors
integration mode, 48
semiconductor devices, 46–51
Deutsches Elektronen Synchrotron (DESY), 60
reviews, 58t
Diadema setosum, spine slices, 131f
Differential attenuation equations, 21
Diffraction, 18, 19f
Diffraction-enhanced imaging (DEI), phase contrast, 63f, 63–64
Diffraction-enhanced radiography, cartilage, 152
Digital-phase mammography, 152
Dilation, 124, 125f
Direct space reconstruction, 29
Distance transformation, 121, 122, 123f
Distance transformation, 3D combined, 136
Dorr, A., 315
Dover, S. D., 4
Drosophilia, neurons, 155
Drug tablet dissolution, 253
Drugs, bone studies, 186

E

Eating, and bone changes, 187–188, 314
Eberhardt, C. N., 216
Echinoderms. *See also* Sea urchins
teeth, 158, 159f
Echinothrix diadema, spine slices, 131f
Egermann, M., 261

Elastic modulus/properties
FEM modeling, 150
trabecular bone, 191, 194
vertebral bone, 187
Electron backscattering diffraction (EBSD), cracking, 286
Elliott, J. C., 4, 40, 42, 46, 47, 50, 87, 100, 177, 255, 256, 295
Emission efficiency, scintillator property, 48
Engineered network solids, 215–217
Environmental interactions, 248–249
biological structures degradation, 253–257
construction materials, 251–252
geological applications, 249–251
metal corrosion, 257–260
Epiphyseal growth plates, 185
Equations
Bragg's law, 18
convolution, 17
differential attenuation, 21
Fourier transforms, 29, 30, 31
linear attenuation coefficient variance, 32
x-ray attenuation, 13
Erosions, 124, 125f
Eswaran, S. K., 192
Etched column-filled scintillators, 49
European Synchrotron Radiation Facility (ESRF), 4, 60
reviews, 58t
Evolution
environmental interactions, 248–249
materials processing, 237–239
microstructures, 237
structure tracking, 128–129
Experiments
design, 115–117
planning for synchrotron, 71

F

Fan-beam geometry
rotate-only, 40f, 40–41
transformation, 42, 43f, 44
tube-based microCT, 23, 52
Fatigue crack
closure, 283

Index

opening, 284–285
path, 309, 310f
Fe nanocatalyst spatial distribution, 68
Feature visibility, 101, 102f
Feldkamp, L. A., 41, 46, 183
Ferrie, E., 288
Fiber-optic scintillatory glasses, 49
Fiber-reinforced composites, cracking, 294–299
Fiberboard studies, 215–216, 216f
Fibrous network solids, 215
Field of view (FOV)
 local tomography, 33–34, 35f
 synchrotron micrcoCT system, 59
Filter, x-ray tube, 10f, 18
Filtered back-projection algorithm, 17
Filtration transport, porosity, 222
Final assembly verification, 3
Finite element modeling (FEM), 150
 cellular solids, 177
Fish skeletons, biomineralization models, 161
Flat field correction, 50
Fluorescence microCT, 67–68
Foam specimens, phase shift, 66
Focal plane tomography, 2
Follet, H., 192
Fontainbleau sandstone, porosity, 220–221
Fossil cleavage-stage embruos, 153, 154f
Four-dimensional (4D) rendering
 description, 129
 packed glass beads, 218f
Fourier components, square wave, 29
Fourier slice theorem, 31
Fourier transform
 inverse, 31
 spatial/frequency domain, 28f
 two-dimensional, 30
Fourier transform infrared spectroscopy (FTIR), 313, 314
Fractional contrast, 16
Frequency space, projections, 31f
Fresnel microzone plates, 62f
Fresnel optics, 19
Fu, X., 247

G

Gallium (Ga) liquid embrittlement, 258, 260, 286
Gastrectomy, trabecular bone loss studies, 186
Gaussian fields, data analysis, 126
Geandier, G. 150
Gender and age, bone studies, 186
Generational analysis, 126
Geological materials, phase distribution, 146–147
Geometric magnification, x-ray tube imaging, 17, 18f
Geometry, in accuracy assessment, 96–98
Gibson, L. J., 171
Glass foams, insulation, 176
Grafts, vascular, 264
Grain size distribution, structure evolution, 239
Grain structure, wood, 173
Granulometry, 124
Grating-enhanced imaging, phase contrast, 63f, 64, 64–65
Grodzins, L., 4
Gualda, G. A. R., 96, 146, 147
Guinea pig model, trabecular morpholoyg, 183
Guldberg, R. E., 184
Gureyev, T. E., 98
Guvenilir, A., 124, 134, 284, 285

H

Halmshaw, R., 13
Haversian canal systems, 223, 224f, 227, 262, 263
Heart valves, 137, 138f, 264
Heggli, M., 150
High definition, 116
High-contrast feature detection, 94–96
 enhancement, 103–104
 sensitivity, 99–101
High-resolution data collection, 3
High-resolution x-ray CT, 1
Hildebrantdt, G., 1, 182, 191
Hip/knee replacements, 260
Hirai, T. H., 11, 45
Hirano, T. K., 4, 149, 294, 295, 298

Hofmann, S., 196
Holotomography, 63
Hot tear damage, chill cast Al-Mg, 239–240
Hounsfield, G. N., 2
Hydroxyapatite scaffolds, 196, 197

I

Image texture analysis, 127
Imaging, 15–17
Implants, bone formation, 194–196, 260
In situ composite studies, structure evolution, 238, 239
In situ loading, fracture processes, 281–282, 284
In vivo microCT systems, 4
Incoherent x-ray scattering, 14
Indentation damage, carbon-fiber-reinforced plastic, 295
Industrial CT, applications, 2
Industrial metrology, 153
Industrial paper studies, 216
Inherent imaging imperfection, 16
Insects, x-ray phase contrast, 154–155
Insertion devices, synchrotron, 11, 11f
Insulinlike growth factor-1 (IGF-1), 313–314
Interferometry, phase contrast, 63f, 64
Internal structure
 reconstruction, 22
 recovery, 1–2
Invertebrates, microCT studies, 154–158, 155f, 158f, 159t
Iron cracking study, 288
Isotropic microstructure, 118
Iterative opening/closing 3D, 124–125, 125f
Iterative reconstruction technique, 24
Itokawa, H., 195

J

Jin, H., 314
Jones, A. S., 223
Jones, J. R., 197
Jones, K. W., 222, 250

K

Kaestner, A., 217

Kang, Z., 242
Kim, C. H., 127, 191
Kinney, J. H., 49, 99, 100, 149, 177, 178, 183, 190, 191, 240, 241, 255, 256, 294, 295, 297
Kirkpatrick-Baez optics, 62
Kiss, M. Z., 64
Knackstedt, M., 219
Komlev, V. S., 196

L

Labeling, data analysis, 126
Laminography, 2, 35–36, 36f, 60
Li, X., 223
Limestone, CO_2 attack, 249
Lin, C. L., 223, 247
Lindquist, W., 121, 122
Linear attenuation coefficient variance equation, 32
Linear attenuation coefficients
 enamel de/remineralization, 256
 high-contrast sensitivity, 99–101
Liquid foams
 evolution, 121, 176
 reconstruction accuracy, 98
Liquid metal embrittlement, 258, 260
Liver microvessels, 226
Local tomography, 33–35, 35f, 60
Lu, S., 251, 261
Ludwig, O., 238, 258, 286
Lung models
 gas volume studies, 228, 228f
 vasculature corrosion casting, 227, 228
Luo, G., 190
Lux, J., 216
Lytechniunus variegatus (pink pin cushion sea urchin)
 endoskeleton, 90f
 intrambulacral plate, 133f
 ossicle imaging, 128
 spine slices, 131f
 teeth, 92f, 157–158, 158f, 307–308

M

M x M object slice, 31, 32
Madi, K., 180

Magnesium (Mg), corrosion studies, 258
Maire, E., 172, 176, 181, 299
Malpighian tubes, silkworms, 155, 155f
Mandible, and teeth, 194
Manufactured composites, phase distribution, 147–151
Martin, C. F., 243
Martin, E. A., 185
Mass attenuation coefficients, 13, 14f
Materials processing, 237–239
Materials processing
   particle packing and sintering, 247–248
   plastic forming, 243, 245–247
   solidification, 238–240
   vapor phase, 240–243
McDonald, S. A., 245, 291, 299, 310, 311
Mean intercept length (MIL), cellular solids, 173
Mean window technique, data analysis, 126
Mechanically induced damage, 281
Medical CT, constraints, 2
Medium definition, 116
Medline, microCT citations, 5, 5f
Menopause, and bone loss, 187
Mesh map, 3D crack position, 136, 136f
Mesotexture, fatigue crack path, 309
Metal corrosion studies, 257–260
Metal fibrous network studies, 217
Metal oxide silicon (MOS) CCD array, 46
Metrology
   definition, 145
   industrial, 153
Microbeam Laue pattern mapping, 309
Microcallus formations, and healing, 184
Microchannel plates, 50
MicroComputed tomography (microCT)
   absorption methods, 39–44
   accuracy assessment, 96–98
   color imaging, 137
   commercial systems, 54t–55t
   data acquisition, 104–105
   data analysis, 117–129
   detectors, 46–51

   fluorescence, 67–68
   future trends, 106–108
   multimodal, 307
   phase contrast, 62–67
   reconstruction, 21
   scatter, 69
   as single modality, 6
   spatial resolution, 1, 4, 45
   websites, 138
   x-ray sources, 44–45
Micropore evolution, chill cast Al, 239
Microstructural feature study, 3
Microstructure measurement, 118–119
Microstructure studies, sea urchin, 128, 133f, 156–157
Microtomography, 1
Miller, J. D., 222, 247, 250
Mineral location, 250–251
Mineralization, bone, 188, 189f
   density studies, 262
Mineralized tissue. *See also* Bone, Sea Urchins, Teeth
   adaptation, 260–264
   deformation studies, 291
   high-contrast sensitivity, 99, 100–101, 102f
   microCT studies, 181–182
Modulation transfer function (MTF), 16, 17
Momose, A., 64, 67, 152, 224
Monochromatic radiation, 18
Monochromators, 18, 19f
Monofilament composites
   multimode studies, 310–312
   silicon carbon, 294, 295
   titanium, 295–296
Monolithic materials, cracks, 283–289
Montel, X., 226
Motion artifacts, 85–86, 87f
Movies, slice stacks, 70, 129, 138
Mulder, L., 188, 193
Müller, B., 225
Müller, R., 195
Multilayer monochromators, 19
Multimode studies, 307
   aluminum alloys, 309, 310f
   bone, 312–315
   copper alloy, 309
   monofilament composites, 310–312

network solids, 315–316
Portland cement, 308
sea urchin teeth, 307–308
Multiscale modeling, 221
Murine trabecular bone
binary segmentation, 120f
erosion/dilation segmented image, 125f
3D rendering, 130, 132f
Myocardial infarct size, mouse model, 265
Myocardial microvasculature, 225

N

Naik, N. N., 308
Nalla, R. K., 293
Namati, E., 228
NanoCT systems, 4, 42, 45, 61–62
detectors, 48
future trends, 106–108
Nanoindentation, 291
Natural gas hydrate, porosity, 222
Network studies
brain blood vessels, 315–316
circulatory system, 223–227, 224f, 227f
engineered solids, 215–217, 216f
pores, 217–223, 219f, 220f
respiratory system, 227–228, 228f
New product development, 2, 3
Nicalon/SiC composites, channel width, 240–242
Nielsen, R. B., 161
Nondestructive evaluation (NDE), 36
Nonidealities, 85
correction for, 93
Nuzzo, S., 185, 313

O

Odgard, A., 172, 173, 183
Ohuchi, H., 225
One-dimensional photodiode arrays, 47f
Op Den Buijs, J., 226
Open-cell aluminum foams, 174–175, 175f
Optical cross-talk, 49
Optical lens systems, 50

Optical lenses, synchrotron micrcoCT system, 59
Organelles, microCT differentiation, 158, 160f
Osteoarthritis
canine study, 183
human bone, 187
murine bone, 61f, 265
Osteointegration, scaffolds, 195
Osteolytic bone loss, 265
Osteon clustering, 263
Osteopontin (OPN), 261
Osteoporosis
and bone studies, 182
postmenopausal, 187
Ovarietomy (OVX), trabecular bone loss studies, 183, 186, 187, 189–190, 192

P

Packing systems, 218, 219, 219f
Paleontology, microCT studies, 153
Parallel beam
geometry, 40f, 41
synchrotron micrcoCT system, 57
Partap, S., 195
Partial view reconstruction, 35
Partial volume effects, 93f, 93–94
Particle displacement, 245, 246f
Particle packing, 247–248
Pencil beam geometry, 39–40, 40f
Penumbral blurring
cone beam reconstruction, 41
x-ray tube imaging, 17, 18f
Peters, O. A., 161
Peyrin, F., 184
Pharmaceuticals, phase distribution, 146
Phase contrast, 62–63
artifacts, 91–92, 92f
diffraction-enhanced imaging (DEI), 63f, 63–64
grating-enhanced imaging, 63f, 64, 64–65
interferometry, 63f, 64
propagation method, 63, 63f, 66, 67f
Phase distribution, 146
geological materials, 146–146

Index

pharmaceuticals, 146
Phase stepping interferograms, 65f
Phosphor powders, 49
Phosphor-optical lens-detector systems, 47
Photon energy, 10f
   vs. synchrotron, 12f
Phylogeny, structural motif quantification, 145
Physical sectioning, microCT comparison, 96
"Pink" beam, 19, 45
Pixel (picture element), 16
Plant roots, 217
Plastic forming, 243, 245–247
Plougonven, E., 178
Pneumatization, birds, 161, 184
Point spread function (PSF), 16, 17
Point x-ray sources, absorption methods, 42
Polychromatic radiation, 19, 45
Polyethylene (PE), mass attenuation coefficient, 13, 14f
Polymer blends, phase distribution, 147
Polymer foams, oil spill absorption, 222–223
Polymeric scaffolds, bone formation, 197
Polypterus, feeding study, 314
Polystyrene (PS)/polymethyl methacrylate (PMMA), inferometer contrast, 67
Polystyrene spheres, interferograms, 65f
Polyurethane foams, oil uptake, 222–223
Polyvinylchloride foam, porosity, 176
Porosity, 134, 173, 217
   deer antler, 184
   and failure, 282
   network studies, 217–223
   propylene fumarate scaffolds, 196
   texture analysis, 127, 223
Porosity formation, epoxy-based composites, 240
Portland cement
   laser micromachining, 247
   microCT study, 116–117

multimode studies, 308
   sulfate attack, 252, 253f, 254f
   3-D rendering, 136, 137f
Preuss, M., 310
Primatology, microCT studies, 153
Printed wiring boards, laminography, 2
Propagation method, phase contrast, 63, 63f
Propylene fumarate scaffolds, porosity, 196
Pulmonary fibrosis, mouse model, 265
Pulmonary wall, 225
Purpose-built filter, ring artifacts, 86
Pyrometric cones, 3
Pyun, A., 148
Pyzalla, A., 309

Q

Quantification, simplicity, 145
Quasi-local tomography, 98

R

R cure behavior, 282, 283
Rabbit femoral bone, 3D rendering, 130, 132f
Radiographic stereo pairs, 33
Radiological imaging, 1
Radio-opaque polymers, 148
Radon, J., 2
Rajagopalan, S., 121, 124
Recentering algorithm, center errors, 87
Reciprocal space representations, 29
Reconstruction
   algebraic, 23–24
   back-projection, 24–29
   basic concepts, 21–22
   fourier-based, 29–31
   performance, 31–33
   precision and accuracy, 92–93
   sample positioning, 51–52
   sinograms, 33, 34f
   x-ray phase contrast, 62–67
Reconstruction artifacts, 85
   angular undersampling, 98, 99f
   beam hardening, 18, 89–90, 90f
   center errors, 87, 88f, 89f

mechanical imperfections, 87–88
motion, 85–86, 8rf
nonphysical streaks, 91, 91f
phase contrast, 91–92, 92f
ring, 86–87, 87f
Region-of-interest (ROI) tomography, 33, 60, 123f
Reinforced scattering, 14, 18, 19f
Ren, W., 260
Renal microvasculature, 226
Replicates analysis, 126
Repositioning, 51
Representative elemental volumes (REV), 126
Resin flow, liquid composite molding, 245, 247
Respiratory system, microCT studies, 227–228, 228f
Restorative dentistry/orthopedics, 238
Reticon devices, 46
Rheumatoid arthritis, bone erosion, 265
Richard, P., 247
Ring artifacts, 86–87
Ritman, E. L., 223, 227, 262
Rivers, M., 96, 146, 147
Robb, R. A., 121
Rolland Du Roscoat, S., 216
Röntgen, W., 1, 13
Root action in soil, 250
Root canal geometry, 161
Rotation axis wobble, 52, 87–88
Ruimerman, R., 190

S

S(uper)P(hoton) ring-8, 4, 60
 reviews, 58t
Samples
 positioning, 51–52
 reconstruction projections, 32
 thickness, 17
Saturation, porous media, 218, 219f
SAXS x-ray scattering, 62, 69, 240, 245, 312, 315
Scaffold structures
 bone ingrowth, 196–197
 bone replacement, 195–196
Scaffolds, 171

Scanco MicroCT-40 system, 56–57
Scanning electron microscopy (SEM), 4, 45, 61, 227f, 310f
Scattered x-radiation, trabecular bone, 190. *See also* X-ray scattering
Schizosaccharomyces pombe, organelle differentiation, 160f
Sciatic neurectomy, trabecular bone loss studies, 186
Scintillator-CCD coupling, 47f
Scintillators
 characteristics, 49t
 properties, 48
Sea urchins. *See also Asthenosoma varium, Echinothrix diadema, Lytechniunus variegatus*
 calcite stereom, 182
 ossicles, 128, 156–157
 spine, 131f, 156
 teeth, 103f, 157, 158f, 307–308
Seeds, cellular structure, 173
Segmentation data analysis
 active contour, 122
 binary, 119–120, 120f
 watershed, 122
Selective laser sintering, polymer powders, 191
Sharpening filter, back-projection, 27, 28f
Sheppard, A. P., 219
Shutters, 46
Silicate wastes, 176
Silicon carbon (SiC)
 monofilament studies, 294, 295
 particles in crack visibility, 290
Silicon carbon (SiC) cloths
 channel width, 240–242, 242f, 243f, 244f
 microstructural variation, 130, 134f
Silicon nitride/silicon carbide, 148–149
Silkworms, Malpighian tubes, 155, 155f
Silva, M. D., 133, 294
Sinclair, R., 311
Single slice microstructure, 118, 119f
Sinograms
 reconstruction projections, 33, 34f
 ring artifacts, 86
Sintering/cementation, 248

Index

Skeletonization, 125–126
Slices, fan-beam geometry, 40–41
Snail shells, 155
"Snake" contour boundary detection, 122
Snow to ice evolution, 249
Soft tissue. *See also* Circulatory system, Respiratory system
   adaptation, 264–265
   studies, 152
Software, microCT, 69–70
Soil composition, 250
Solder bump integrity, 60
Solid-phase/void-phse burn, 124
Solidification studies, 238–240
Solids, x-ray refraction, 62
Spatial coherence, 14–15
Spatial resolution, 1, 4, 16, 45
   design, 115
   light-emitting scintillators, 51
Specimen characteristics, design, 116
Specimen rotator, 51
Spiral tomography, 33
Spowage, A. C., 122
Square wave, 29, 30f
Stainless steel wire, crack propagation, 257, 259f
Starch/polycaprolactone biodegradable scaffolds, 195
Stent implantation, PCI, 256–257
Stereometry, 36
Steroids, trabecular bone loss studies, 186
Storage rings, synchrotron, 11
Strain localization, 281
Strain measurement, trabecular bone, 192, 193, 193f
Streak artifacts, 91, 91f
Stress fractures, cracking, 294
Strontium ranelate, 187
Structural connectivity, 126
Structure model index (SMI), 118
   cellular solids, 172, 173
Swiss Light Source (SLS), 60
   reviews, 58t
Synchrotron
   beam time, 70–71
   facilities missions, 60
   monochromators, 19
   nanoCT reconstructions, 62
   phase imaging, 63, 66
   radiation, 10, 11f, 12f
   radiation source, 45
   research, 4, 6
   system specifications, 69–71
   systems, 57–61
   systems reviews, 58t
   tube-base reconstruction comparison, 61, 61f, 70

T

Tabeculae strain displacement studies, 128–129
Tabletop synchrotron radiation, 11, 45
Tafforeau, P., 153, 254
Talbot interferometry, 64
Tanaka, E., 188
Tatahouine meteriorite, mapping, 68
Teeth
   echinoderms, 92f, 103f, 157–158, 158f, 159f
   human, 161, 162f
   microstructure studies, 254–256
Textiles, network studies, 216–217
Texture, trabecular microarchitecture comparison, 190
Thermal barrier coatings, solidification, 240
Thicknesses, determination, 122
Thin film scintillators, 48
Thompson, A. C., 4
Thompson, K. E., 247
Three orthogonal plane rendering, 130, 131f
Three-dimensional (3D) imaging, 4
   internal structure recovery, 1–2
Three-dimensional (3D) numerical model, developing bone, 190
Three-dimensional (3D) reconstruction, 29
Three-dimensional (3D) rendering, 130, 131f, 133
   cellular solids, 172
   crack openings, 134, 136f
   microstructure evolution, 121, 237
   porosity system, 219, 220f
Thresholding, 119

Throats and pores, porosity system, 219, 220f
Time delay integration, 42
Tissue engineering scaffolds, 5
Tissue-engineered bones, 196
Titanium (Ti)
  mass attenuation coefficient, 13, 14f
  monofilament studies, 295–296
Titanium dioxide-polymer composites, 150
Titanium-silicon carbon (Ti-SiC), cracking, 299, 310, 311
Toda, H., 148, 286
Tomography, definition, 2
Tomosynthesis, 35–36, 36f, 60
Tortuosity, cracks, 134
Trabecular bone
  binary segmentation, 120f
  damage, 191–194
  erosion/dilation segmented image, 125f
  growth and aging, 184–190
  reconstruction accuracy, 98
  studies, 159, 182–184
  3-D rendering, 130, 132f
Tree growth study, 3
Tumor formation, 265
Turnkey commercial systems, 53, 56, 56t
Two-dimensional (2D) detector systems, 46–47
Two-dimensional (2D) rendering, 129
  crack openings, 134, 135f, 136
  SiC cloths, 134, 136

U

Ultrasonic analysis, bone analysis, 314–315
Unidirectional fiber composite fracture, 295

V

Vacumn, x-ray tube, 10f
Vapor deposition, 19
Vapor phase processing studies, 240–243
Vertebrates, microCT studies, 159, 161, 162f

Vessel system
  replicates analysis, 126
  studies, 159, 315–316
Views, reconstruction projections, 32, 33
Viggiani, G., 281
Voids, 3
Volcanic silicate foams, bubble-size distributions, 175
Volkman canals, 262
Volumetric data collection, 4
von Kossa staining, 313, 314
Voxel-value interpretation, 128
Voxel-value-based thresholding, 122
Voxels (volume elements), 4, 21, 22f, 53
  synchrotron micrcoCT system, 57, 59

W

Walther, T., 215, 216
Wan, S. Y., 224, 227
Wastewater sludge, 222
Water distribution, porous media, 218–219
Water-immiscible liquids, porosity, 221–222
Watershed segmentation, 122
Watson, I. G., 238
Wave-length emission, scintillator property, 48
WAXS x-ray scattering, 62, 313, 315
Weiss, P., 260
Weitkamp, T. C., 59, 64, 65, 66, 88
White field/dark field correction, 50–51
Wilkinson, A. P., 308
Williams, Jr., J. C., 152
Withers, P. J., 310, 312
Wobble, rotation axis, 52, 87–88
Wong, F. S. L., 255
Wood
  cellular structure, 173–174
  degradation, 252
  orthogonal sections, 174f
Wu, J. P., 176

X

X-radiation, 9

attenuation equation, 13, 17–18
contrast and imaging, 17–20
generation, 9–15, 10f, 11f, 12f
optics, 19
X-ray brightness imaging, 11
X-ray computed tomography (CT), 1
X-ray fluorescence, 68
X-ray projections, reconstructions, 21–23, 22f
X-ray scattering
loaded bone/teeth, 312–313
SAXS, 62, 69, 240, 245, 312, 315
WAXS, 62, 313, 315
X-ray sources, 44–45
X-ray to light converters, 46, 48, 50
X-ray tube, 9, 10f
commercial systems, 52–57
imaging, 17, 18f
photon energy, 12f
sources, 44
synchrotron reconstruction comparison, 61, 61f, 70

Y

Yoder, G. R., 309
Young's modulus (elastic properties), 220
bone minerals, 128
cell walls, 176
trabecular bone, 192
yarn, 217

Z

Zang, H., 288
Zeolite catalyst growth, 174
Zettler, R., 245
Zhu, X. Y., 225